THE LOEB CLASSICAL LIBRARY
FOUNDED BY JAMES LOEB 1911

EDITED BY
JEFFREY HENDERSON

ARISTOTLE
IV

LCL 228

ARISTOTLE

THE PHYSICS

BOOKS I–IV

WITH AN ENGLISH TRANSLATION BY

PHILIP H. WICKSTEED

AND

FRANCIS M. CORNFORD

HARVARD UNIVERSITY PRESS
CAMBRIDGE, MASSACHUSETTS
LONDON, ENGLAND

First published 1929
Revised and reprinted 1957
Reprinted 1963, 1970, 1980, 1993, 1996, 2005

ISBN 0-674-99251-2

Printed on acid-free paper and bound by Edwards Brothers, Ann Arbor, Michigan

CONTENTS

PREFACE

A CHARACTERISTIC of the work of Philip Henry Wicksteed was his deep interest in the sources of any literature he sought to understand, and his long study of the Middle Ages, represented by such minds as those of Aquinas and Dante, constantly led him back to the classics and eventually to a serious study of Aristotle as the great master of Mediaeval Philosophy.

But though he considered Aristotle was for various reasons more intelligible to the mediaeval scholar than to us, he did not believe that his importance to the world had decreased. His own thought ranged from man's urgent material needs, represented by Economics, to his deepest spiritual needs, represented by Religion. He would not have considered it an exaggerated claim that Aristotle's practical genius working upon the Greek love of thought for its own sake, made his mind and method of vital importance in almost all our modern problems : a touchstone clear of the structure and débris of ages we stand on, and of the utilitarian element so difficult, if not impossible, to escape to-day, and yet so practical a danger to the value we call truth.

When he left London to live in the country some thirty years ago, he conceived the project of translating the *Physics*, and after a severe illness in 1918 which compelled him to give up his lecturing, it became more and more his principal literary occupation. In 1924 the work was sufficiently advanced to be submitted to the directors of the Loeb Classics, who accepted it for publication in their series.

Both the scholarly and the popular character of this series made a strong appeal to him, and one of its features—the inclusion of the Greek text—seemed of particular importance in a work intended to be an interpretative paraphrase. With Aristotle's text at hand for reference it would seem justifiable to express the same Greek word by different English ones and different Greek words by the same English one, whenever Aristotle's intention seemed subtly different or subtly akin.

In these general principles and in the whole conception and purpose of the work, he was supported by the sympathy of Professor Gilbert Murray, who, in the midst of his many activities, never failed to show a practical and enlightening interest in the work. Grateful acknowledgements have at the same time to be made to Professor E. H. Neville for criticisms and suggestions on certain points submitted to him at the translator's request, and on others which have arisen in course of preparation for the press; also to Sir Thomas Heath and to several others for generous help and advice.

When the first draft of the paraphrase was finished, the writing of the Arguments and expository notes and of the various introductions was begun, while at the same time a careful and systematic revision of the translation was proceeding. Of this the only part brought to something like its present form was the first few chapters of Book IV.

PREFACE

It was at this critical stage of the work and with only a few months' warning that Philip Wicksteed received the final summons to cease labour. Rapidly increasing infirmity made it impossible for him to do without secretarial assistance, and one of his grandchildren (the grandchild also of his friend and master, the late Abraham Kuenen of Leiden) suspended her own studies in order to come and help him. Shortly before his death he dictated the following for inclusion in the preface of the book :

> " I should not have been able to carry on the work at all when I became unable to manipulate my books and find references, without the help of my granddaughter, Emily Kuenen. She was a highly intelligent amanuensis, capable of finding all the relevant passages in the Greek, Mediaeval and Renaissance commentaries, with complete understanding of my purpose, and submitting them to my judgement with the utmost economy of effort on my part."

It soon became evident that it was a question of weeks, not of months, and when he saw that there was no possibility of being able to get the book ready for the press himself he sent an urgent call to his old friend, Mr. J. V. Saunders : " Saunders," he said, " will know when I have made a mistake, and when I have said it on purpose." Mr. Saunders came immediately, and, with much diffidence, agreed to shoulder the responsibility, and they conferred together whenever it was possible to do so.

But as every writer knows, the last stages of a work tend to lengthen out, and the labourer's tale of days was progressively drawing in. It was my privilege to be with him for several days in his last week and for a few hours at the very end. His spirit never flagged, and he greeted me on my arrival one evening with radiant delight at having completed an exceptionally good day's work, that had taken him to within a few pages of that part of the revision which he still hoped to complete. Later in the evening Mr. Francis Cornford, who was intimate with some members of the family and whom my father had expressed a desire to see, arrived from Cambridge. My father saw him for a few moments and explained his plan of action. They had only once met, many years before, but the two now passed from the status of fellow scholars to that of intimate friends. The next day a few more pages of the revision were done, but there remained some five or six which he was never able to attempt. For two or three days it was still occasionally possible to dictate notes, and Mr. Cornford was able to get some valuable points clearly set forth. But from the hour of his arrival the future of the book could no longer be a matter of anxiety to my father himself, to Mr. Saunders, or to any other. Mr. Cornford has shown himself continuously anxious not only to spare no pains in perfecting the work, but, in constant consultation with Mr. Saunders and with my sister, Rebecca, to preserve everything possible of my father's spirit in it.

During my father's last years my sister was probably more intimate than anyone else with his mind and work, and has therefore been able to supply many links between his written and spoken word. She has prepared from MS. notes left by my father the section, in the General Introduction, on the principles of translation adopted in the work, and she is mainly responsible for the Note on Mathematical Conceptions.

Mr. Cornford sends the following for inclusion in the Preface :

" It is due to Dr. Wicksteed's memory and to the reader that I should explain, as exactly as I can, what I have done towards the preparation of this book for the press.

" Although I had looked forward to the appearance of this translation, I had never seen any part of it when, within a week of Dr. Wicksteed's death, I was summoned to Childrey. I came without any idea that it would be my privilege to share in the inheritance of his work. He was still able, by an heroic effort of will, to dictate notes bearing upon certain points that were on his mind ; but, not having read the translation, I could not ask those questions to which we have since had to find an answer. Mr. Saunders welcomed my help with the generosity of a brother in scholarship, and it was agreed that he and I, in collaboration with Miss Rebecca Wicksteed, should do what remained to be done. Since each of us lives at a long distance from the others, it became obvious that one of us should revise the text and translation and submit the result to his colleagues. As I had some special knowledge of Greek philosophy, this task fell to me.

" When I had taken the manuscript away and examined it, I found that, while some parts had been re-written several times, others had not been thoroughly revised. Only a part of Book IV (the Introduction and Chapters I-V) had gone to the press and been corrected in proof by Dr. Wicksteed. The rest of the translation still contained such inaccuracies, and even positive mis-renderings, as any scholar, translating so difficult a work, would be likely to leave uncorrected in his earlier drafts. I conceived it to be my task to bring the translation closer to the Greek, in so far as this could be done without spoiling its intended character.

" It was a delicate matter to retouch the work of a master whose approach to any subject was marked by a singular independence. My difficulties were the greater because the translator had not prepared a Greek text beyond Book I, and I was not seldom in doubt as to the reading he had adopted. He seems to have worked sometimes with the Berlin edition, sometimes with the Tauchnitz text of 1881, whose obsolete readings and grotesque punctuation here and there misled him. He did not, so far as I know, possess Prantl's edition of 1879 (Teubner), which is out of print and hard to obtain. M. Carteron's text and translation (Paris, 1926) appeared too late to be of much use.

" It was not my business to re-edit the text of the *Physics* ; on the other hand, I was bound to print what seemed, upon the evidence available, most likely to have been written by Aristotle. I have no knowledge of the MSS. beyond what can be learnt from the critical apparatus of the abovenamed editions, from Diels, *Zur Textgeschichte der aristotelischen Physik* (Abhandl. d. k. pr. Ak. der Wiss., Berlin, 1882), and from Torstrik (*Philologus*, xii. 494 ff.). I have constantly consulted the commentaries of Themistius, Philoponus, and Simplicius.

Apart from Book IV. Chapters I-V, the text I have printed is the best I could construct with these materials. It differs from the current texts chiefly in paragraphing, in the accentuation of enclitics, and in punctuation. No two editors will ever agree upon the punctuation of Aristotle, and my only guide has been the desire to make the structure and meaning of each sentence, as I understood it, as clear as possible to an eye accustomed to the punctuation of contemporary English (rather than German) books. Where the reading and interpretation are doubtful, I have sometimes been unable to find out how the translator arrived at his paraphrase, which at such points naturally tends to be freer than elsewhere ; and I have printed what appears to me to be the best authenticated text, adding a literal translation in a note. As a rule, I have given a critical note only where the reading has not the authority of any of Bekker's MSS., but is either found in the Greek commentaries or due to conjecture. Professor Ross of Oriel College has kindly helped me with his judgement upon a few very difficult passages.

" The additions I have made to the Arguments and notes are enclosed in square brackets with my initial [C.]. The large number of minor changes and corrections in the translation could not be marked in any way without making the book unreadable. For these alterations, though they have been submitted to Miss Wicksteed and Mr. Saunders, I must bear the responsibility. The editors of the Series were good enough to allow me an entirely free hand.

" The final result might be described as an unprincipled compromise. I must have made many changes which Dr. Wicksteed would have rejected with characteristic vehemence. On the other hand, I have let some interpretations stand, which strike me as doubtful, only in some cases offering an alternative rendering in a note. So I may claim that, if I should not have satisfied him, neither have I satisfied myself.

" I cannot end without looking back to the experience of that last week at Childrey. It was as if we were witnessing a conflict between the spirit of a man entirely fearless and Death himself. With a force of will that touched the limit of human power, he seemed to be holding the phantom at arm's length until he was assured that his work would be finished. He left us like Heracles refusing to linger with Admetus and the restored Alcestis because he was bent upon some new labour that lay beyond his defeat of Death.

F. M. Cornford."

Mr. Cornford left Childrey only when the master's mind at last began to flag, holding still undimmed its courage. Then, on the second evening following, as the sunset faded and the stars shone in through the open window, Philip Wicksteed, conscious to the last and in deep peace of spirit, joined the company of his masters.

Joseph Wicksteed.

Childrey, *January* 1929.

GENERAL INTRODUCTION

I. What to expect from the Physics

The title 'Physics' is misleading, and the reader must expect to find little or nothing that it suggests in this treatise. 'Lectures on Nature,' the alternative title found in editions of the Greek text, is more enlightening. But 'Principles of Natural Philosophy' (as the term would have been understood in the eighteenth and earlier nineteenth centuries) would be better still.

The realm of Nature, for Aristotle, includes all things that move or change, or that come and go, either in the sense of passing from 'here' to 'there,' or in the more extended sense of passing from 'this' to 'that,' which latter phrase is equivalent to 'becoming something that it was not'—a solid becoming a liquid, or a hot thing becoming cold, for instance.

Thus anything that 'becomes' this or that (substantively or qualitatively), any concrete thing to which, as such, an inception of being or a cessation from being can be assigned, belongs to the realm of Nature; but so also do things eternal if, and in so far as, they move or otherwise change. Thus the ultimate 'matter' which, according to Aristotle, underlies all the elementary substances must be studied, *in its changes* at least, by the Natural Philosopher. And so must the eternal heavenly spheres of the Aristotelian philosophy, in so far as they themselves move or are the causes of motion in the sublunary world.

Thus all things of which motion, change, or becoming (*i.e.* the inception of, and cessation from, 'being this or that') can be predicated are the subject matter of the study of Nature.

What is excluded, then? Not Psychology. For although it may be a disputable point whether the 'being' of the (human) *psyche* (or 'soul') had any inception or will have any cessation, it is at least certain that it passes from one state or act of consciousness to another, whereas if there are (as Aristotle believed there were) immaterial beings, and one supreme being, not subject to conditions of time, place, matter, dimension, or change of consciousness, the study of them, if possible at all, must lie outside the realm of Nature. But yet the study of Nature may point beyond itself to the necessity and possibility of such a Theology[a]; and it is one of Aristotle's directive purposes throughout the *Physics* to prove that it actually does so.

For the purposes of this Introduction it is unnecessary to pursue the subject of the division of studies any further. Enough has been said to show that the domain of Aristotle's Natural Philosopher extends beyond what we call (or used to call) Natural Philosophy.

But the *Physics* only covers the *principia* or fundamental and elementary problems of Natural Philo-

[a] One of the titles given by Aristotle to the work called by us, but never by him, the *Metaphysics*.

sophy. It was actually known to Dante and the Schoolmen as the 'First Book' on 'Nature,' and in Avicenna's paraphrase of Aristotle's works the Psychology was enumerated as the 'Sixth Book' of the 'Things of Nature.'

What, then, are the elements or principles of the study of Nature? Hardly a trace of what we understand by Physics will be found amongst them. They are concerned with the analysis of the conceptions of change, becoming, and so forth. What is 'motion'? What, if anything, are we to say to those who deny that there is such a thing? If motion means change of place, what does 'place' mean? If movements take time, what does 'time' mean? What is the meaning of an acorn being potentially but not actually an oak, or of cold water being potentially hot water? What, indeed, is the difference between the potentiality of anything 'becoming' one of several things and the actuality of its being the one thing it is? And why are potentialities at once so varied and so strictly limited? Or again, when things change, why do they change? And what do I mean by the cause or causes of change? Are the changes, so far as they are natural, purposeful? Have they an aim? And if so, do they always hit the mark they aim at?

These, and such as these, are the fundamental questions and concepts with which Aristotle deals in the *Physics*, and since he was neither an experimenter nor a close observer of falling, moving, floating, rising or sinking, expanding or contracting bodies, and does not know anything of what we understand by Chemistry, we shall not find in the *Physics* anything analogous to the close observation of sea animals, for instance, or the insight into comparative anatomy and embryology, which excite the admiration of biologists in Aristotle's biological treatises.

Nor again shall we find anything corresponding to the insight into the processes of the human mind and principles of human conduct which make some direct or indirect acquaintance with Aristotle's logical treatises, his *Psychology*, his *Ethics*, his *Politics*, or his aesthetical treatises, the common property of all students and the still living inspiration or direction of so much of their thinking.

Nor, yet again, shall we find in the *Physics* what we might more reasonably expect, namely a sheaf of those inspired guesses that students of the 'Pre-Socratics' delight in culling and that cause such wonder and delight in the mind of the general reader—a heliocentric cosmos, for instance, or an elaborated atomic theory of matter, or the belief that life on the dry land was developed out of life in the sea.

Indeed, in the most important of these matters Aristotle was on what is generally called the 'wrong side,' though it does not follow that he was not on the 'right path.' For he generally had good reasons for his wrong opinions.[a]

Aristotle himself did not undervalue the 'sagacity' that foresees a conclusion that it has not reached, but he also very well knew the difference between shooting at a mark though you miss it, and shooting at a venture though you make a valuable hit; and

[a] Even for his (admittedly erroneous) belief that the heart was the seat of mental function, which Plato (rightly) assigned to the brain.

it has been said of the ‘ Ancients ’ at large that ‘ they said everything, but proved nothing.’

What, then, are we to expect from the *Physics* ? Something that is still of philosophical interest ; very much that is of historic interest and that has entered deeply into the texture of our language ; much of purely intellectual interest and bracing gymnastic ; but also much that is of vital significance in relation to that borderland between physical and metaphysical thought where mathematics and philosophy meet.

The present translation has been inspired by a conviction that contemporary philosophic thought and discussion is seriously handicapped by the fact that Aristotle's *Physics* is practically unread, and without such help as is here attempted must remain unreadable except to a few special students.

It must be in the progress of the work itself, if at all, that this conviction must be justified ; but this section of the Introduction may be concluded by a bare indication of the way in which the elementary philosophy of mathematics enters into the tissue of the *Physics*, and of the points on which Aristotle insists.

Many of the Catholic theologians insist on the services rendered by heretics in compelling the Church to bring out points that gave precision and support to the faith but could hardly have been understood except when the fatal effects of overlooking them had been developed by the heretics. So Aristotle found that the metaphysicians who denied that motion or change could possibly exist—the Paradox of Achilles and the Tortoise, by which Zeno tried to prove that the quicker could never overtake the slower, being the stock instance—

compelled him to go outside his domain of Nature in order to rid his readers of a bewildered obsession by the thought that, after all, the study of moving and changing things must be all in vain, since great thinkers had proved that there was no such thing as change or motion.

This led him to the following conclusions, the bearing of which on thought it will be our task to develop.

(1) The terms 'finite' and 'infinite' ('limited' and 'illimitable') belong to quantity and are only incidentally concerned with the other categories. They are attributive, *i.e.* adjectival or adverbial, not substantial. (2) Time is the non-spatial dimension of 'motion' and 'station.' (3) 'Motion' is not made up of 'stations,' nor a lapse of time of 'nows,' nor a line of points. There is no 'next possible' station to a given station, nor a next 'now' to a given 'now,' no 'next point' to a given point. (4) What is complete and whole presupposes limitation and exists within the limits which define it. (5) Any definition whether factual or conceptual automatically limits what it defines. (6) What is not limitable is not complete or whole and therefore is not definable. Thus the indefinable exists in the definable, and not the other way round : the illimitable is inherent in the limited, not the limited in the illimitable. (7) There is no limit to the factual process of subdividing a limited magnitude into a limited number of parts, therefore it can never be completed ; but it may be arrested whenever we choose, so as to embrace only a definite (limited) number of divisions. (8) Arresting a process does not alter its nature, and the nature of an illimitable process may be apprehended by analysing a limited stretch of it, within our own experience. (9) Every potentiality is *capable* of actualization (otherwise it would have no meaning) ; if it is inexhaustible so also is the corresponding actuality. There is no distinction between potential and actual in things eternal.

II. Aristotle's Philosophy

1. *Mind and Matter, and their meeting-ground*

Aristotle has been claimed by many as the philosopher of common sense, for his fundamental assumptions are frequently the very ones that underlie such thinking as the plain man cares to indulge in and excuse him from thinking metaphysically at all.

For instance, Aristotle never questions the reality and objectivity of the material world revealed to us by the senses. Our senses seem to assure us that this world has an existence of its own, independently of our consciousness of it, and Aristotle accepts this assurance at its face value. The material world is to him neither an illusion nor a mere creation or manifestation of mind.

Further, he regards it as evident, upon reflection, that our mind, consciousness, or awareness of things (whatever name or names we give it) is not itself of a material or tangible nature. Body, colour, or warmth can be touched, seen, or felt. Thought, or that in us which thinks, can be neither touched, seen, nor felt.

Matter, then, is not mind, nor is mind matter. But the organs of sense, through which news of the material world reaches our consciousness, are themselves material. So, though mind and matter cannot be resolved into each other either way, there are nevertheless close and intimate reactions between them. Experience makes us familiar with these reactions themselves ; but (unless we are content with facile and unconvincing 'explanations') it leaves them unexplained and permanently mysterious, though it reveals some kind of kinship and correspondence between the worlds of matter and of mind. The fact, at least, remains that material things announce themselves to our consciousness through our senses, and consciousness can act upon matter, immediately by causing the limbs of the conscious organism to move, or mediately by directing those limbs to move other things.

So far Aristotle is quite content to accompany the plain man and to adopt his assumptions in their entirety. But he is very far from being content to leave these assumptions in the vague and undeveloped state in which the plain man holds them. For as soon as we begin to reflect upon them and try to connect and systematize them, they seem to reveal confusions and contradictions ; and the plain man who has not been accustomed to hard thinking can easily be misled by thinkers who have themselves gone astray or by sophists who take pleasure in bewildering him.

So Aristotle attempts to give precision to the plain man's conceptions, to release him from confusions into which the ambiguities and imperfections of language may have betrayed him, to teach him to hold the thread of a continuous argument, to grapple with familiar but baffling concepts such as 'time,' 'cause,' 'infinity,' until he sees them steadily, and to detect the underlying agreements that are sometimes concealed under the mutual hostilities of rival schools of thought.

The *Physics* is full of examples and illustrations of all this ; and we shall find that Aristotle, when revealing the plain man to himself and justifying him—sometimes against the philosophers—in his fundamental convictions, always keeps him, the plain man himself, in control of the ultimate court of appeal, as judge of the final conclusions to be accepted. He must by no means be outraged, and

Aristotle is always anxious to show that his conclusions are just what the plain man has always been feeling after and can now rest in and adopt as his own.

And yet, after all, it is one of the main purposes of the *Physics* to lead the plain man beyond his own range into a region bordering upon that of the mystic, by showing him that natural science points beyond itself to a ‘ divine ’ world, for which the very savage already has a dim and instinctive sense, though the sage in his highest moments can but touch upon its borders and can never make it his own. In his sense of a divine mode of being that transcends human experience (though it is by contact with it that man becomes most human) Aristotle betrays the overmastering influence of his great teacher, Plato ; and the question arises how far he succeeds in showing that his own characteristic line of thought—sharply contrasted as it is in so many ways with Plato's—really leads him, on its own principles, to retain and confirm as much of the Platonic outlook upon the world of conclusive truth and reality as he tries to justify by it. On this question the careful student of the *Physics* and of portions of the *De caelo* and the *Metaphysics* will form his own conclusions, and we shall touch upon it again in dealing with Aristotle's Cosmography and Theology in this Introduction. But meanwhile we must return to the fundamental problems of the nature of mind and matter and of the connexion between them.

Aristotle accepts the teaching of Empedocles that earth, water, air, and fire are the four elementary substances with which the physicist has to deal ;

but he differs from him in two respects, for he adds to these four a fifth element of an essentially different character (of which more hereafter), and he holds that the four generally acknowledged elements are capable of changing into each other, which implies that there must be a common something underlying them all and preserving its own individuality while assuming, for instance, now the watery and now the aerial forms and qualities, though never manifesting itself nakedly.

This common underlying something is what Aristotle means by ' matter,' the ultimate ' material ' out of which all tangible substances are built. He never gives us an express account of the nature of matter, and indeed tells us that we can only get at it at all by inference and analogy, not by direct access through the senses ; but nevertheless he incidentally reveals to us that in his own mind, as in that of Locke, matter must have certain primary attributes of its own which it retains under all its metamorphoses, in addition to the attributes which distinguish the elements from each other. Can we say what they are ?

To begin with, Aristotle constantly falls back upon the axiom that ' two bodies cannot occupy the same place.' He never defends or examines this belief, or seems to think that anyone can question it. This, then, gives us at once Locke's ' extension ' (dimensionality) and ' impenetrability ' as primary attributes of matter.

We shall be able, as we go along, to add more to our conception of matter, but we have enough already to show us that to Aristotle it is matter that constitutes the ' thingness ' of things. ' Things ' '*have*' such qualities as weight, colour, shape, density ; but they '*are*' something more and other than aggregates of distinguishable and distinguishing qualities—and that in virtue of their objective 'materiality.'

Now all this is profoundly unsatisfactory to the metaphysician, who will allow no privileged position to the supposed 'primary attributes' of matter ; but the plain man (in spite of the daily shocks he now receives from Science herself) still stubbornly continues to hold that things 'have' attributes but that the attributes do not constitute the 'thingness' of the things that 'have' them, since attributes must be the attributes of something and cannot themselves be the basis that they presuppose. And this is precisely Aristotle's position.

Passing now from 'matter' itself to the four elements, which are the simplest forms under which it manifests itself, we find Aristotle thinking of them as characterized and differentiated by two 'couples' giving four combinations : Earth is dry and cold, Water cold and moist, Air moist and warm, Fire warm and dry. Add to these expansion produced by heat and contraction produced by cold and we have a rough conception of Aristotle's analysis of the objective transformations that the elements themselves (and their mixtures and combinations) undergo, independently of the states of consciousness they produce in us. But note that we are now adding another attribute of matter : it can be contracted or expanded. This, unlike 'dimensionality' and 'impenetrability,' is not tacitly assumed but deliberately claimed, and defended against a rival opinion, by Aristotle, to whose system, as we shall see, it is integral. Note also that it is not one of Locke's 'primary attributes.'

The careful reader of Aristotle, however, will soon notice an uncertainty, and at least an incipient uneasiness, in his treatment of fire. He sometimes seems to be on the very verge of discovering that it is not a substance or element at all, but the extreme manifestation of the quality or state of 'heat.' And he certainly does not regard fire as the sole cause of heat. Of this more under the heading of Cosmography. But meanwhile we can trace something analogous to this uncertainty or absence of 'definition' (in the photographic sense) in the conception of the other elements also.

If we were to say that Aristotle and his contemporaries meant by 'air' anything that is gaseous, by 'water' anything that is liquid, and by 'earth' anything that is solid, much of what startles us now would become quite natural. We should see why Aristotle is so confident that air and water can pass into each other, for instance, and why he regards metals as prevailingly 'aqueous' (for they can be almost completely liquefied), but partially 'earthy,' perhaps because, as well as being solid at ordinary temperatures, they always seem to leave a slag. But to substitute 'gas,' 'liquid,' and 'solid,' for 'air,' 'water,' and 'earth,' would be to make the ancients too definitely modern. We can only say that they regard the 'elements' as types or norms, to the 'nature' of which (pure or mingled) all other material substances approximate.

We have next to observe that there are members of our own bodies by which we feel things to be hot or cold to the touch, and these members themselves become hotter or colder to the touch in the process of feeling the things they touch to be so ; but nevertheless, inasmuch as inorganic nature is not conceived by Aristotle or by the plain man as having any consciousness at all, the hot water, for instance, does not itself ' feel hot.' The physical condition of water or iron that we call ' being hot ' is, therefore, a different thing from the state of consciousness which we call ' feeling hot.'

As we pass from the sense of touch through those of taste, smell, hearing, and sight, the distinction between the physical cause and the mental effect becomes more and more obvious, even to the thoughtless, though it can hardly become more real to the thoughtful ; and Aristotle keeps vigilant guard against vague and fanciful analogies—such as, that ' smooth ' things will naturally taste sweet, and ' sharp-cornered ' things bitter—which profess to give ' explanations ' of the correspondences between states of sense-consciousness and the physical properties of the things that provoke them. We shall meet with this problem again when we come to speak of the emotions and the intellect ; but, so far as the senses are concerned, we may note at once that Aristotle accepts, registers, and examines these correspondences between states of matter and states of consciousness rather than tries to ' explain ' them.

But this is not to say that he has nothing enlightening to say about them. On the contrary, an examination of his conceptions on this subject will lead us straight to the heart of one of the most fundamental and luminous of his philosophical principles. He is perpetually referring to the distinction between potentiality and actuality, and he often assures us that where one thing acts upon another the potentialities of the ' agent ' (which is the cause) and of the patient (that experiences the effect) are quite distinct as potentialities, but coincide when actualized. And, moreover, this coincidence is realized in the ' patient,' not in the ' agent.'

Now, the most telling illustration of this is to be found in the magnet and the iron filing. Aristotle himself does not give it (though he was acquainted with the properties of the magnet), but his medieval disciples never did him and his readers a better service than when they hit upon it. The magnet has the power of drawing an iron filing (call it the ' needle ') to itself, and the needle has the property (which a chip of wood, for instance, has not) of experiencing and answering to the pull of the magnet. As potentialities, these two properties are distinct and inert. But when magnet and needle are brought into suitable relations of distance with each other they both actualize their potentialities simultaneously in the single event of the motion of the needle, the effect caused by the magnet-agent being realized in the needle-patient.

Now take the case of vision. The consciousness (in virtue of its association with an organized body possessing eyes) has, even in the dark, the potentiality of seeing. A neighbouring apple, in virtue of the form and colour of its surface, has the potentiality of being seen, *i.e.* of producing an effect upon the consciousness. These two potentialities are distinct; but let there be light, and the two potentialities coalesce as a single actuality in the experience of ' seeing,' which is realized in the consciousness.

As regards inorganic nature little more need here

be said. Aristotle understood that combinations of elements had properties that were not mere mixtures of the qualities of the elements themselves; but his presuppositions as to matter and the material things and states that provoke our sense-experiences, generally assume a greater simplicity and continuity in them than can be detected in the impressions they produce on our senses. In the same way we moderns think of the different colour-effects as caused by different wave-lengths which could be measured off on a linear scale as a simple case of 'more' or 'less' of the same thing, though no one could suppose that the sensation of 'red' is simply a 'more' or 'less' of the sensation of 'yellow.'

2. *The Scale of Life and Consciousness*

If we 'take stock' at this point, we have as items of Aristotle's creed, expressed or implied: (1) Consciousness is not material. (2) In a 'sensation' of any kind a potentiality of a given material object or attribute and a potentiality of a given consciousness meet and are actualized in a psychic experience. (3) We are to accept as authentic our impression that matter exists in different and changing states independently of our consciousness of it, and would so exist if no man were conscious of it. The relation of these three articles of faith with each other awaits further examination.

It will be natural to inquire first whether Aristotle believes that consciousness too, on its side, has an independent existence apart from matter and would exist if there were not a material universe at all.

As to this, we note at once that it is the dominating

purpose of a great part at least of the *Physics* to prove that there are immaterial beings, and there is one supreme immaterial being subject in no way to time, place, or change, and that all these beings live a 'divine' life transcending the highest experience of fruition known to man.

But this statement must be accompanied by two warnings. For (1) it must not be taken as implying any doctrine of creation. Aristotle never represents the supreme being as the creator of the material universe. On the contrary, he believes in the eternity of matter-in-motion.

And (2) he never assumes that we have a direct and not-to-be-challenged assurance of immaterial existence, corresponding to that which he supposes us to have of the objective existence of material things. On the contrary, he holds that our belief in the immaterial and divine being of 'theology' must rest on inference, since we have no direct cognisance of mentality or of any form of conscious experience or activity except in close (however mysterious) association with matter,—and that matter 'organized' and 'unified' in some determined and 'vital' organism.

These terms must be more closely examined. 'Organs' are differentiated parts of a single whole (the 'organism') so related mutually that their several functionings are subservient to the functionings of the organism as a whole. These functionings of an organism as a whole are something more than the mere sum of the several functionings of the organs. Health consists in the perfect balance of the activities of the several organs, considered as subservient to the 'life' of the organism itself.

But what is 'life'? It is a kind of functioning only found in organisms and always found in them. But it varies in its scope. At its lowest it is found in plants, which 'grow' (in virtue of a principle intrinsic to themselves) by assimilation and dissimilation. No individual plant lives if these processes cease; and in no case do they go on indefinitely. All individuals, therefore, sooner or later die. But if some or all of these individuals reproduce their own kind, then that 'kind' may be propagated in an unlimited succession of growing and decaying individuals. 'Nutrition' and 'reproduction,' then, constitute life in its lowest terms.

Now this 'life' strikes Aristotle as so marvellous that, though he has been content to consider that changes in the proportions and modes of combining of the four elements may be a sufficient objective basis for the varying impressions which material objects make upon the senses of sight, hearing, and so forth, yet when it comes to matter itself exhibiting 'vital' phenomena, he seems to feel the necessity of enlarging this material basis. In his extant works there is only a single passage [a] which gives us direct light on this matter, and it is not where we should naturally have expected to find it. But I am not aware that its authenticity has ever been challenged, and I am far from being disposed to challenge it myself. In the passage in question Aristotle declares that into the composition of all organisms (that is to say all living things) there enters, in addition to the four terrestrial elements, a fifth and 'celestial' element, which element is no other than the substance of which the revolving firmament

[a] [*De gen. anim.* ii. 3, 736 b 29.—C.]

and the heavenly bodies are composed.[a] This fifth element shares in the dimensionality and impenetrability of the other four, but not in their contractility. Moreover, its potentiality of motion differs from theirs.

But plants, according to Aristotle, do not feel ; so we have not yet reached the proper meeting-ground of matter and consciousness. Animals, however, in addition to sharing the nutritive and (in all cases that we need consider) the reproductive life of plants, add to these powers susceptibility to sensations and (in most cases) powers of locomotion. Here, then, we encounter consciousness in association with matter ; for a sensation, we have seen, is a psychic experience, not a material event.

We must resist the temptation to linger over the many fascinating subjects discussed by Aristotle in relation to the special senses and their organs. He knows, for instance, that 'touch' is common to all animals, and that some have no other sense. Also that under the denomination of 'touch' we really include more than one distinct sense. He is interested in the growing refinement and, as it were, tenuity of the physical impacts on the sense organs as we pass from touch up through taste, smell, hearing, and sight. And he notes the significance of the material composition of the organs of sense, in relation to their functions. Further, he attempts to supplement and define our imperfect and confused vocabulary, which uses 'sight,' for instance, to mean indifferently what we see, our power of seeing it, and the organs through which we exercise this power. But all this and much else we must pass over as not

[a] [*De caelo*, i. 2-3.—C.]

strictly necessary to our general line of advance. We will, therefore, go on at once to a special doctrine of Aristotle's that is of central importance and also in great need of clear exposition.

How do we distinguish between the sensations we receive through different organs, and how do we combine them? If we know that honey is yellow and sweet and that gall is yellow and bitter, this information is not given us by the taste, which takes no cognisance of colour, nor by sight, which takes no cognisance of savour. Yet these distinctions and combinations pertain to sense. There must, then, be a general or 'common' sense which receives, adopts as its own, and assimilates the reports of all the special senses, and combines these reports, as well as distinguishing between them.

It is of the greatest importance to note that the special senses already discriminate, each on its own ground. This discrimination rests upon the contrasts of change or 'passing' from this to that. If there were no change, we should note and distinguish nothing. This 'passing' or change, then, of which change of place or position is the simplest form, is noted by all the senses. Colours, tastes, etc., all change, and the several senses discriminate between that from which and that to which they pass. These changes may be in quality (from red to yellow, or sweet to bitter) or they may be in intensity. In this latter case, therefore, each sense discriminates between the more-or-less-ness of its special percepts. This may be conceived as their great-or-small-ness, magnitude, or quantity. Again, the sensation, even if of uniform quality and intensity, may be intermittent or continuous, by whatever sense discerned.

Thus the strokes of a clock are discrete and intermittent, but each one of them is continuous, though perhaps varying, as long as it dwells upon the ear. And so too each other sense discriminates on its own ground between unity and plurality and between the discrete and the continuous. And lastly each sense can distinguish between different orders of succession, or arrangement and position, as the case may be.

Thus the 'common sensibles'—the things of which each of the senses makes us aware on its own ground—are typified by movement, magnitude, figure (*i.e.* position and arrangement of parts), plurality and unity.

The 'common sense' has no *special* relation to the 'common sensibles' (though many expositors, betrayed by the terms, have supposed that it has), for its office is identical with respect to the 'common' and to the special or 'proper' sensibles (colour, sound, etc.), namely to lay them side by side, and to associate and combine them while distinguishing between them, but without *adding* to them. The common sensibles are common to all the senses ; and the 'common' sense, therefore, has access to them through all the senses, but through the senses only ; just as it has access to the special 'sensible,' (say) colour, by vision, but by vision only.

Thus, in the case of special sensibles, sight may say 'there is yellow,' touch 'there is stickiness,' and taste 'there is sweetness.' But these senses cannot make these reports *to each other*, and so far as they are concerned there is no relation or combination between them possible. But when they all report to the 'common' sense, that sense can at once distinguish between them and pronounce that they co-exist in associated combination. It functions in exactly the same way in relation to the 'common sensibles.' Sight may report either changes of intensity (*i.e.* more-or-less-ness) or continuities and discontinuities (*i.e.* one-or-more-than-one-ness) in its impressions. And the like with taste and touch. But only the common sense can report coincident more-or-less-nesses or coincident one-or-more-than-one-nesses, which are just as unlike each other as colour and sound. The common sense, then, gives us direct perception of a new kind of oneness, the oneness of an *associated combination* of different orders of sensation. On this Aristotle lays express and emphatic stress.

Add to this that the common sense can tell us of a combination in which *some* of the factors hold together in continuous association, while others change, and we shall see something, which Aristotle does not (I think) expressly tell us, but which we can see very clearly for ourselves, namely that it is the common sense (not the special senses) that objectifies the world for us; for it incidentally asserts the existence of a number of concrete entities outside ourselves, which would be there though we were not. When sight says 'there is yellow,' in the last resort it means only 'there is a certain sensation.' But the common sense, which can take cognisance of other sense impressions, says 'there is *something* that *has* colour, but, without the reports of the other special senses, I don't know what it tastes or feels like.' Metaphysicians may reasonably quarrel with Aristotle for taking what either the common sense or the special senses tell us as a direct perception of ultimate truth, but I think it is quite clear that it is to his 'common' or synthetic sense that he would have to trace the direct and inevitable impression of objectivity and of concrete objects.

This will explain his frequent assertion that the proper senses never err. This is only intelligible if we understand their testimony to be 'there is this sensation,' and no more than that. Whereas the common sense can err, because the report of common sense is really an assertion, though it does not appear to be so. For when the common sense receives the visual impression of whiteness and the combined visual and tactual impression of fine granulation, it spontaneously objectifies a 'something' that 'has' these qualities, and it may hastily announce that it will taste sweet and call it sugar, whereas, if it had waited for the special report of taste, it would have called it salt.

We are now prepared to understand how, on the basis of the synthetic sense (as we shall henceforth call it, to avoid ambiguity), we may trace not only changes of sensation but changes and developments of things ; for if we can trace any sufficient continuity in the group of associated sensible qualities of a defined object, we think of 'it' as retaining its identity while changing, at once or successively, one or more of its qualities. An acorn that we have seen develop into root, stem, and leaves, while suspended in a glass vessel, and have then planted out and watched till it becomes a robust sapling, retains its identity for us throughout. Only when we cease to be able to trace *any* continuous 'association of attributes' at all do we say that a thing has altogether 'ceased to be.' Though even so we may believe, as Aristotle did, that the ultimate 'matter' of the thing really has a continuous existence still, so that in strict truth nothing ever begins to exist or ceases to exist in its entirety.

But we are anticipating ; and we must go back, to ask whether the synthetic sense has an organ of its own. The modern reader will at once suggest the brain, and that was Plato's answer. Aristotle too says in one passage that 'some people' think it is the brain. In another passage he treats it as indifferent to the special point he is discussing whether it is the brain or the heart. But in yet a third passage he himself pronounces definitely for the heart. Naturally this has been regarded as showing great perversity or want of intelligence on his part by those 'moderns' who judge of the intelligence of an 'ancient' by the measure in which his conclusions coincide with those of modern science or diverge from them, rather than by the way in which he handled the materials that were at his command. From this latter point of view it is interesting to note that the late Sir D'Arcy Thompson tells us that, in the then state of anatomical and embryological science, Aristotle could give a very good account of the grounds of his belief on this matter.

Be this as it may, the further examination of the relations of the special and the synthetic senses to each other will throw much light on Aristotle's general philosophy ; but before we try to develop it we must take a rapid survey of the psychic superstructure which, according to Aristotle, is built up in animals (to different heights in different cases) on the basis of the life of sense.

The modern reader will find it best to regard these

developments as taking place in direct connexion with the organ of the synthetic sense (to us the brain), considered as that to which all the others make their reports—remembering, however, that it is entirely dependent upon these special senses for the materials which it combines.

Sensations, then, are provoked by external objects. But as a fact some kind of impression is apt to remain upon the consciousness when these objects are withdrawn ; and such impressions, even when no longer present to the consciousness, can sometimes be recalled. This is to say (though not to explain) that the life of some animals includes memory and recollection and, further, a power of forming 'images' of absent objects and of the sensations they might provoke and have provoked when present. 'Image-forming' (whence our 'imagination') is, however, a misleading term, in so far as it has felt the 'tyranny' (as it has been called) of the visual sense, which in virtue of its vividness asserts a kind of primacy in determining our language. It must be understood, then, that to recall or 'conjure up' sounds, or tastes, or other sense-impressions, is as much to be included in our conception of image-forming as if these impressions were visual.

But further, sensations are pleasant or unpleasant. Also our image-forming may be anticipatory as well as retrospective. Hence desires, loves, fears, and so forth.

Moreover we can move, and by moving can often avoid or secure painful or pleasant experiences.

We are now already in the region of desires, purposes, and even passions ; and it will repay us to re-examine, with reference to the richer psychic life we are now dealing with, the relations (mysterious as they must always remain) into which the physical and psychic factors of existence enter with each other in a vital organism.

In the case of mere sensations we recognized the object perceived as the cause and the sensation as the effect, while guarding against the blunder of supposing that the vibrating gong, or the vibrating air outside or inside the ear, *is* the sensation.

When we come to the passions and affections, we can, at any rate in some cases, trace the associated physical and psychic events and describe them each in terms appropriate to its nature ; but the difference between them becomes easier to see, and their causal relations far more complex, than in the case of the senses.

Aristotle tells us [a] that anger might be defined by the biologist as ' the seething of the blood, or heat, in the region of the heart,' and by the psychologist as ' a desire to hit back ' ; but he leaves us in no doubt that, in his opinion, the psychologist really is trying to tell us what anger *is*, and the physiologist only to describe physical events that are concomitant with it. But can we now say that the physical event is the cause of the psychic experience ? Might we not equally well say that my being angry is what makes the blood seethe ? Neither statement is all the truth. I may be angry because a man has used insulting language to me, and that feeling may literally stir my blood. Or my blood may be stirred by some physical process, and that may make me (if not actually angry) so irritable that the most trivial provocation makes me dangerous. Here we have

[a] [*De anim.* A 1, 403 a 29.—C.]

two causal chains of succession, one physical and one psychic, each of which may be traced along its own line irrespective of the other, but which are constantly reacting both ways (not only one way, as with the senses) upon each other.[a] Fear (also cited by Aristotle) is, perhaps, a still better example than anger of psychical causes producing physical effects that seem to have no physically causal antecedents, or physical causes producing psychical effects that seem to have no psychically causal antecedents.

There is another aspect of the reactions we are considering, which has only been touched on as yet incidentally. We men, and all the higher animals, can move our limbs. Thereby we can change the positions of things amongst themselves and so modify their reactions upon each other, and we can also change our own position with reference to them and so modify the reactions between ourselves and them. All we can say about this mysterious power is that it does not appear to be of the order of physical impacts by which a body already in motion sets or keeps another body in motion, but is rather a case of a psychic fact, of the order of desires or the like, which produces a physical effect subservient in its turn to the gratification of some psychic impulse or desire. All the higher animals, then, can adapt their actions to the satisfaction of their impulses and desires.

But there is a region which we shall, perhaps, still be safe in claiming, as Aristotle claimed it, as

[a] [Using 'cause' in Aristotle's sense, which covers the internal constituents of a thing as well as its antecedents, anger can be called the *formal*, seething of the blood the *material* 'cause' (or aspect) of a single event.—C.]

accessible to man alone of earthly beings. Does anyone claim for dog, elephant, or ape a love of truth in the abstract? Aristotle (somewhat rashly) claimed it for every human being. 'All men by nature desire to know,' he says, and by 'knowing' he meant not 'being familiar with' or 'knowing how to deal with,' but 'understanding,' to the limit of human capacity, the nature and truth of things, for the sake of understanding it, apart from any material utilities it may bring with it. This is what I mean by love of 'truth in the abstract'; and the power of seeking it Aristotle regarded as an aspect, or factor, or activity of human 'aliveness' not shared by any other terrestrial animal. We shall not be far from Aristotle's own terminology if we call it 'mind' or 'mentality.'[a]

The stress laid on it by Aristotle is characteristically Greek, and he is so deeply impressed by its significance that as he demanded a fifth and celestial element when he came to life, so here he demands, not another element but a contact of man, and man alone, with a divine mind or mentality that already enjoys in actuality all, and more than all, the potentialities which constitute the human mind. On this subject we have but a few broken and mysterious phrases, which Aristotle's immediate disciples could but imperfectly interpret; they will be best treated under the heading of Cosmography. But as to the nature, as distinct from the origin, of this mental faculty in man, we are perfectly well informed. It lies in the word 'abstract,' already used, and we must approach it historically.

The early Greek thinkers had felt that truth must

[a] See below, p. lxxviii.

in its nature be always the same and always true, whereas things and people, and all that our senses and sense-experiences reveal to us, are perpetually changing ; and since our minds can get at nothing outside themselves except through the senses, how can we extract the abiding and changeless from the fleeting and variable, or build the permanent truth upon passing shows ? One answer was : We cannot. The very permanence of truth warns us against looking for it in passing shows. If it is anywhere, it is in the mind. Whatever is or appears to be changing (that is to say, all the world of sense) is untrue and illusory. Another was : We cannot, because there is nothing permanent, and the mind that vaunts its superiority over the senses is absolutely dependent upon them and, if they are false, is doubly false itself.

The great philosophers of the fifth and fourth centuries B.C. inherited this problem and found both answers unsatisfactory. According to Aristotle, it was Socrates who first put us on the right track by calling our attention to *classes* or *kinds* of actions and qualities which we group together because of their resemblances in spite of their manifest differences. Their differences are observed, their resemblances are felt, and in virtue of those resemblances we give them collective names. Wherein exactly does that resemblance consist ?

If answers could be found to the questions thus raised, we should have in the goodness of things good, in the beauty of things beautiful, in the convincingness of convincing arguments, in the materiality of material things, in the straightness of straight things, in the equality of equal things, in the virtue of virtuous acts and the rightness of right dispositions, possibly also in the tableness of tables and the vitality of things living, something that does not change and that can give us the feeling that we are not only acquainted with beautiful things and so on, but understand why they are beautiful, or whatever it may be. This would be true knowledge of permanent and abiding truth reached by the mind, even if the material out of which the mind extracted it were all of it fleeting.

It was in the field of the moral, social, and intellectual life that Socrates sought for the abiding and essential truth, but it was under the stimulus of his search that the general conception of ' classes ' or ' kinds ' in which there was a group resemblance amid individual differences first distinctly emerged. It has ever since remained at the basis of science and philosophy alike, for it is the principle of generalization and classification.

The many-sided genius of Plato inherited this problem from Socrates, and according to some recent writers he also inherited from him the attempted solution known as the ' philosophy of Ideas.' But in this exposition I follow, with great confidence, the express testimony of Aristotle, who assigns the philosophic system to Plato himself.

The word ' Idea,' in this connexion, is open to legitimate challenge, and easily lends itself to misconception. In any case it seems clear that the ' Ideas ' sank into the background of Plato's mind in the later and perhaps most characteristic phase of his thought. Moreover the dialogue, which is Plato's favourite literary form, makes it difficult definitely to commit him to the utterances of any interlocutor, however strongly we may suspect such utterances to be the unqualified expression of his own belief; and the iridescent beauty of his suggestive and elusive handling of the ineffable makes it the extreme of rashness to translate his vision into dogma. But this is exactly what Aristotle is accused (and I think justly accused) of doing; and as our main business is with Aristotle, and Aristotle's polemic against Plato is generally directed specifically against the doctrine of Ideas, I must try to give such an account of that doctrine as will enable us to understand Aristotle's dissent from it. Such an account must not pretend to adequacy or completeness, but it must aim at complete fidelity as far as it goes.

I think it is safe to say that, according to the doctrine of Ideas, there is an order of reality higher than that of the phenomenal world of things and concrete experiences. In this order truth is nothing but Truth's self, beauty nothing but Beauty's self, and so forth, whereas the beautiful things we know are beautiful things, or poems, or thoughts, that is to say they are poems, or whatever it may be, that participate in real and absolute Beauty and are beautiful in virtue of such participation, but are not Beauty's self.

But how do we know that they participate in Beauty if we never saw or knew Beauty herself? At least two suggestions of an answer are made by Plato in different dialogues.

Had we, before we were born into this material order of things, actually experienced a life of converse with absolute Truth, Beauty and the rest, and did birth bring with it forgetfulness, but yet leave us a reminiscent instinct by which we can recognize its traces, or its more or less distorted reflections, in the phenomenal world, as recalling some far-away we-know-not-what that claims our delighted allegiance ? This would make a mathematical demonstration, for instance, less an acquiring of fresh knowledge than a recalling to distinctness of knowledge that had been obscured.

Or should we rather say that the things we call beautiful are really beautiful only because they partake of Beauty's self, and they really attract us in virtue of this participation ; but in our material and immature state it is only when embodied in comparatively gross things that Beauty can attract us at all ? And is there a discipline, intellectual and moral, that can teach us to apprehend beauty in ever more refined and noble forms till at last Beauty herself reveals herself to us unassociated with anything meaner than herself and we are rid at last of illusions and touch upon the eternal and changeless reality ?

Now I shall try to show presently that Aristotle had been profoundly and permanently influenced by Plato's vision of life in the realm of Ideas. But his own specific contribution to philosophy consisted, in his own mind, in a demonstration that the Ideas have no such absolute existence as Plato claims for them, that the hypothesis of their existence is quite gratuitous and explains nothing, and that what they were vainly designed to find is within our grasp without them.

What then is his own teaching on the subject of access to the permanent truth in a world of constant changes ? It is very simple, and the plain man will probably adopt it as his own as soon as he understands it and will wonder why anything so obvious ever needed to be ' found out ' at all.

Let us go back to the common or synthetic sense and the special senses. In a remarkable passage, after he has introduced us to the synthetic sense, Aristotle emphasizes the value of the special senses in spontaneously, and in a way objectively, separating out and disengaging from their concomitants certain sense-impressions. Colour is seen, not felt, smelt, tasted, or touched. Colour does not and cannot exist except as the colour of something, but we are aware of it in itself as distinct from all the other sensations that the ' something ' may or does provoke in us. Now we have noticed already that, though Aristotle assumes (without proving) the objectivity of things, yet, in assigning to the common sense the direct perception of a concrete ' oneness ' that he does not find revealed by any of the special senses, he comes very near to declaring objectivity to be a direct datum of the common sense. It would seem, then, that, though we are tempted to call the common sense ' synthetic ' because we have been introduced to it as the putting together of the data of the special senses, we shall probably be nearer Aristotle's fundamental thought if we regard the common sense as giving us direct cognisance of concrete objects, and the special senses as barring out all the orders of impressions which the object is capable of provoking except one, simply because it is incapable of being provoked by them. Thus the special senses are spontaneously and inevitably analytical. Now Aristotle finds in the human ' mind,' which has faculties not granted to the highest of the brutes, a power to isolate in thought what is never isolated in fact, analogous to the necessary action of a special sense in isolating to the experience of the consciousness what is never isolated in the challenge of the object. This is the power of 'abstraction' by which we can think apart of that which never exists apart. The eye enables—nay, compels us—to 'abstract' or draw away from all coloured things a vast range and variety of impressions which all have, amid their differences, a something in common, distinguishing them as a group from all other impressions. The word 'colour' is the record of this sense-abstraction. All names, or 'nouns,' are in like manner records of spontaneous 'abstractions,' from concrete individual examples, of something that exists in all the individuals to which the name applies, but never by itself in separation from them all.

But 'colour' is the name of a group of sensations, and its isolation from other groups is done for us by our senses. In the mental world we must learn to isolate in thought likenesses or identities of which there is no sense-register, and which are never found in fact or experience isolated from unlikenesses and diversities.

If we can do this we shall have found the realm of abiding realities and shall 'understand,' so far as man can understand, what it is, in virtue of which beautiful things are beautiful, living things alive, convincing arguments convincing, and all the rest. We shall then command a realm of abiding truth, but we shall find it not outside and above the change and flux of things and experiences but at their very heart, separable from them in thought but never in reality. Such ' abstractions ' exist *apart and sejunct* from ' concrete ' things and beings only conceptually ; but their existence *in* existing things and experiences is as real as those things and experiences themselves, and it is only by isolating or abstracting the conceptual that we can have any grasp of the permanent and unchanging amidst the flux and change of sense-experiences, which we mean by true ' knowledge ' as distinct from cognitive familiarity.

This is Aristotle's solution of the old problem as to where we are to find the permanent and changeless when we have no direct access to anything but the evanescent and varying. He places it in sharp contrast to the Platonic doctrine of Ideas ; and in his first attempt to arrest and define it before he has gained the complete mastery of it to which he ultimately attained, we catch him in one of those rare moments in which pure intellect kindles on his lips into passion as he feels himself to be in the very act of bringing the unceasing flux under the control of an intelligible order—like one who has arrested the rout of a bewildered and panic-stricken army and is rallying and inspiring it for victory.[a]

Thus Aristotle died, as he had lived, a passionate Platonist, under the abiding illusion that he was building up on his own rational basis of abstraction a complete reinforcement of his beloved master's teaching, and letting himself go straight for the central truth which his master never approached directly, but always in some oblique manner. So his love burned undimmed and steady, though to us

[a] [In *Anal. Post.* ii. 19, 100 a 11, Aristotle uses this image to illustrate the power of the mind to pass from particulars to universals.—C.]

he seems to be always attacking the doctrine of Ideas, which he held to be superfluous irrelevancies, and always fretting against certain misunderstood passages in Plato's *Timaeus*, of which indeed there does not appear to be any real explanation—there are only hints—but Aristotle thought he understood them and rejected them.

With this clue in his hand Aristotle examines language, the treasure-house and record-office of all the spontaneous activities of the faculty of 'abstraction' that has not yet recognized itself. There are 'kinds' of words, as well as words: verb and noun, for instance. What is it exactly in which all verbs are alike each other and unlike nouns? And what is the difference between words, which merely present something to our minds, and sentences, which perhaps assert connexions between things (Grammar)? And what are the 'kinds' of assertion we can make? What is the difference between telling what a thing *is*, and telling what qualities it *has*, or how much of it there is, or in what position it is, or in what relations to other things? Is it possible to make an exhaustive list of possible kinds of assertion, to one and only one of which every assertion must belong (the Categories)? In such inquiries we are trying to isolate in thought some *function* or *characteristic* of a group of words or propositions; the characteristic, namely, which makes them what they are, and enables us to understand why they are a group, which is indeed the very essence and truth of their being, but which never and nowhere exists except in them, and in them is never objectively isolated.

3. *The Categories*

This brings us to the crucial distinction between the *aestheta* and the *noëta* in Greek, the *sensibilia* and the *intelligibilia* in Latin, the *concrete* and the *abstract* or the *particular* and the *general* in English.

But unfortunately the English terms do not adequately emphasize the point that a concrete and ' particular ' stone or tree, for instance, is perceived by the senses as having an objective existence of its own, whereas the ' generals,' that is to say the intrinsic characteristics that constitute stones or trees as distinguishable groups or classes of things, while existing in and to the intelligence in separation from the ' particulars,' have no objective existence except in them. These ' generals,' then, are *abstracted* or ' drawn off ' from the concrete and particularized individuals by the intelligence, and, as contemplated apart from the particulars, exist only in and to the intelligence.

The contrast would be well marked by ' sensible ' and ' intelligible,' if the words were free enough from other associations to be safe depositories of the meanings : ' accessible to cognition by the senses,' and ' accessible to cognition by the intelligence ' ; and perhaps with due precautions we may be able sometimes to speak of the concrete objects of sense-perception simply as ' sensible ' ; but for ' intelligible ' we shall have to substitute the word ' conceptual.' It is harsh and, though true, not very illuminating, to say that the objects of sense are ' unintelligible,' but not harsh to say that they are ' not conceptual.' It is quite misleading to say (as you may in Latin) that we can only reach the intelligible through the unintelligible, but to say that we can only reach the conceptual by distilling, sublimating, or ' abstracting ' it from the non-conceptual and concrete, that is to say from the ' sensible,' is (it may be hoped) illuminating.

Let it be understood, then, that ' the conceptual ' (and sometimes ' a concept ') is to represent the Latin *intelligibile* and the Greek *noëton*. It is no more than a variant of ' abstract,' which stresses the fact that the abstract exists objectively nowhere but in the concrete, and also that it is only (according to Aristotle) from the concrete that the intelligence can draw it off.

We can now go forward with more rapid steps. ' Abstraction ' consists in grouping, and we may group the concrete sensibles with which we start as our primary data just according to our convenience or fancy ; but the ' common nouns ' of our language are the register of a series of spontaneous abstractions and groupings which have established themselves as serviceable. Science attempts to systematize these groupings, to give them precision, to fill in gaps, and so forth ; but wherever there are common nouns, there are ready-made ' generals,' ' abstractions,' or ' concepts ' ; for language is a deposit and a reservoir of thought.

The simplest illustrations may be drawn from the consideration of such a scale as this : ' horses ' and ' men ' are two groups included in the wider group of ' animals ' ; animals and plants are two groups included in the wider group of ' living organism ' ; living organisms and inorganic substances are two groups included in the wider group of ' material entities.' Is there any yet higher group which includes material entities and some other group?

The attempt to answer this question will take us deep into the heart of our subject, but before embarking upon it we may note that it is usual to call any group a *genus* with respect to the groups it includes, and a *species* with reference to the *genus* that includes it. Thus, 'animal' would stand in our illustrative scale for the genus of which 'horse' and 'man' are species, and for a species of the genus 'vital organism.'

Our question then is: Do 'material entities' constitute a supreme genus which cannot be ranked as a species under any genus wider than itself?

If we take 'material entity' to mean anything that the senses, and pre-eminently the common sense, announce to us as existing objectively, independently of our knowledge or perception of it, but distinguishable by us from other things like or unlike itself, then we shall say that such 'entities,' perceived by the common sense, are always concrete. A substance such as air, or a stone, a tree, a man, are concrete substances, things, or beings, revealed (the metaphysical inadequacy of all this range of conceptions has been sufficiently emphasized) to us by the senses as existing both in themselves and independently of our cognitions. But the colour, weight, size, or extension of these things, even if announced by the senses, is not revealed to us as existing 'in itself,' but only as existing *in* the concrete thing that is coloured, heavy, or extended. Yet *in* the things they *do* exist, and (in some cases at least) exist to the senses.

Colour only exists as the colour of something,

smell as the smell of something, shape as the shape of something, weight as the weight of something, size as the size of something ; but the apple which has its own colour, smell, shape, weight and size, is not the apple of something, but the concrete apple itself, perceived as such by the senses. Colour, shape, and size are not ' things ' ; the apple is.

' Existences,' then, are of various kinds, and if we take ' entity ' to mean the concretely existent that exists in itself and not in something else, then ' entities ' are a ' kind ' or class of ' existences '—those namely which are concrete and exist in themselves—and ' qualities ' and ' quantities ' are other kinds of existents, alike in that neither quality nor quantity can exist except in concrete entities.

' Entity,' then, is a genus which may include other than material entities ; for there may be (and Aristotle was profoundly convinced that there are) conscious beings that exist in themselves and have mental activities or attributes, but are wholly immaterial.

Material entities, then, may be a species of entities. But entities themselves constitute a genus indeed (for theirs is one ' kind ' of existence, distinguishable from other kinds), but they form a *supreme* genus, and are not subject as a species to any higher genus or ' kind.' For ' existence ' in the largest sense, the only term that can embrace entities and other kinds of existents, is not a ' kind ' at all, but embraces all that (in any sense of the word) ' is.'

This brings us to Aristotle's celebrated and fundamental doctrine of the ' Categories ' or *summa genera*, each of which is a genus but not a species. We have here encountered three of them, entity, quality, quantity. There are more such categories, 'relation' of one thing with another, for instance, actions and effects of actions, 'place where' and 'time when' (ten in all); but it is entity, quality, and quantity, almost exclusively, that we are concerned with in the *Physics*, and we are concerned with them from the beginning to the end.

Once again, then, sense presents us with concrete stones, apples, trees, animals, each of which is perceived as an individual 'something' distinguishable from other 'things.' Also there are substances or materials such as iron, clay, water, or air, which are distinguishable from each other, but can be divided into parts in a different way from that in which an apple can be divided; for if you divide an apple in half, it becomes two half apples, but not two apples, whereas if you divide a mass of iron in half, it becomes two masses of iron, each of which is just as much iron as the whole mass was, each existing in itself. Again, when you divide an apple into two half apples, you divide one coloured thing into two coloured things, but neither the colour of the one nor the colour of the two exists 'in itself.' The substance iron, then, is an entity, but colour is not.

From these things, organisms and substances, we must start. They 'exist' in the primary sense, and we regard them as existing independently of our thought of them. When we *think* of them, we analyze them into the things they are and the qualities and so forth that they have, but we never *perceive* a thing apart from the qualities that it has, nor qualities apart from the thing that has them. The concrete or *compositum*, therefore, is the immediate datum of the senses. The distinction between the 'matter' that constitutes the objectivity of all material things in common, and the special qualities that distinguish this thing from that, is purely mental, for these inferred 'components' of the *compositum* are never met with apart except in the conceptual world of thought; and since these conceptuals are distilled by the intelligence out of the perceived concretes, it follows that qualities and quantities, for example, cannot be thought of as existing unless the *composita* exist. The primacy of existence, then, belongs to the concrete.

Before leaving the Categories we may note that throughout the *Physics* Aristotle likes to begin his examination of any concept that presents difficulties or obscurities, such as 'time,' 'place,' or 'movement,' for example, by trying to determine to which of the categories it is to be referred, and whether it is a potentiality or an actuality. We are now prepared to understand the relevance of these inquiries.

4. *Abstract thinking and Self-consciousness*

We shall be much concerned in the *Physics* with another set of abstractions—surface, point, distance or dimension, motion, for example—which cannot even be conceived as existing apart from bodies, but which can nevertheless be thought of (though not 'imagined') apart from them and only so 'understood' (Mathematics and Physics).

This last set of concepts we shall naturally encounter again as we come to Aristotle's detailed treatment of them in the *Physics*, but we can already see how in Logic and Mathematics Aristotle conceived himself to have entered a conceptual or noëtic world of abiding and changeless truths.[a] These sciences became to him the standards of precision and certainty in knowledge; and we can discern through all his work a continuous effort to link his conclusions on to truths above them as well as building them up upon facts beneath them. But he is well aware that in this attempt we cannot always be equally successful and that we must be content in each case with such degree of precision and certainty as that case allows. For instance, on the ground explored by Socrates, we note that every man recognizes a type of *character* and *conduct* (as distinct or at least differing from *ability*, for instance) that he admires and would like to see prevalent. We may ask what exactly the general characteristic is that distinguishes this type of human excellence from other kinds of human excellence, or excellence in general, as of an animal or a tool; and what minor groups and distinctions are included in the main group (Ethics). Our conclusions in this field may be enlightening in a high degree, but they will

[a] Unfortunately, we possess no works of Aristotle expressly devoted to mathematical science, but his grasp of fundamental mathematical conceptions is apparent everywhere, and it is clear that, next to logic, he regards mathematical science as presenting us with the highest type of generality and certainly of thought. Moreover, though mathematics is the most abstract of the special sciences, and so the remotest from materiality, it is the most universally and most precisely applicable to material things, for its concepts inhere in them all, and so its inferences, while exceptionally rigorous on the conceptual side, are exceptionally capable of verification on the 'sensible' or practical side.

hardly give us the 'paramount belief' [a] that Aristotle found, and the plain man still finds, in Mathematics.

We will not follow Aristotle further in the vast survey covered by his extant works and indicated by the fragments and reports of those that are lost, but there is one all-important field which we cannot ignore.

In all these mental analyses and abstractions the mind itself is active, and the mind is conscious and can examine its own mentality. Neither the eye nor the visual sense can see or visualize its vision, nor can the hearing ear hear its hearing ; but the thinking mind can think upon its thought. Sensation is conscious, but mind alone is self-conscious. And so when we come to the human mind, of which the power of abstraction is a function, we have come to something more directly accessible to our consciousness than the material world itself, and we have here an announcement of reality more authentic and unchallengeable than that announcement of an objective world existing apart from our consciousness, which Aristotle and the plain man have allowed to pass unchallenged. Into this region too Aristotle carries us. Indeed we have been in it all along. Thought is a function of our vitality, and it was by thought that we distinguished, first, the inorganic from the organic or vital world, next, the unconscious life of plants from the conscious life of animals, and, lastly, the notes on the ascending scale of animal consciousness in sensation and in the traces left by sensation (recalling, image-making, and the rest) from

[a] The *phrase* is Wordsworth's, not Aristotle's. But the poet and the philosopher both express the same *thought* in reference to Mathematics.

each other and from the crowning self-conscious mentality of man (Psychology).

It need hardly be pointed out that in all these inquiries mental 'defining' (which is only to say 'distinguishing') must be perpetually aimed at, and that verbal 'definitions' will be of frequent use. Only by their aid can the concepts of the plain man be developed, clarified, and systematized for him.

It is by thinking, then, that we have arrived at the conception of thought itself as abstracted from all other vital functionings. But more than that, if we look back upon the processes involved in thinking, we find one that has been always present but of which we have had little direct occasion to speak. What is 'inference,' and when is it valid? All thinking involves the belief that one thing 'follows upon' another, or that, if two connected assertions are true, they warrant a third. To show that the conclusion really does follow is the office of the teacher; to ascertain it, of the discoverer; to make it appear to follow when it does not really do so is the trick of the sophist. To discover, therefore, the general and distinctive quality which gives an inference the right to demand our assent is a condition *sine qua non* of fruitful thinking (Logic).

The results reached in this region of speculation, high up as they come in our progress from matter to mind, must in a systematic exposition come first of all, preceding even our examination of language; for it is of fundamental importance to distinguish between the *truth* of an alleged inference and the *validity* of the inference. If we argue: 'Men are animals, and mammals are animals, therefore men are mammals,' the conclusion is true, but the 'therefore' false. Otherwise it would be equally true to say that because men are mammals and squirrels are mammals men are squirrels.

In studying the laws of inference we are studying the way in which the mind works. But there is one more step to make. There is no point on which Aristotle is more insistent than that inference alone cannot furnish the mind. Everything cannot be proved by inference, for inference must start from something already accepted. The direct data of the senses are not inferred; they are known without inference. And in the mind itself there are truths which are not inferred or arrived at, but are directly known. One of Aristotle's favourite examples is that, as soon as we know what 'whole' and 'part' mean, we know that the whole is greater than the part. But the most fundamental of them all states that unqualified assertion and negation of the same thing cannot both be true. If 'A is B' and 'A is not B' are unqualified assertions, they cannot both be true. Axiomatic truths are truths that the human mind, by its very constitution, cannot do other than accept.

The constitution of the human mind, then, (1) gives it certain truths that it cannot deny or doubt, and (2) lays down certain paths of inference on which it can securely tread, and (3) gives it powers by which it can extract the permanent and universal out of the transient and the particular.

All certainty is based upon these certainties and can be no more certain than they. But Aristotle accepts them as absolutely valid. And these axiomatic certainties, together with the processes of our minds in dealing with them, we can 'understand.' The data of our senses we must accept.

But since self-consciousness in man develops only under the stimulus of consciousness of an objective world other than itself, the impressions of the senses furnish all the material on which the mind can work till it rises upon material steps to the recognition of immaterial being—from the region of sensible facts to the region of intelligible truths.

And here, according to Aristotle, the study of Nature ends and the study that he sometimes calls First Philosophy and sometimes Theology begins.

5. *Cosmography and Theology*

In a connected series of works (the *Physica*, the *De caelo*, and the *De generatione et corruptione*[a]) Aristotle sets out, and attempts to justify, his theories as to the alternations and successions of physical phenomena which constitute the order of Nature.

He bases his theories on the data of the senses and the inferences that they appear to him to justify.

Among the most important of these inferences are :

1. That there cannot ever have been a time in the past when the order of Nature, as we know it, was not in actual operation ; nor will there ever be a time when it ceases to operate.

2. That although the material universe and the processes of Nature must thus be regarded as everlasting, without beginning or end, yet they cannot be regarded as a self-contained and self-sustaining system, but point beyond themselves to an immaterial

[a] On the exclusion of the *Meteorologica* from this list see note on pp. lxxv f.

order of existence, on which they depend for all their functionings.

The first of these conclusions explains why Aristotle, who is keenly interested in the origins and history of concrete human communities and institutions, never attempts to give us any kind of cosmogony, or account of the origin of cosmos, such as we find in the mythologies of all nations and, in some shape or another, in the speculations of many philosophers.

The second conclusion shows us how it is that Aristotle's cosmography leads us to his theology, and even in a certain measure *into* it. This theology (Aristotle's own term) is dealt with expressly in the work which Aristotle himself describes as ' Theology,' or sometimes under the more comprehensive title of First Philosophy. This work is the treatise which we call the *Metaphysics*, because in the systematized Aristotelian canon it came ' after ' (*meta*) the group of physical treatises. But though theology is most expressly dealt with there, we have glimpses of it, not only in the *De caelo*, but also in the *Physics*, into which indeed it is inseparably mortised.

Many details of Aristotle's cosmography will meet us in their places as we go through the *Physics* and will be dealt with as they arise. But it will be well to start with some general conception of the universe as it presented itself to his mind.

Aristotle's universe, then, was a bounded plenum. That is to say, it was bounded by a nest of transparent spheres, which carried the stars and other heavenly bodies, all composed of the fifth or celestial element, within which the four-elemental world, with the earth at its centre, was contained. Throughout this whole universe, from the outmost sphere to the inmost core, there was no vacant space (or spaces) anywhere.

The sphericity of the earth herself was already firmly established in the science of Aristotle's day, and this carried with it a recognition of the relativity of the conceptions of ' up and down.' To us ' up ' means in the direction from our feet to our heads. To the antipodes it means the direction from *their* feet to their heads—which coincides with our ' down.' And every direction between these opposites is someone's ' up ' and someone else's ' down.' True to his general principle, Aristotle asks what is it in which all these opposites and variations agree, and he finds it in the fact that what ' down ' means to everyone everywhere is ' on a straight line from wherever they are to the centre of the earth.' ' Up ' means ' along the same line in the other direction ' ; and this definition holds everywhere in the universe.[a]

Heavy things have a trend towards the centre—

[a] But the universe rotates on its axis, and movement (whether of rotation or translation) presupposes direction. Hence the six elemental directions must be in some sense applicable to the universe as a whole. Now it is axiomatic with Aristotle that motion starts from the right. This is illustrated by the action of a man hurling a stone or javelin or by our right-handed bowler in modern times. Motion begins by a swing forward of the right side counterclockwise. Therefore from right to left is the natural movement. And it is with the power eternally and without effort to move thus from right to left that the eternal matter of the heavenly spheres is endowed, though with different degrees of intensity (apparently) according to the sphere or the place in the sphere which that particular matter occupies.

If I in this hemisphere regard myself as part of a solid sphere which is that of the fixed stars, and (ignoring for the

that is what being ' heavy ' means—and therefore the question why the mass of the heavy earth does not ' fall down through space ' is futile. ' Falling ' means ' moving towards the centre,' and nothing else. The earth is at the centre and has nowhere to fall to. It will be seen how very near Aristotle is to the modern conception of gravitation. If the mass of earth shapes itself as a sphere by mutual pressure,

moment the secondary movements) take the sun as marking a particular place in that moving sphere, if I face the sunrise and turn my body always to face the sun I should be moving with the movement of that sphere, and that is from left to right, which is clockwise. But that is cosmically wrong. If, however, I were standing on my head or were standing up on my feet at the Antipodes, that same movement from East to West would have been from right to left or counter-clockwise, which is cosmically right. Therefore the south pole is the top of the universe; the East is to the right and the West to the left. So in Lucan (*Phars.* iii. 248) the Arab troops of Pompey are spoken of as

Umbras mirati nemorum non ire sinistras

—wondering that, as they look on their own shadows, they did not find themselves moving counterclockwise, from right to left, as they had been used to in the southern hemisphere.

And contrariwise Dante on the mount of Purgatory having seen the sun rise and noting its place some hours later is bewildered to find that it has moved from right to left counterclockwise, instead of from left to right clockwise as would be the case in the northern hemisphere (*Purg.* iv. 56 ff.).

Plato too speaks of the East lying to the right when he is talking about the earth, but as lying to the left when he supposes the observer to be contemplating the heavens. (*Laws* 760 D τὸ δ' ἐπὶ δεξιὰ γιγνέσθω τὸ πρὸς ἕω, *Epinomis* 987 B τρεῖς δ' ἔτι φορὰς λέγωμεν ἐπὶ δεξιὰ πορευομένων μετὰ σελήνης τε καὶ ἡλίου.) So to this day in our modern maps the East lies to the right hand in the terrestial map and to the left in the celestial map.

See *De Caelo* ii. 2.

and if every several clod of earth, when separated from the mass, has a natural trend to join it, is not this the very formulating of the fact of masses gravitating to each other? No, for Aristotle expressly tells us that a clod falls to the centre not because it is the centre of the *earth* but because it is the centre of the spherical vault of heaven; and that, if the mass of earth were forcibly removed to the position of the moon and held there, a free clod would fall not to it but to the centre of the universe. Thus he attributes to the 'centre' a kind of active power of drawing heavy things (which are potentially subject to this power) towards itself; and unfortunately he attributes a corresponding active power to the inner celestial verge, by which it too attracts to itself the buoyant things which by potentiality are subject to its attraction.

Thus Aristotle commits himself to gravity and levity as two distinct qualities, both of them positive. Fire has levity alone, earth gravity alone; air and water have both levity and gravity in different degrees. This preconception prevents Aristotle from accepting the belief (already held by some other thinkers) that all the (non-celestial) elements have some gravity, though that which has least will rise above all the rest. Aristotle himself is constantly pushed in this direction by his own speculations, so much so as to be occasionally betrayed into inconsistencies with his own avowed theory, which nevertheless he vigorously and expressly defends.

The earth, with its swathings of water, air, and fire, Aristotle held to be, as a whole, stationary, as the central body of the universe, though its parts were (or were capable of being) in continuous change of relative position, by falling, rising, or crossing. This doctrine of the 'stationary earth' (as we may call it for convenience) was not unchallenged in Aristotle's day, and he himself was well aware that it might be conceived either as 'rolling' or 'spinning,' and he implies that, so far as appearances go, a revolving earth under a stationary sky would not reveal itself as differing from a stationary earth under a revolving sky. But (in addition to less valid considerations) he made, as we shall see, a truly scientific attempt to find an independent standard of reference by which to judge between the two hypotheses, and he decided in favour of the unmoving earth.

But if the four elements naturally arrange themselves in the stated order (counting inwards) fire, air, water, earth, how comes this order ever to be disturbed, and why is anything so out of its natural place as to manifest its trend towards it? Aristotle's answer to this question, though explicit enough as far as it goes, is not worked out with the completeness and detail we could desire. Broadly, Aristotle's answer is closely akin to that of modern science. Evaporation and condensation are continuously going on, and they are typical of all the transformations which, by changing the bulk of things, tend to change their relative gravity or levity; and so, for instance, air (vapour) and water (liquids) establish a circulation under the influence of accessions or withdrawals of heat, for it is a cardinal doctrine with Aristotle that the elements can be transmuted into each other, and that the same ultimate matter occupies more or less room according as it is heated or chilled. But here the hesitancy we have already noted as to the relation of fire, as an element, to heat, as a state, obscures the connexion between the sometimes isolated data to be found up and down in Aristotle's physical treatise, or incidentally in his other works.

Under the reserve implied in these warnings, we must attempt to reconstruct Aristotle's system. It is certain that the celestial 'fifth element' is not fire and has not its properties ; and since the heavenly bodies are composed exclusively of this element (being, apparently, in some sense concentrations or intensifications of it) they are not and cannot be the direct sources of heat in the sense in which an earthly fire is. Yet it is obvious (and Aristotle lays stress on the fact) that changes of heat do accompany the presence or absence of the sun's rays, and are affected by the directness or obliquity with which they strike things. There is only one passage in his extant works in which Aristotle gives us a direct clue to the solution of the problem thus suggested. We know that friction is a potent cause of heat, potent enough to melt a leaden bullet as it passes through the air. We may suppose, then, that the friction of the inner surface of the rotating celestial shell of the universe heats the upper air in contact with it into actual fire, more especially where its potency is highest, that is to say in its concentrated presence in the sun.[a] The heat thus generated descends by conduction and reaches the other elemental regions with diminished force. When the sun is withdrawn cooling succeeds to heating.[b]

[a] *Cf.* Vol. II. p. 315, ll. 4 ff.

[b] In his biological treatises Aristotle appears, at least, to treat the heart as an independent source or agent of heating, and the moist brain as a corresponding refrigerator.

This account of the inorganic or mechanical movements of the elements must close with a repetition of the warning that though all the principal data on which it is based are authenticated by explicit texts, yet it is impossible not to have some misgiving as to their power, in every case, to bear the weight that an attempted reconstruction of the whole system necessarily lays upon them.

Under our next step, on the other hand, we shall find the ground perfectly firm. With inexhaustible earnestness, perseverance, and ingenuity Aristotle urges throughout the *Physics* that no mechanical movement is self-initiated by matter. Earth falls indeed, and fire rises, in virtue of an intrinsic ' principle of movement.' But that principle is a passive capacity for being moved, and not a capacity for self-initiated movement. The falling clod does not move itself, though it is in itself capable of a natural and unforced movement if suitably acted upon.

Now in the inorganic world movement in a body is always caused by some other body that is already in motion. This will land us in an *ad infinitum* chase for the ultimate origin of movement unless we can find some cause of movement that is not itself in motion.

The cosmic and ' theological ' conclusions to which this leads us close the *Physics* ; but our next step towards them must be the observation that living things, and especially animals, unlike inorganic things, do seem to initiate movements. This

We may perhaps connect this with the belief that some infusion of the actual celestial element was an invariable concomitant of the phenomena of life. But the links of this connexion (if it exists) are lost.

Aristotle admits, but points out with elaborate care the danger of expressing this fact by saying that animals 'move themselves.' An animal's free movements are probably *prompted* in every case by some change in its physical organism or environment, but it is always some kind of *desire* (however prompted or suggested) that actually makes the animal move. Now a desire is not a material body or substance that can impinge upon another body and cleave or thrust it. Yet we see that as a fact—a primary datum of observation and experience that we must accept whether we can or cannot 'explain' it—an animal's 'desire' can and does move its body. And since a desire is not capable, directly and on its own account, of moving locally (for only a 'body' can so move), we seem already to have reached our 'unmoving source of motion.'

But this conclusion would be too hasty. With the sole exception of the human intelligence, with its natural 'desire to know' for the sake of knowing, all the desires of animals are prompted by physical and sense-perceived changes, and therefore postulate them as antecedents and cannot be their prime source. And even intellectual desires, though Aristotle thought they could rise into independence of all sense organs, can draw the materials on which they so rise only from material objects through the organs of sense.

But let us look from earth to heaven. Aristotle's astronomy will require our attention in due course, but for our present purpose we need only know that he believed the whole ethereal shell of the universe to consist of a number of concentric spheres, one inside another, each being in frictionless contact with the next within it, until the inmost rubs against the sublunary or ' four-elemental ' world.

Perhaps the homely illustration of the coats of an onion will best serve us here. Replace the core of the onion by some imaginary ' stuffing ' of unlike substance ; Aristotle's concentric spheres will then be represented by the remaining coats of the onion, and the imagination will easily rise to the idea that they can slip about within each other in immediate but frictionless contact. This is what Aristotle's concentric ethereal spheres (or shells) do. Each of them has a ' proper ' movement of its own, but, with the exception of the outmost, each of them also sympathetically obeys the movement of some one or more of the spheres above it. It is thus that the spiral course of the sun throughout the year, of the moon throughout the month, of the planets (with their retrocessions) throughout their periods, and the simple circular movement of the fixed stars (in Aristotle's time the ' precession of the equinoxes ' had not been observed) were accounted for by a single circular movement in one case, or a ' proper ' circular movement imposed upon one or more derivative movements (all circular) in the other cases. The fundamental principle of this analysis and synthesis still survives as the basis of our modern astronomy.

But our immediate concern is with the ' proper ' movement of each several ' heaven,' or ethereal coat, shell, or envelope. Each such heaven was seen by Aristotle as ' moving itself,' just as an animal does, in addition to any movement it may receive from other moving spheres. Such a heaven, then, manifests the phenomenon which Aristotle regarded as peculiar to animated beings, the power, namely, of initiating bodily movement in virtue of an inherent vital principle, not by impact but by the impulse of desire.

Each heaven, then, is an animated being. But the desires of sublunary animals are stirred, directly or indirectly, by sense-impressions or physical changes in the organs of sense, and (except in the case of the human intelligence) are directed towards, or at least accompanied by, actual 'sensations' of a material order; whereas the heavens or celestial animals have no organs of sense, and the substance that constitutes their bodies is incapable of any modification except that of (rotatory) movement. What, then, can be the nature and the object of the 'desire' that their motion implies?

The highest desire of the highest of sublunary animals and the least dependent on material sensations is the human 'desire to know.' Lift that into a higher potency, and make it continuous instead of intermittent, and regard it as continuously gratifying itself by fruition, and you will have the best conception that man can form of the blessed and eternal joy of the divine animals.

But what are the objects of their desires? These must needs be higher than themselves. Moreover, since they all move, and we have seen that the ultimate source of movement must itself be motionless, and we now know that it must be higher than even the celestial animals with celestial bodies, we must take it that the object of each desire that originates the proper movement of a celestial animal or 'heaven' must be an immaterial consciousness free from movement, in blessed possession of 'truth' its very self, wherein desire is quenched in eternal fruition.

Such, in the supreme degree, is the life of the supreme being, described by Aristotle as the 'principle on which all heaven and nature depend.' Such is the direct object to which the desire of the supreme heaven, expressed in its continuous and simple movement, is directed. The proper movement of each of the other heavens is exercised in independence, indeed, but in a certain subordination to that of the supreme or outmost heaven; and the divine and immaterial object of the desire that inspires it must be conceived (though Aristotle does not expressly develop this line of inference) as standing in a corresponding relation of subordination to the Supreme Principle 'on which heaven and nature depend.'

The Poet Laureate may well call the attention of his readers to the 'extreme interest' of the chief passage in which this 'theological' doctrine is expounded. For it is, as he declares, 'the one original foundation of the Christian doctrine on the subject.'[a] More especially it is the basis on which all the elaborate Angelology of the Schoolmen is built, and that because it gave the Neoplatonists, Christian and ethnic alike, and the Arabian Aristotelians of the tenth to twelfth centuries, the point of attachment for their synthesis of the Platonic and the Aristotelian philosophies.

Indeed it must be evident to the most careless observer that in all this theological construction Aristotle betrays himself as still dominated in his highest spiritual life by Plato's teaching. While he

[a] Robert Bridges, *The Spirit of Man*, extract 39.

is honestly persuaded that he is building up his system towards a conclusion that only reveals itself as we approach and reach it, it is obvious that in reality his biological predispositions, his reverence for popular instinctive and traditional beliefs, and his own instruments and methods of thought, are all directed into the channel they take in this argument by the influence of the pre-established goal to which they lead us.

Plato had taught Aristotle—a willing pupil—that the supreme life is a life of direct perception of immaterial and eternal realities, in which truth and beauty are one. And though Aristotle on his own ground broke away from his master, yet his ultimate conception of the divine and ideal life not only remained Platonic but continued to influence him in all his thought.

We shall never understand the under-currents of Aristotle's thought till we understand this subtle influence. It is at the root of his 'teleology,' that is to say his doctrine that the perfect and the actualized is presupposed in the potential and the undeveloped and at once directs and explains its course. It is that that makes him regard even the up-and-down-ness that observation establishes as a fact in the movement of the four elements as finding its reason and explanation in the consideration that it is the nearest approximation that the incomplete linear movement can make to the 'circumlation' of the complete and self-re-entrant circular movement.

On the same principle he explains why animals and plants that are condemned to mortality by the divergent trends of their component elements nevertheless strive after immortality by leaving beings like themselves behind them to make the collective life of their kind eternal.

It is the same impulse that makes him—though he defines Nature as the objective sum of things, attributes, and functionings that are subject to change—nevertheless treat her and feel towards her as a divine (impersonal, it is true) agent of impulse, purposeful and aspiring.

But in one respect Aristotle has the advantage of Plato. It must be particularly obvious to the reader who compares the *Timaeus* with Aristotle's physical and metaphysical treatises. Plato, starting with the perfect and absolute, but unable to get rid of the world in which we live, gives us a saddening, sometimes a horrifying, description of the badness of the 'second best' which was all that a material creation could possibly aspire to, and our prevailing impression may well be of the degradation involved in our material nature. Aristotle, starting from the world in which we live, finds it full of intensest interest in itself, and ennobled, from centre to circumference, by upliftings and aspirations towards that which is above itself. Even its failures are not-yet-nesses, or not-quite-nesses. There is no degradation or confinement in our having bodies, but there is a divine glory in our having intelligence, and joy in our having disinterested affections. And if there is a region of divine life to which we mortals cannot rise, it is inspiring to push out the limits of our thinking to the utmost of our power and to dwell upon the thought that there are diviner lives in which those limits are broken.

It remains that the common, if not the universal,

verdict pronounces that Plato's power of effectually bringing us into an ideal world transcends Aristotle's. But if to the mystic and the metaphysician Aristotle can never be more than the philosopher of the plain man, yet to the plain man himself he may bring moments in which he forgets that he is not a mystic.

III. Note on the *Meteorologica*

As to the Aristotelian canon, I am content in the main to accept the authority of Ross (*Aristotle*, pp. 9 *sqq*.) as to the *spuria* (such as the *De coloribus*, the *Mechanica*, the *Problemata*, parts of the *Historia animalium*, etc.) and the *dubia*. But I am compelled to reject the *Meteorologica* in its entirety, although it is classed by Ross as 'undoubtedly genuine' (p. 11); and as it might naturally be expected that its data would be taken into account in expounding Aristotle's cosmography, I must at least indicate the grounds for rejecting it, though I cannot here enter upon a detailed argument.

To begin with, the *Meteorologica* is written in conspicuously pleasanter and more flowing Greek than is at all frequently to be met with in Aristotle's genuine works. This was noticed by earlier commentators. On the other hand the argumentation is indefinitely looser than Aristotle tolerates even when he is least severe, as, for instance, in parts of his biological treatises. Even where Aristotle's reasoning lags farthest behind his observation, and where he is readiest to adopt any explanation that seems to serve him for the moment, he never approaches the random and facile opportunism of the author of the *Meteorologica*.

Apart from these general characteristics, the whole system of the *Meteorologica* differs from that of the *Physica*, the *De caelo*, and the *De generatione et corruptione*. In these genuine treatises, for instance, there is no trace of the doctrine of ' dry and moist exhalations ' which in the *Meteorologica* is so prominent as almost to oust the doctrine of the ' four elements,' nor is this doctrine of exhalations itself anywhere clearly and fundamentally expounded.

Again, in the *Meteorologica* the revolving heaven carries the sphere of fire and the sphere of air with it in its circulation, down to the region in which the mountains, by their barriers, make as it were pools of still air; whereas in the *De caelo* it seems to be the friction of the heavenly sphere against the stationary atmosphere that heats the upper air into fire; and in any case the idea of the rotating atmosphere would bring confusion into the whole mechanism of the natural up-and-down-ness of the motion of the elements.

I may add that the traits in medieval science drawn specifically from the *Meteorologica* have never appeared to me either to connect up with the general system or to have any definiteness and consistency of their own. I cannot but think that other students must have shared this impression.

IV. Note on the Principles of Translation here adopted [a]

"The frankly paraphrastic and expository character of the translation of the *Physics* here submitted to

[a] Compiled by R. W. The sentences between inverted commas were written as they stand by the translator; the rest is from his notes.

students was adopted under the conviction that a so-called literal translation would be wholly unintelligible in many passages to the English reader, and would be of very little value to the student of the Greek text.

" The writer's object was primarily to make what he understood Aristotle to *mean* clear to the English reader who knows no Greek, but is not afraid of hard thinking ; and secondarily (if he could) to make what Aristotle *says* intelligible to the reader of Greek who has made no previous study of the book or of the Greek and mediaeval commentators who have thrown so much light upon it."

The translation, therefore, must not be regarded as an introduction to Aristotle's technical terms, but as an attempt to express his thoughts in terms with which the English reader is familiar. " Such an attempt must necessarily be tentative and must at best be open to legitimate challenge at many points, at worst be convictable of positive and frequent error. I should not have had the audacity to enter upon so high and perilous an emprise, had I not been able to lean at every step on the ample Greek and Mediaeval commentaries, without which the *Physics* must have remained to me a sealed book almost from the first page to the last, and which the reader to whom I hope to be of service cannot be supposed to have studied."

A ' literal translation ' " would have necessitated the constant use of English words in meanings that they cannot bear in English writing, a practice which is particularly dangerous and misleading when the words in question carry strong and definite associations of their own. No insistency of warning can suffice to prevent such words from drawing the reader after them in their natural and customary suggestions." Therefore the principle of never using a well-known English word in a sense alien from that in which it is used in current literary English speech or writing has been carried as far as possible.[a]

"For instance, to translate ψυχή by 'soul' (as distinguished from ζωή as 'life') would raise false but inexpugnable associations from which no effort, however firm and vigilant, could protect the reader." Both ψυχή and ζωή mean 'life' in exactly the same sense—but ζωή is 'life,' ψυχή is '*a* life.' There is no indefinite article in the classical languages, but ψυχή has the individualized quality of the indefinite article already in itself.[b] Therefore ψυχή will be found

[a] Some of his readers will think that he has carried it too far, a few may think that he has not carried it far enough, as, for instance, when he uses 'air' where we should use 'vapour.' But what we mean by 'air' was to Aristotle the elemental vapour, as 'water' was the elemental liquid, and to substitute 'vapour' for 'air' here would be substituting not merely a current English word, but a modern conception.

[b] "In the middle ages there is an *Unum quod* (*unus qui*) *convertitur cum ente*; it is the hypercategorical *unum*, and it plays an enormous and very perplexing part in Mediaeval physics. It is an *unum* which is not numerical and therefore is not opposed to plurality, and there is no contradiction in saying that things which are numerically distinct may be hypercategorically *unum*. This *unum* is simply the indefinite article. There is, so far as I am aware, no one of the Western group of the Indo-European languages, except English and Anglo-Saxon, that have a distinctive form for this *unum quod convertitur cum ente*."

Even in English, though we have an indefinite article, 'one' is sometimes used in its place, *e.g.*, "I am now wanting *a*

translated as ' a life ' instead of ' a soul,' for we may speak with propriety of a celestial spirit and a plant as each having *a life* (ψυχή) of its own, but it would be quite another thing to say that a celestial spirit and a plant each had a *soul* of its own.

There is, however, no one English word which will always be the equivalent of ψυχή, and in order to make clear what Aristotle is thinking of at the moment, it has been translated variously by—a life principle, a terminable life (which may be cut short by an accident), a reproductive or digestive function, a thought, mentality or mind—but it is always an individual something, never (as ζωή is) just life, or thought in general.

Again ' substance ' has been rejected in favour of 'being' or ' concrete entity ' as the English equivalent of οὐσία " chiefly because ' substance ' irresistibly suggests materiality ; and to raise the question whether the soul is a substance, or to assert that the immaterial and non-spatial, but conscious beings of Aristotle's theology on the one hand, or individual men, stones and trees on the other, are ' substances ' is too great a violation of English usage to be permissible except in professedly technical compositions." ' Concrete ' means *compositum* in the Latin sense, *i.e.* the complex of body and form, which everything is that has a primary existence of its own : everything that would be there whether we existed or not, and this constitutes the first category, *οὐσία*.

---

book on such and such a subject ; can you recommend *one* ?" means just the same as " Can you recommend *a* book on such and such a subject ? I am now wanting *one*." And there is no intention here to set the limit at a single one.

Yet again : *shortage* is substituted for *privation* (the established translation of στέρησις), for *privation* implies that something has been taken away, and *shortage* merely that something has not been attained. For instance, the seals that might under suitable circumstances have developed the full form of quadruped, or the water that might have become ' air,' though they are both *short* of something which they might have had, are neither of them *deprived* of anything which they have had.

" Illustrations could of course be multiplied, but these will suffice as apologies to the students of Greek and of philosophy, and as warnings to the English reader that he must never suppose himself to be in possession of Aristotle's terminology in an English dress, however faithfully I may hope to convey his meaning.

" On the other hand I have to crave indulgence for the use of uncouth words, some of them ignored and others expressly disallowed by the dictionaries, such as ' aliveness,' ' awareness,' ' dimensionality,' ' quantitive ' and ' qualitive ' ; and of innovations or archaisms such as ' station ' as the correlative of ' motion ' or ' continent ' as the correlative of ' content,' [a] and of some frank barbarisms in the way of hyphened composites which are in no danger of causing confusion of thought however much they may offend propriety."

[a] " I say ' a motion interrupted by *stations* ' ; ' a motion is not made up of stations, any more than a line is made up of points.' And I call a vessel that contains ' the vessel *continent* ' or even ' the continent ' in contradistinction to the ' content,' as meaning in fact the vessel that contains the content."

V. Note on certain Mathematical Conceptions

The following is an attempt to give the unmathematical reader some idea of the mathematical conceptions which permeate the *Physics*. No detailed knowledge of mathematics is necessary, but the meaning of certain abstractions such as magnitude, dimension, divisibility, surface, line, point and so forth must be firmly grasped in order fully to understand the significance of Aristotle's reasoning.

Divisibility.—The question of divisibility is a crucial one, and we will begin by considering that.

We see that a collection of individuals can be broken up into groups, and these can again be broken into smaller groups, and so on and so on ; all these divisions will, roughly speaking, be alike, and this can go on until each individual stands alone. The process of division can still be carried on, but if an *individu-al* is broken up it thereby loses its *individu-ality* For instance, if a chair is broken up, the parts will not be small chairs but pieces of wood, cane, and so on. The divisions of an orange are not small oranges, nor the parts of a man small men. The divisions of a homogeneous mass on the other hand are all exactly alike. But things that in everyday life we call homogeneous may not really be so, and experiment may show that if they are divided and subdivided again and again the parts will be homogeneous only up to a certain point, after which they too are reduced to individuals, and if the process is continued they will lose their individuality.[a]

[a] Observe that the term ' division ' is here, as elsewhere, used in different senses.

(*a*) The *Process* of dividing a limited quantum into parts is called *division.*

If a lump of gold is cut in two there will be a certain amount of wastage and distortion, the whole of the gold will not remain in the two parts, and they will not fit together perfectly. If the process is continued, the wastage will tend to exhaust the gold (it does not destroy it, but it makes it unavailable); but this is due to the limitations of actual conditions. Wastage and distortion are imperfections, not functions of dividing. The perfect 'cut' would neither waste nor distort: the two parts would comprise the whole of the gold, and they would fit together so as to leave no spaces between them and would take up no more room than the uncut lump. There would then be no trace of the 'cut,' but it would still 'mark off' the parts, and if the pressure which held them together were removed it would determine where they fell apart and how much of the gold was in each part. So though the cut takes up no room, and is no part either of the gold it divides or of anything else, it has a definite position and extension.

(*b*) The *Parts* into which a limited quantum is divided are called *divisions*. These divisions when added together make up the whole of the quantum of which they are parts, and if they are successively removed there will be less and less of the original quantum left.

There is yet another sense in which the term is used, viz.:

(*c*) The *Boundaries* whose function it is to mark off the parts are also called *divisions*. In practice these divisions 'use up' some of the quantum they divide. For instance, the hedges which mark out fields, occupy some of the ground, and if there were enough of them they would eventually fill up the whole of the ground; but this is not part of the function of a boundary; it is an imperfection due to physical conditions. The perfect boundary would take up no room, and removing it would neither increase nor diminish the available space.

Its existence, though not material, is none the less real; it is in fact what is meant by a mathematical surface.

Now suppose a lump of gold to be divided (by 'perfect' cuts) into layers. These layers may be again divided into thinner layers, and so on and so on. The layers are parts of the gold and when added together build up the whole of it. The 'cuts' or 'surfaces,' on the other hand, are not parts of the gold; and as we have seen they have no material existence.

There is no limit to the conceptual possibility of thus slicing up a solid continuum, for however long we go on there must always be a layer between one 'cut' and the next, which layer, however thin, must occupy *some* room and is conceptually divisible into yet thinner layers. Thus we can never say that any two cuts (surfaces) are the nearest conceivable to each other, for they are always separated by a conceptually divisible layer.

A surface is conceptually divisible into less extensive surfaces, which exactly fit into each other, and together make up the whole of the original surface. The conceptual 'boundaries' by which it is divided up are not themselves parts of the surface, and therefore do not tend to exhaust it; they are in fact mathematical lines which are no more parts of a surface than a surface is part of a solid.

Two lines running side by side may be as close together as we choose to imagine them, but they will always be 'spaced': there is always a conceptually divisible strip of surface between them, and a strip, however narrow, is still a part of the surface, and the surface can be built up from and divided into these strips, but the dividing lines are no part of the surface; they simply mark off its parts.

Yet a line, though it has no material existence, has length and position. It therefore does exist, though not materially. Its length can be divided into shorter lengths, the sum of which is identical with the whole length. The 'boundaries' of a line are not parts of it; they simply serve to mark it off into parts. They are in fact mathematical points, which are no more parts of a line than a line is part of a surface or a surface part of a solid.

Two points may be as close together as we choose to imagine them, but they must always be spaced; there is always a conceptually divisible length of line between them, and the line can be built up from, and divided into these lengths. But the dividing points are no part of the line; they simply mark off its parts.

To sum up:

A solid body which has a material existence is conceptually divisible by surfaces.

A surface has no material existence; it exists in virtue of having extension and position, and it is conceptually divisible by lines.

A line exists in virtue of having length and position, and it is conceptually divisible by points.

A point exists in virtue of having position; it is indivisible.

Magnitude can be conceived as 'that which is divisible'; that which is indivisible is not a magnitude. The terms 'divisible' and 'indivisible' belong to the class of opposites which he calls contradictories; they are mutually exclusive and there can be no intermediate state.[a]

[a] Some opposites are not contradictories: hot and cold, for instance, or wet and dry, are opposites but they are not contradictories. They are extremes between which there are

Points are indivisible and so are not magnitudes.

Lines, surfaces, and solids are all divisible and are therefore all magnitudes. But they do not all belong to the same 'order' of magnitude.

The 'order' or 'dimensions' of a magnitude depends on the further divisibility of the boundaries by which it can be divided up.

A line is a one-dimensional magnitude, being divisible by *in-divisible boundaries* (points).

A surface is a two-dimensional magnitude, being divisible by *one-dimensional boundaries* (lines).

A solid is a three-dimensional magnitude, being divisible by *two-dimensional boundaries* (surfaces).

Now as there is no bridge between divisibles and indivisibles, neither can there be any bridge between magnitudes of different dimensions. For if a one-dimensional magnitude could become two-dimensional, its indivisible boundaries would have to become divisible, which is impossible, as we have seen; neither can a two-dimensional be built up from one-dimensional magnitudes, for adding magnitudes together does not alter the dimensions of their boundaries. It is easy to see that transformation is equally impossible the other way round.

The same reasoning will demonstrate that there is no bridge between two-dimensional and three-dimensional magnitudes.

Now, owing to our inability to visualize abstractions

different degrees of hotness and coldness, or of wetness and dryness. The same thing may vary from time to time between the opposite extremes, tending now to become hotter now colder. The actuality of one extreme implies the potentiality of the other in the same subject. But with contradictory opposites the actuality of the one is the absolute negation both of the actuality and of the potentiality of the other, in the same subject. See Bk. V. ch. iii.

and to the difficulty of following a demonstration without some actual construction, we habitually represent points by dots, lines by thin strokes, and surfaces by thin laminae, and are thereby subjected to the insidious danger of regarding these representations as at any rate approximations to mathematical abstractions, and if we argue from constructions rather than from definitions, we may easily make false assumptions without knowing it. There is no safe receipt for avoiding this danger, but the test of divisibility is an invaluable check. An assumption is convictably false if it implies that a magnitude is of a different order of divisibility from that which defines it. But many false assumptions pass unconvicted through want of real grasp of Aristotle's teaching in this matter.

Measurement.—For purposes of measurement and comparison it is convenient to divide up quanta into standard *units*, and to express them as aggregates of these units.

A ' unit ' is not a ' unity ' in the sense of being indivisible, for it may be either an individual, or a group, or a magnitude, or anything else which for the purpose of manipulation, measurement, and comparison it is convenient to treat as a single whole. Any unit suitable for the purpose in hand may be taken as the *standard unit*, and different units will be required according to the quanta and the conditions to be dealt with.

Sometimes the unit will be a specific magnitude, such as a yard, a ton, a load or a cupful ; sometimes it will be a kind of individual irrespective of magnitude, such as a man or a chair ; or it may be a group of individuals, such as a constellation of stars or a grove of trees ; or again it may be such a thing as a dream, or a taste. It may in fact be anything which can be counted.

Units can of course only measure quanta of their own kind : units-of-length measure lengths, units-of-time measure time. An aggregate of units-of-weight is itself a weight, an aggregate of units-of-length is itself a length.

The immaterial ' boundaries ' which tell off the successive aggregates are their ' numbers,' and the aggregates themselves are named after their specific numbers ; so that a given number stands both for a numerating boundary and a numeral aggregate.

The system of aggregation applies to units *qua units*, not *qua* parts of specific quanta, and the science of arithmetic depends on the system of aggregation, and is applicable to aggregates of any kind of unit. It is not therefore necessary to specify the nature of the unit in order to investigate the properties of numbers, and when this principle has been grasped it becomes possible to experiment with aggregates of easily accessible units, and to apply the conclusions to those not susceptible of direct manipulation. Thus having discovered experimentally that an aggregate of three pebbles and an aggregate of four pebbles together make an aggregate of seven pebbles, we can deduce the fact that three men and four men make seven men ; three hours and four hours make seven hours ; three dreams and four dreams make seven dreams, and in general, that three and four units together make seven units.

But if we want to measure specific quanta we must choose specific units. If we are told that a distance is ' three ' we do not know how long it is, but if we are told that it is three feet or three inches we do know. If we are told that a weight is 'three' we do not know how heavy it is, but if we are told that it is three pounds we do.

The question now arises of the measurability of a given continuum by a given unit. A unit which *will* measure the continuum can of course be found, by dividing the continuum itself into equal parts and taking these parts as units. The continuum will then contain an exact number of the chosen units. But can we expect that any unit will be an exact fit, if it has been chosen without reference to the continuum to be measured?

Take, for instance, a length OA and measure off successive units on the same line.

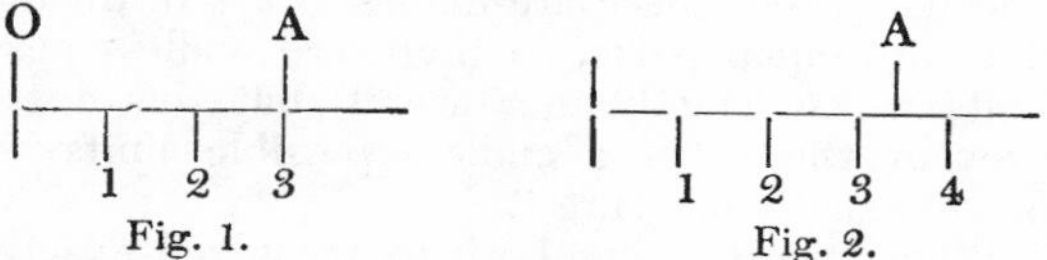

Fig. 1. Fig. 2.

If one of the boundaries which mark off the successive units coincides with the point A which marks the end of the line OA (as in Fig. 1), then the unit will be an exact measure of OA and OA will contain an exact number of these units. But if the point A falls between two successive boundaries (as in Fig. 2), then the unit is not an exact measure of OA; OA exceeds one number of units and falls short of the next.

Now it would appear at first sight that we could be sure of an exact fit by taking a small enough unit, for the smaller the unit the greater is the number of boundaries in a given space. But smallness alone gives no ground for the presumption that the fit will be exact. For however numerous the boundaries may be, they must always be 'spaced,' and the point A may fall between two consecutive ones. We cannot prevent this from happening by inserting still more boundaries, for however close these may be, there will always be *some* interval between them, and as long as there is *any* interval the point A may fall within it, and then the unit is not an exact measure of the line.

The measurability of a given line by a chosen unit, then, depends not on whether the unit is large or small, but on an exact adjustment which is a pure accident, in Aristotle's sense of this word, unless it has been secured by constructing the unit from the line itself. As there are endless ways of dividing a line into equal parts, so there are endless possible units which exactly measure it, but this does not prevent there being endless possible units which do *not* exactly measure it.

When two lines can both be measured exactly by the same unit they are said to be *commensurable.*

We cannot assume that any two lines OA and OB taken at random will be commensurable, for there is no ground for supposing that any unit constructed from OA (by dividing OA into equal parts) will be an exact measure of OB, or that a unit constructed from OB will be an exact measure of OA.

When the ratio (*i.e.* relation of magnitude) of two lines is such that they cannot be measured exactly by the same unit they are said to be *incommensurable.* But this does not imply any vagueness or instability in their relationship, but simply that it cannot be expressed numerically, for the lines are not both aggregates of the same unit. We are constantly dealing with ratios that are neither vague nor unstable though they are incommensurable. For instance, the ratio of the side of a square to its diagonal is incommensurable, and so is the ratio of the circumference of a circle to its diameter.[a]

[a] The atomists contended that all quanta were reducible to indivisible particles or 'monads,' and that these constituted the ultimate units from which all quanta are built up. In their view, therefore, all ratios were commensurable. Aristotle, on the other hand, held that there was no limit to the potential divisibility of a continuum. In his view, therefore, there would be no 'ultimate units' of a continuum, and therefore there must be incommensurable ratios.

ARISTOTLE'S PHYSICS

BOOK I

INTRODUCTION

A favourite method of Aristotle's, in investigating any subject, is to begin by surveying the opinions of previous thinkers (or, in lack of such, the current opinions and usages of speech) and endeavouring to discover some common trend in which they agree, perhaps even when they suppose themselves to be in flat contradiction. If he succeeds so far, he endeavours to define the direction of this common trend, to push it forward to its natural goal, and there to formulate it. If the result so reached enables him to reveal some aspect or fragment of truth underlying each thinker's system, and at the same time to show why his imperfect formulating of it exposed him to error, we may then feel that we have a reasonable guarantee of its soundness.

This is the method pursued in Book I., but it seems, at first sight, to bring us in a very slender harvest; for at the end of a discussion which puts a very severe strain upon our attention we find ourselves rewarded by nothing more startling than the assurance that some changes are possible and others not, and that, where there is any change at all, there is something that undergoes the change and some change that it undergoes, so that something is there at the end that was not there at the beginning. A good deal of trouble is spent in considering how to formulate this statement most effectively, but the statement itself appears not only to be obvious but to be irrelevant to the questions with which the Book opens, and as to which the opinions of the earlier thinkers are collected; for we begin by asking, out of how many elemental bodies the world of Nature is physically compounded, and we end by trying to determine into how many factors it is conceptually convenient to resolve the process of 'changing.'

This is all so, and Aristotle will deliberately have it so. But all the same this Book is a very serious piece of work and in its way a model of discussion. A short introductory chapter gives us the key, if we can but find the skill to turn it through the wards.

In this chapter Aristotle lays it down that investigation should begin at the beginning and make for the end, which means beginning where we are and making for where we want to be, and this again (in the case of Nature at any rate) means beginning with the concrete and changing phenomena of the world revealed to the senses, and making for abstract and stable theoretical generalizations that will introduce order, connexion, and sequence into our intellectual apprehension of them; in a word, our progress must be from the phenomena accessible to the senses to the conceptions that satisfy the intelligence.

Now, according to Aristotle, all his precursors took the opposite course. They began by laying down a general proposition that satisfied, or was even demanded by, their intelligence, and then tried to arrange the facts in harmony with it. The consequence was that their starting-point had no firm attachments to the real subject of investigation, and their deductions led them to more or less unfruitful and insecure results. But no one knows better than Aristotle that it is half the battle to know what you want to get at and to 'ask the right questions' about it; and it will therefore be a good day's work if he can show his readers that, in so far as his precursors began at the wrong end and were true to their own method, they disagreed with each other and arrived at comparatively little of primary value; but in so far as they allowed Nature herself to force them, surreptitiously as it were, to follow the true method, they all manifest a common trend and are found to be *groping for the right question*, which emerges at the end of the Book, instead of the wrong one, suggested by their own axioms, which meets us at the beginning.

Perhaps it may be well at this point to introduce parenthetically a few illustrations, from our own circle of familiar ideas, of the way in which an accepted ' axiom ' of reason may entangle thought and perplex observation.

The conception that a thing cannot act and produce an effect except where it is (*i.e.* the denial of the possibility of *actio in distans*) has led to dissatisfaction with the law, or even the observed fact, of gravitation, because the statement of it treats distant bodies as affecting each other. The same conception has also led to the hypothesis of a luminiferous ether, the existence of which has obstinately evaded all tests, and the constitution of which defies consistent formulation.

Or again, Jevons, in his attempt to unify the principles of science, was long baffled by finding that, whereas in arithmetic one and one make two, in logic one and one make one. At last he discovered that in arithmetic too one and one make one if they are both the same one. It is one and *another* that make two.

Compare Bergson's assertion: that the axiom 'two bodies cannot occupy the same space' is contradicted by science all along the line, but still holds its own, because, though not a property of matter, it is a property of two-ness.

These latter examples will serve to illustrate the further point (of great significance in connexion with Aristotle's refutations) that any speculative axiom we jump at has to be expressed in language; and language—which has been evolved by processes that are anything but rigidly logical, systematic, and consistent—embodies many ambiguities and conceals many obscurities; so that, if apparently axiomatic statements are suddenly arrested and petrified, without adequate scrutiny and analysis, the deductions drawn from them may lead us far indeed astray.

To return to Aristotle. The particular axiom which he believes to have misled and enslaved his precursors is variously expressed as ' nothing can come out of nothing or ' the existent cannot come out of the non-existent, or the non-existent out of the existent.' This axiom, as understood by those who accepted it, Aristotle regards as false. The deductions they made from it, so far as they were seriously accepted, flatly contradicted Nature, and also led to internally self-contradictory conclusions ; or at the very least, tended to divert attention from the true problem and from fruitful attempts to deal with it.

Parmenides and his disciples, according to Aristotle, clearly saw that their axiom amounted to a blank contradiction of all change and movement, and so to a conception of Nature as unitary and rigidly unchanging and undifferentiated. They granted that ' appearances ' contradict this ; that is so much the worse for the appearances. They contradict the truth and therefore do not exist at all, but only ' appear ' to do so.

Others were less consistent, but as they too held it to be impossible for anything to come out of nothing, the question that chiefly interested them was not ' How are we to regard change, and how investigate changes ? ' but ' What are we to suppose it is that is *always there* in reality, while it appears as if everything were passing in and out of existence ? ' It is really all air, said one ; all water, said another ; the four elements, said a third ; every known substance or tissue out of which things are visibly built up by addition and interlacing, said another. Thus they were all of them looking for the ' existent,' *i.e.* the reality (as distinct from appearance) which never comes to be or ceases to be, but is ' always there.' This, however, was not examining the fact of change (which is the distinctive characteristic of what we mean by Nature) at all. It simply diverted the mind from it. And, obscurely feeling this in spite of themselves, all the thinkers (even including Parmenides himself, when he deigns to look at ' appearances ') had recourse to certain agents of change, which they brought in under the guise of ' contrasted principles,' such as ' hot and cold,' or ' more and less,' and so forth, between which, as between two poles, change takes place

—or seems to do so. Obviously, then, it is here that we must look for the tracks of that driving force of the truth that secretly diverted them from the lines they were trying to follow, and lured them into an undesigned accord with Nature, and so with each other, even where they were most conscious of mutual opposition.

Thus at last we come to the right questions, and ask: What conceptual analysis of the observed process of change will give us the best starting-point for clear thinking and fruitful observation? What and how many are the factors or constituents of change? What technical terms give best promise of precision and accuracy?

Aristotle tries a variety of formulae and compares their merits, always keeping steadily in view the passage of the changing thing from one state to another. He notes a certain goalfulness in Nature, analogous to our own purposeful actions. He sees that, if we follow any definite series of changes, such as that from the acorn to the oak, we find that at every stage of progress the changing thing *lacks something* of the thing it is to be, and this 'shortage' is an essential characteristic of what, at the moment, it already is: and this provokes interesting reflections as to the relation of the 'existent or non-existent' philosophy to these 'absences' or 'shortages,' which are negative yet definite. Are *they* existent or non-existent? What *is* this 'not-yet-there-ness' (which, under a more positive aspect, we call a 'potentiality') that goes shares with 'already-there-ness' in the ceaseless 'becomings' of Nature?

ΑΡΙΣΤΟΤΕΛΟΥΣ ΦΥΣΙΚΗΣ

Α

CHAPTER I

ARGUMENT

By 'understanding' things we mean getting at their causes and constituent elements, and apprehending first principles. In the order of nature we think of things being built up from their elements, of effects following causes, and of movements and changes as obeying principles (184 a 10–16).

But our experience, on the other hand, introduces us first and most directly to concrete and individual objects and their behaviour, and we have to work back from them to general principles or ultimate elements. Our progress, then, in studying nature, must be from what is familiar to our senses to what is luminous to our intelligence, or from what is first *or most accessible* to us, *to what is* first *or most fundamental* in nature (a 16–26).

Names and definitions are related to synthetic and analytic concepts respectively (a 26–b 14). (*Aristotle uses a single word* (γνώριμον) *alike for what can be directly recognized by the senses as fact, and for what is demanded or accepted by the mind as lying behind it and explaining it. The translation makes no attempt to follow him in this.*)

[*This prefatory chapter is to 'clear up the question of the beginnings of the science of Nature'* (*l.* 15). *The term 'beginnings'* (ἀρχαί) *is loosely used without clearly dis-*

ARISTOTLE'S PHYSICS

BOOK I

CHAPTER I

ARGUMENT (*continued*)

tinguishing three senses: (A) *the* primary elements of natural things (ὅθεν πρῶτον γίγνεται ἐνυπάρχοντος, Met. 1013 a 4); (B) *the* starting-points of a science. *In a systematic science,* e.g. *geometry, these are* (i) *the* premisses *or* basic truths (ὅθεν γνωστὸν τὸ πρᾶγμα πρῶτον, 1013 a 14), *which are ' true, primary, immediate, and intrinsically more intelligible than the conclusion and prior to it,' and are apprehended by intuition* (νοῦς). Anal. Post. 71 b 20, 100 b 12. (ii) *But the best* starting-point for inquiry *or learning* (μαθήσεως οὐκ ἀπὸ τοῦ πρώτου καὶ τῆς τοῦ πράγματος ἀρχῆς ἐνίοτε ἀρκτέον, ἀλλ' ὅθεν ῥᾷστ' ἂν μάθοι, 1013 a 2) *must be ' things more immediately cognizable to us ' by sense-perception.*

The path to knowledge starts from (B ii) *the indistinct* (συγκεχυμένον, *cf.* Poet. 1450 b 37) *unanalysed whole* (ὅλον) *given by sense-perception, which has for its object the concrete individual thing* (*a man*), *but dimly discerns in this the universal* (*Man*), Anal. Post. 100 a 16. *The analysis of this universal into its constituent parts* (μέρη, τὰ καθ' ἕκαστα) *carries us up towards the primary concepts and basic truths* (B i), *just as we understand the nature of a complex thing when we have analysed it into its elements* (A).

This might lead us to expect that Aristotle would go on to establish, by induction from the data of sense-observation

ARGUMENT (*continued*)

(B ii), *the premisses or basic truths* (B i) *of Physics. But the* ἀρχαί *which he proceeds to discuss in Chapter II. are not sense-data, but the alleged elements of natural things* (A). *The connexion between the chapters may lie in the un-*

Ἐπειδὴ τὸ εἰδέναι καὶ τὸ ἐπίστασθαι συμβαίνει, περὶ πάσας τὰς μεθόδους ὧν εἴσιν ἀρχαὶ ἢ αἴτια ἢ στοιχεῖα, ἐκ τοῦ ταῦτα γνωρίζειν—τότε γὰρ οἰόμεθα γινώσκειν ἕκαστον, ὅταν τὰ αἴτια γνωρίσωμεν τὰ πρῶτα καὶ τὰς ἀρχὰς τὰς πρώτας καὶ μέχρι τῶν στοιχείων—, δῆλον ὅτι καὶ τῆς περὶ φύσεως ἐπιστήμης πειρατέον διορίσασθαι πρῶτον τὰ περὶ τὰς ἀρχάς.

Πέφυκε δὲ ἐκ τῶν γνωριμωτέρων ἡμῖν ἡ ὁδὸς καὶ σαφεστέρων ἐπὶ τὰ σαφέστερα τῇ φύσει καὶ γνωριμώτερα· οὐ γὰρ ταὐτὰ ἡμῖν τε γνώριμα καὶ ἁπλῶς. διόπερ ἀνάγκη τὸν τρόπον τοῦτον προάγειν ἐκ τῶν ἀσαφεστέρων μὲν τῇ φύσει ἡμῖν δὲ σαφέστερων ἐπὶ τὰ σαφέστερα τῇ φύσει καὶ γνωριμώτερα.

Ἔστι δ' ἡμῖν πρῶτον δῆλα καὶ σαφῆ τὰ συγκεχυμένα μᾶλλον· ὕστερον δ' ἐκ τούτων γίνεται γνώριμα τὰ στοιχεῖα καὶ αἱ ἀρχαὶ διαιροῦσι ταῦτα. διὸ ἐκ τῶν καθόλου ἐπὶ τὰ καθ' ἕκαστα δεῖ προϊέναι· τὸ γὰρ ὅλον κατὰ τὴν αἴσθησιν γνωρι-

[a] καθόλου and καθ' ἕκαστα, when contrasted, usually mean 'general' and 'particular'; but here they are used in the other and less frequent sense of 'concrete whole' and 'constituent factors'; as in *De gen. et cor.* ii. 4 (331 a 20), where 'air,' 'water,' etc., are regarded as each of them constituting a single πᾶν or καθόλου, and their attributes 'moist,' 'warm,' etc., as each constituting a καθ' ἕκαστον. Were the words taken in their more usual sense there would not only be an insuperable difficulty here, but also a flat contradiction between this passage and the τὸ μὲν γὰρ

ARGUMENT (*continued*)

expressed thought that the starting-point (B ii) *of our present inquiry must be the confused notions of our predecessors about* (A), *which we proceed to analyse, with a view to arriving at the true fundamental notions.*—C.]

In all sciences that are concerned with principles or causes or elements, it is acquaintance with these that constitutes knowledge or understanding. For we conceive ourselves to know about a thing when we are acquainted with its ultimate causes and first principles, and have got down to its elements. Obviously, then, in the study of Nature too, our first object must be to establish principles.

Now the path of investigation must lie from what is more immediately cognizable and clear to us, to what is clearer and more intimately cognizable in its own nature; for it is not the same thing to be directly accessible to our cognition and to be intrinsically intelligible. Hence, in advancing to that which is intrinsically more luminous and by its nature accessible to deeper knowledge, we must needs start from what is more immediately within our cognition, though in its own nature less fully accessible to understanding.

Now the things most obvious and immediately cognizable by us are concrete and particular, rather than abstract and general; whereas elements and principles are only accessible to us afterwards, as derived from the concrete data when we have analysed them. So we must advance from the concrete whole to the several constituents which it embraces;[a] for it is the concrete whole that is the

καθόλου κατὰ τὸν λόγον γνώριμον, τὸ δὲ καθ' ἕκαστον κατὰ τὴν αἴσθησιν, of chap. v. (189 a 5 *sqq*. and 72 a 1 *sqq*.).

184 a μώτερον, τὸ δὲ καθόλου ὅλον τί ἐστιν· πολλὰ γὰρ περιλαμβάνει ὥσπερ μέρη τὸ καθόλου.

184 b 10 Πέπονθε δὲ ταὐτὸ τοῦτο τρόπον τινὰ καὶ τὰ ὀνόματα πρὸς τὸν λόγον· ὅλον γάρ τι καὶ ἀδιορίστως σημαίνει (οἷον ὁ κύκλος), ὁ δὲ ὁρισμὸς αὐτοῦ διαιρεῖ εἰς τὰ καθ' ἕκαστα. καὶ τὰ παιδία τὸ μὲν πρῶτον προσαγορεύει πάντας τοὺς ἄνδρας πατέρας καὶ μητέρας τὰς γυναῖκας, ὕστερον δὲ διορίζει τούτων ἑκάτερον.

CHAPTER II

ARGUMENT

How many primary constituents of things are there? A few elements or principles, or infinity of atoms? (184 b 15–25).

The dogma of Parmenides that all Nature is one and rigid, amounts to a denial that any science of 'changing material things' (which is what we mean by Nature) can exist at all; and therefore the discussion of it belongs to First Philosophy, not to Physics. Nevertheless, since it has a philosophical interest, and incidentally raises questions that concern the Physicist, we will not stand on our right to ignore it (b 25–185 a 20).

To begin with, then (since there are different ways of 'existing'), does the assertion 'everything that exists is one' refer to substantive, or quantitive, or qualitive existence? If to all, or to more than one of them, then there is more than one 'existence,' and Parmenides is wrong. If only to one, then that one must be substantive; for the other categories exist only in relation to the substantively existent. But to say either with Melissus that the existent is unlimited, or with Parmenides that it is limited, is to make it quantitive as well as substantive (a 20–b 5).

more readily cognizable by the senses. And by calling the concrete a 'whole' I mean that it embraces in a single complex a diversity of constituent elements, factors, or properties.

The relation of names to definitions will throw some light on this point; for the name gives an unanalysed indication of the thing ('circle,' for instance), but the definition analyses out some characteristic property or properties. A variant of the same thing may be noted in children, who begin by calling every man 'father' and every woman 'mother,' till they learn to sever out the special relation to which the terms properly apply.

CHAPTER II

ARGUMENT (*continued*)

Again, 'one' is just as ambiguous as 'existent.' It has three meanings: (1) *a continuous line may be one, though divisible infinitely;* (2) *there may be two names for one thing;* (3) *or an individual* (i.e. *an 'individable' unit*) *that ceases to exist as a unit when broken up, may be one. But in none of these senses can 'everything' be 'one'* (b 5–25).

Some later thinkers also—not understanding that the concrete oneness of a single subject may embrace a multitude of attributes and relations, and also believing (*through failure to distinguish between the copula and the verb substantive*) *that to say 'a man* is *pale-complexioned' would imply that 'pale-complexioned' existed as an entity—wrenched language in a vain attempt to avoid the* (*really quite unimpeachable*) *attribution of unity and multiplicity to one and the same subject* (b 25–186 a 3).

184b Ἀνάγκη δ' ἤτοι μίαν εἶναι τὴν ἀρχὴν ἢ πλείους· καὶ εἰ μίαν, ἤτοι ἀκίνητον (ὥς φησι Παρμενίδης καὶ Μέλισσος) ἢ κινουμένην (ὥσπερ οἱ φυσικοί, οἱ μὲν ἀέρα φάσκοντες εἶναι οἱ δ' ὕδωρ τὴν πρώτην ἀρχήν)· εἰ δὲ πλείους, ἢ πεπερασμένας ἢ ἀπείρους· καὶ εἰ πεπερασμένας πλείους δὲ μιᾶς, ἢ δύο ἢ τρεῖς ἢ τέτταρας ἢ ἄλλον τινὰ ἀριθμόν· καὶ εἰ ἀπείρους, ἢ οὕτως ὥσπερ Δημόκριτος, τὸ γένος ἓν σχήματι δὲ ἢ εἴδει διαφερούσας, ἢ καὶ ἐναντίας.

Ὁμοίως δὲ ζητοῦσι καὶ οἱ τὰ ὄντα ζητοῦντες πόσα· ἐξ ὧν γὰρ τὰ ὄντα ἐστί, πρῶτον ζητοῦσι ταῦτα πότερον ἓν ἢ πολλά, καὶ εἰ πολλά, πεπερασμένα ἢ ἄπειρα· ὥστε τὴν ἀρχὴν καὶ τὸ στοιχεῖον ζητοῦσι πότερον ἓν ἢ πολλά.

Τὸ μὲν οὖν εἰ ἓν καὶ ἀκίνητον τὸ ὂν σκοπεῖν οὐ
185a περὶ φύσεώς ἐστι σκοπεῖν· ὥσπερ γὰρ καὶ τῷ γεωμέτρῃ οὐκέτι λόγος ἐστὶ πρὸς τὸν ἀνελόντα τὰς ἀρχάς, ἀλλ' ἤτοι ἑτέρας ἐπιστήμης ἢ πασῶν κοινῆς, οὕτως οὐδὲ τῷ περὶ ἀρχῶν· οὐ γὰρ ἔτι ἀρχή ἐστιν, εἰ ἓν μόνον καὶ οὕτως ἕν ἐστιν· ἡ

---

[a] ἢ καὶ ἐναντίας, not only 'differing in form,' but also 'of contrasted qualities' (particles of bone, flesh, etc.). The reference, apparently, is to Anaxagoras. *Cf.* I. iv, p. 43.

[b] [For πρῶτον Bonitz conjectured πρώτων, *i.e.* 'primary constituents.' Similarly Plato (*Soph.* 242 c), after discussing Parmenides, reviews the other philosophers who have undertaken to decide the number and nature of τὰ ὄντα (ultimately real constituents of things): are they two or three, one or many, or both one and many, etc. ?—C.]

Well, then, there must be either one principle of Nature or more than one. And if only one, it must be either rigid, as Parmenides and Melissus say, or modifiable, as the Physicists say, some declaring air to be the first principle, and others water. If, on the other hand, there are more principles than one, they must be either limited or unlimited in number. And if limited, though more than one, they must be two or three or four, or some other definite number. And if they are unlimited, they must either be, as Democritus held, all of the same kind generically, though differing in shape and sub-characteristics, or of contrasted nature as well.[a]

The thinkers who inquire into the number of 'absolute entities,' again, follow the same line. For their first[b] question is whether the constituents of which things are composed are one or more than one; and, if more than one, are they limited or unlimited? So they, too, are inquiring whether there is one principle or ultimate constituent, or many.

Now, as to the contention that all existence is one and is rigidly unchanging, we might say that it does not really concern the student of Nature, and he need not investigate it; any more than it concerns the geometer to argue with one who denies the geometrical axioms. Such questions must be dealt with either by some other special science or by a fundamental discipline that underlies all the sciences. And so it is in this matter of the unity of the natural principle, for if it is only one, and one in the sense of rigidity, it is not a principle at all; for a principle must be the principle of some thing

γὰρ ἀρχὴ τινὸς ἢ τινῶν. ὅμοιον δὴ τὸ σκοπεῖν εἰ οὕτως ἓν καὶ πρὸς ἄλλην θέσιν ὁποιανοῦν διαλέγεσθαι τῶν λόγου ἕνεκα λεγομένων (οἷον τὴν Ἡρακλείτειον, ἢ εἴ τις φαίη ἄνθρωπον ἕνα τὸ ὂν εἶναι) [ἢ λύειν λόγον ἐριστικόν, ὅπερ ἀμφότεροι μὲν ἔχουσιν οἱ λόγοι, καὶ ὁ Μελίσσου καὶ ὁ Παρμενίδου—καὶ γὰρ ψευδῆ λαμβάνουσι καὶ ἀσυλλόγιστοί εἰσιν—μᾶλλον δ' ὁ Μελίσσου φορτικὸς καὶ οὐκ ἔχων ἀπορίαν]. ἀλλ' ἑνὸς ἀτόπου δοθέντος τὰ ἄλλα συμβαίνει· τοῦτο δὲ οὐδὲν χαλεπόν.[1] ἡμῖν δ' ὑποκείσθω τὰ φύσει ἢ πάντα ἢ ἔνια κινούμενα εἶναι· δῆλον δ' ἐκ τῆς ἐπαγωγῆς. ἅμα δ' οὐδὲ λύειν ἅπαντα προσήκει, ἀλλ' ἢ ὅσα ἐκ τῶν ἀρχῶν τις ἐπιδεικνὺς ψεύδεται, ὅσα δὲ μή, οὔ (οἷον τὸν τετραγωνισμὸν τὸν μὲν διὰ τῶν τμημάτων γεωμετρικοῦ διαλῦσαι, τὸν δ' Ἀντιφῶντος οὐ γεωμε-

[1] [Bekker proposed to excise 8 ὅπερ ἀμφότεροι . . . 12 χαλεπόν, comparing 186 a 6–10. W. bracketed ἢ λύειν λόγον ἐριστικόν as 'an editorial interpolation' and καὶ γὰρ ψευδῆ . . . ἀπορίαν (*sic*) 'as a stray from 186 a 7.' But he must have meant to excise also ὅπερ ἀμφότεροι . . . Παρμενίδου, which cannot stand alone and is not translated. 9 ἀλλ' ἑνὸς . . . 12 χαλεπόν is bracketed at 186 a 9. W.'s note is :

'The whole of the passage which appears in Chapters II. and III. in the mss. and editions was regarded by Bekker, followed by Prantl, as belonging to the latter context. But if it is taken as a whole it contains a flat contradiction: ἀσυλλόγιστοί εἰσιν . . . ἀλλ' ἑνὸς ἀτόπου δοθέντος τἆλλα συμβαίνει — "Their syllogisms are false . . . but their conclusions follow from their premises." If the passage is divided between the two places, as I suggest, the contradiction

or things other than its naked self. To consider, therefore, whether the natural principle is one in this sense, is like discussing any other paradox that is set up just for the sake of arguing, like the paradox of Heracleitus,[a] or the contention (should anyone advance it) that the totality of existence is one single man. If their fundamental paradox is admitted, there is nothing strange in all their other paradoxes following from it. Let us then start from the datum that things of Nature, or (to put it at the lowest) some of them, do move and change, as is patent to observation ; and let us make a note that we are not bound to answer every kind of objection we may meet, but only such as are erroneously deduced from the accepted principles of the science in question. Thus, it is the geometer's business to refute the squaring of the circle that proceeds by way of equating the segments, but he need not

[a] [That 'opposites' are 'one,' *e.g.* 'good and evil are one,' Heracleitus, *frag.* 58 (57 Byw.). *Cf.* b 20. Heracleitus was understood by some to be denying the maxim of contradiction, *Met.* 1005 b 23.—C.]

---

disappears ; for in Chapter II. Aristotle says that, if you accept the paradoxical *conclusion* that Nature is a rigid unit, all the other paradoxes of the school (as to time and space for instance) inevitably follow. In Chapter III., on the other hand, he starts with the false (because ambiguous) *premise* that "nothing can come out of nothing" and shows that his opponents do not argue soundly even from that. The words ἢ λύειν λόγον ἐριστικόν (a kind of echo of the ἀμφότεροι γὰρ ἐριστικῶς συλλογίζονται) appear to be an unintelligent editorial attempt to make room for both terms of the contradiction.'

The repetition of the illustration from the geometer (185 a 1–3 and 14–17) suggests that this whole paragraph is a conflation of two recensions.—C.]

185 a τρικοῦ). οὐ μὴν ἀλλ᾿ ἐπειδὴ περὶ φύσεως μὲν οὔ, φυσικὰς δὲ ἀπορίας συμβαίνει λέγειν αὐτοῖς, ἴσως ἔχει καλῶς ἐπὶ μικρὸν διαλεχθῆναι περὶ αὐτῶν· ἔχει γὰρ φιλοσοφίαν ἡ σκέψις.

Ἀρχὴ δὲ οἰκειοτάτη πασῶν, ἐπειδὴ πολλαχῶς λέγεται τὸ ὄν,[1] πῶς λέγουσιν οἱ λέγοντες εἶναι ἓν τὰ πάντα, πότερον οὐσίαν τὰ πάντα ἢ ποσὰ ἢ ποιά·

[1] [After τὸ ὂν E has ἰδεῖν.—C.]

---

[a] The problem of squaring the circle as it was originally conceived is to construct, by means of ruler and compasses alone, a rectilinear figure whose area is equal to that of a given circle. And it is to this problem that the three solutions mentioned by Aristotle address themselves.

1. *The solution by segments*: Simplicius tells us that this refers to some very ingenious constructions of *menisci*, or lunes, by Hippocrates of Chios. Simplicius gives us the constructions and explains them at length. The fallacy in this demonstration, it appears, consists in confusing lunes of different forms, and assuming that, because certain kinds of lunes (as Hippocrates showed) can be squared, others, required for the quadrature of the circle, can also be squared, whereas in fact they cannot. It seems unlikely that Hippocrates himself should be guilty of this confusion, but whoever the author of this demonstration was, he purports to have based it on the accepted principles of geometry, and it is therefore, Aristotle says, the business of the geometer to refute it, by pointing out where the error lies.

2. *Antiphon's solution*: Antiphon is only known to us by this passage, but Simplicius informs us that the construction he proposed was to inscribe a square in the circle, and on each of the four sides regarded as chords of the circle to build two chords meeting in the midway point of the circumference above it. We shall then have an octagon inscribed in the circle, but the chords which constitute its sides will still be

consider Antipho's solution.[a] And yet, after all, though their (the Eleatics') contention does not concern the study of Nature, it does incidentally raise points that are of interest to the physicist; so perhaps it will be as well to examine it briefly,—especially as it has a certain philosophical interest of its own.

Now since the term 'existent' is itself ambiguous,[b] it lies at the very heart of the matter to inquire whether (1) they who assert all existing things to be 'one' are thinking of all existing things substantively or quantitively or qualitively. Or (2) are all existing

short in their total periphery of the length of the circumference of the circle. But, says Antiphon, you must continue the process until the sides are so small as to fall coincidently upon their respective sections of the circumference, and then you will have a polygon equal in area to the circle.

To this Aristotle simply remarks that a geometer, as such, has nothing to say, because it assumes that the circumference is already made up of minute straight lines, so that a very small chord may be coincident with a corresponding section of it; this is contrary to the axioms and principles of geometry, and the geometer is only concerned with what professes to be based on those principles.

3. There was also one Bryson, referred to in the *Posterior Analytics*, ix., 75 b 40. He argued that any polygon inscribed in the circle must have a less area, and any polygon circumscribed to it must have a greater area, than the circle. It follows that there must be somewhere a polygon which equals it (observe that the sides of such a polygon would obviously cut the circumference). To this demonstration in itself Aristotle raises no objection, but he says that it is a general or logical, rather than a geometrical demonstration. For the geometer must show you not only that such a polygon is possible, but how you are to construct it. Bryson does not do this.

The question of squaring the circle is further discussed in an excursus, p. 98.

[b] See Gen. Introd. p. liii.

185 a καὶ πάλιν πότερον οὐσίαν μίαν τὰ πάντα (οἷον ἄνθρωπον ἕνα ἢ ἵππον ἕνα ἢ ψυχὴν μίαν) ἢ ποιόν, ἓν δὲ τοῦτο (οἷον λευκὸν ἢ θερμὸν ἢ τῶν ἄλλων τι τῶν τοιούτων)· ταῦτα γὰρ πάντα διαφέρει τε πολὺ καὶ ἀδύνατα λέγειν.

Εἰ μὲν γὰρ ἔσται καὶ οὐσία καὶ ποσὸν καὶ ποιόν —καὶ ταῦτα εἴτ' ἀπολελυμένα ἀπ' ἀλλήλων εἴτε μή—πολλὰ τὰ ὄντα· εἰ δὲ πάντα ποιὸν ἢ ποσόν, εἴτ' οὔσης οὐσίας εἴτε μὴ οὔσης, ἄτοπον, εἰ δεῖ ἄτοπον λέγειν τὸ ἀδύνατον. οὐθὲν γὰρ τῶν ἄλλων χωριστόν ἐστι παρὰ τὴν οὐσίαν· πάντα γὰρ καθ' ὑποκειμένου τῆς οὐσίας λέγεται. Μέλισσος δὲ τὸ ὂν ἄπειρον εἶναί φησιν. ποσὸν ἄρα τι τὸ ὄν· τὸ γὰρ ἄπειρον ἐν τῷ ποσῷ, οὐσίαν δὲ ἄπειρον εἶναι
185 b ἢ ποιότητα ἢ πάθος οὐκ ἐνδέχεται εἰ μὴ κατὰ συμβεβηκός (εἰ ἅμα καὶ ποσὰ ἄττα εἶεν)· ὁ γὰρ τοῦ ἀπείρου λόγος τῷ ποσῷ προσχρῆται, ἀλλ' οὐκ οὐσίᾳ οὐδὲ τῷ ποιῷ. εἰ μὲν τοίνυν καὶ οὐσία ἐστὶ καὶ ποσόν, δύο καὶ οὐχ ἓν τὸ ὄν· εἰ δ' οὐσία μόνον,

---

[a] On the supposition that the 'life' or 'soul' is an independent entity, to which, however, Aristotle must not be supposed to commit himself here.

[b] [I understand Aristotle's main alternative here to be: Do they mean (1) that all things *belong to only one category*: 'all things are Substance,' or 'all things are Quality,' etc., and, if so, to which category? or (2) that all things are *a single thing* belonging to one category: 'all things are a single substance' (a man), or 'a single quality' (white)? —C.]

[c] [This refers to the first alternative. (*a*) If all things belong to one category only, they must choose which category; 'for if all things are to belong to several categories, there must be a plurality.' He then points out (*b*) that the only

things 'one substance,' like 'one man' or 'one horse' or 'one soul'? [a] Or is it that they are all one in quality, (say) all the same 'white,' or 'hot,' or so forth, so that the same qualitive predication can be made of them all? [b] For though all such assertions are impossible, yet they are far from being all identical.

(1) Thus (*a*) if all things are to be one both substantively, quantitively, and qualitively, then (whether we consider these forms of 'being' as objectively separable from each other or not) in any case existences are many and not only one.[c] Whereas (*b*) if it be meant that all things are one *magnitude*, or one *quality*, then, whether there be any substantive existence at all or no, the assertion is absurd—if we may so call the impossible. For none of the categories except 'substance' can exist independently, since all the rest must necessarily be predicated of some substance as their *subjectum*. But Melissus [d] says that the Universe is unlimited, which would make it a magnitude or *quantum*, since 'limited' and 'unlimited' pertain to *quanta* only, so that no substantive being and no quality or affection can, as such, be called unlimited (though it may incidentally be a *quantum* also); for the conception of a *quantum* enters into the definition of 'unlimited,' whereas the conception of substantive existence, or quality, does not. If then (*a*) the 'existent' is both a substance and a magnitude, it is two and not one; whereas (*b*) if it is a substance

possible category is Substance; but then Melissus ought not to say that Being is unlimited; for this implies that it is also a magnitude (quantity).—C.]

[d] [*Frag.* 2.—C.]

οὐκ ἄπειρον, οὐδὲ μέγεθος ἕξει οὐδέν· ποσὸν γάρ τι ἔσται.

Ἔτι ἐπεὶ καὶ αὐτὸ τὸ ἓν πολλαχῶς λέγεται ὥσπερ καὶ τὸ ὄν, σκεπτέον τίνα τρόπον λέγουσιν εἶναι ἓν τὸ πᾶν. λέγεται δ' ἓν ἢ τὸ συνεχὲς ἢ τὸ ἀδιαίρετον ἢ ὧν ὁ λόγος ὁ αὐτὸς καὶ εἷς ὁ τοῦ τί ἦν εἶναι, ὥσπερ μέθυ καὶ οἶνος.

Εἰ μὲν τοίνυν συνεχές, πολλὰ τὸ ἕν· εἰς ἄπειρον γὰρ διαιρετὸν τὸ συνεχές. (ἔχει δ' ἀπορίαν περὶ τοῦ μέρους καὶ τοῦ ὅλου—ἴσως δὲ οὐ πρὸς τὸν λόγον ἀλλ' αὐτὴν καθ' αὑτήν—πότερον ἓν ἢ πλείω τὸ μέρος καὶ τὸ ὅλον, καὶ πῶς ἓν ἢ πλείω· καὶ εἰ πλείω, πῶς πλείω· καὶ περὶ τῶν μερῶν τῶν μὴ συνεχῶν· καὶ εἰ τῷ ὅλῳ ἓν ἑκάτερον ὡς ἀδιαίρετον, ὅτι καὶ αὐτὰ αὑτοῖς.)

Ἀλλὰ μὴν εἰ ὡς ἀδιαίρετον, οὐθὲν ἔσται ποσὸν οὐδὲ ποιόν· οὐδὲ δὴ ἄπειρον τὸ ὄν, ὥσπερ Μέλισσός

---

[a] Συνεχές (continuous) and ἀδιαίρετον (indivisible) each have several meanings. Both words may be used in the strict mathematical sense, or in the popular sense when συνεχές would apply to any simple or complex body, so held together—whether by nature or by art—that by moving a part you move the whole (1016 b 5 *sq.*), and ἀδιαίρετον to anything which (though not actually undividable) is taken as a whole. Here Aristotle (after his wont) is thinking of both words primarily in their mathematical sense, which he regards as typical; the 'continuous' applying to magnitudes of one, two, or three dimensions, and the 'indivisible' to their limiting points, lines, and surfaces; points being indivisible as to all three dimensions, lines as to depth and breadth, surfaces as to depth only. *Cf.* Gen. Introd. p. lxxxiv.

only, it cannot be unlimited, nor indeed can it have any dimension at all, for otherwise it would be quantitive as well as substantive.

(2) Again, since 'one' is itself quite as ambiguous a term as 'existent,' we must ask in what sense the 'All' is said to be *one*. (*a*) Continuity establishes one kind of unity; (*b*) indivisibility another;[a] (*c*) identity of definition and of constituent characteristics yet a third; as for instance (the Greek) *methy* and *oinos* (both of which mean 'wine') are 'one and the same' thing.

So then, (*a*) if a single *continuum* is what is meant by 'one,' it follows that 'the One' is many, for every *continuum* is divisible without limit. (And this suggests[b] the question—not to our present purpose, perhaps, but interesting on its own account: whether the part and the whole of a *continuum* are to be regarded as a unity or as existing severally, and in either case in what sense they are a unity or plurality, and if they are regarded as a plurality in what sense a plurality. And the question arises again with respect to a discontinuous whole, consisting of unlike parts: In what sense do such parts exist, as several from the whole? Or if each is indivisibly one with the whole, are they so with each other?)

Whereas (*b*) if 'one' signifies an 'indivisible,' then it excludes quantity (as well as quality); and therefore, the One Being, not being a *quantum*, cannot be 'unlimited' as Melissus declares it to be; nor indeed

[b] 'And this suggests, etc.' This admittedly irrelevant digression I take to refer to the subject elaborately discussed in IV. iii. apropos of the question whether a body can 'contain itself' or 'be its own place.' A reference to the whole passage will, I think, justify the expansion, in the translation, of this highly elliptical reference.

185 b φησιν· οὐδὲ πεπερασμένον, ὥσπερ Παρμενίδης, τὸ γὰρ πέρας ἀδιαίρετον, οὐ τὸ πεπερασμένον.

Ἀλλὰ μὴν εἰ τῷ λόγῳ ἓν τὰ ὄντα πάντα (ὡς λώπιον καὶ ἱμάτιον), τὸν Ἡρακλείτου λόγον συμβαίνει λέγειν αὐτοῖς· ταὐτὸν γὰρ ἔσται ἀγαθῷ καὶ κακῷ εἶναι καὶ μὴ ἀγαθῷ καὶ ἀγαθῷ—ὥστε τὸ αὐτὸ ἔσται ἀγαθὸν καὶ οὐκ ἀγαθὸν καὶ ἄνθρωπος καὶ ἵππος, καὶ οὐ περὶ τοῦ ἓν εἶναι τὰ ὄντα ὁ λόγος ἔσται αὐτοῖς ἀλλὰ περὶ τοῦ μηδέν—καὶ τὸ τοιῳδὶ εἶναι καὶ τοσῳδὶ ταὐτόν.

Ἐθορυβοῦντο δὲ καὶ οἱ ὕστεροι τῶν ἀρχαίων ὅπως μὴ ἅμα γένηται αὐτοῖς τὸ αὐτὸ ἓν καὶ πολλά. διὸ οἱ μὲν τὸ ἔστιν ἀφεῖλον (ὥσπερ Λυκόφρων), οἱ δὲ τὴν λέξιν μετερρύθμιζον, ὅτι ὁ ἄνθρωπος οὐ λευκός ἐστιν ἀλλὰ λελεύκωται, οὐδὲ βαδίζων ἐστὶν ἀλλὰ βαδίζει, ἵνα μή ποτε τὸ ἔστι προσάπτοντες πολλὰ εἶναι ποιῶσι τὸ ἕν, ὡς μοναχῶς λεγομένου τοῦ ἑνὸς ἢ τοῦ ὄντος. πολλὰ δὲ τὰ ὄντα ἢ λόγῳ (οἷον ἄλλο τὸ λευκῷ εἶναι καὶ μουσικῷ, τὸ δ' αὐτὸ ἄμφω· πολλὰ ἄρα τὸ ἕν) ἢ διαιρέσει, ὥσπερ τὸ ὅλον

---

[a] [*Cf.* 185 a 7. If two contradictory predicates can be simultaneously asserted of the same thing, we may just as well say 'all things are not-one' (*quasi* μηδ' ἕν) as say that they are 'one.'—C.]

[b] [*Cf.* Plato, *Soph.* 251 A, *Philebus* 14 D. Lycophron was a sophist, a pupil of Gorgias (Zeller, I.[6], 1323[3]).—C.]

'limited' as Parmenides has it, for it is the limit only, not the *continuum* which it limits, that complies with the condition of indivisibility.

Lastly (*c*) if the contention is that all things are identically 'one and the same' by definition (as 'clothes' and 'garments' are) we are back again at the Heracleitean paradox; for, in that case, being good and being bad will be the same and being not-good the same as being good—with the consequence that the same thing will be both good and not good, or both a man and a horse, and we shall no longer be maintaining that all existences are one, so much as that none of them is anything [a]—and being of a certain quality will be the same thing as being of a certain magnitude.

Indeed some of the later ancients were themselves disturbed by the danger of finding themselves admitting that the same thing is both one and many; [b] and therefore some of them, like Lycophron, banned the word 'is' altogether; while others did such violence to language as to substitute for 'the man is pale-complexioned' the phrase 'the man has complexion-paled,' or to admit 'the man walks' but not 'the man is walking,' for fear of making the one existent appear many by affixing the word 'is'; which nervousness resulted from their equating 'is' with 'exists' and ignoring the different senses of 'one' or of 'being.' But the fact is that one thing obviously *can* be many, either in the sense of being several conceptually distinct things at once (for though it is one thing to be pale-complexioned and another to be cultivated, yet one and the same man can be both, so that 'one' is 'many'); or in the

186 a καὶ τὰ μέρη. ἐνταῦθα δὲ ἤδη ἠπόρουν καὶ ὡμολόγουν τὸ ἓν πολλὰ εἶναι—ὥσπερ οὐκ ἐνδεχόμενον ταὐτὸν ἕν τε καὶ πολλὰ εἶναι, μὴ τἀντικείμενα δέ· ἔστι γὰρ τὸ ἓν καὶ δυνάμει καὶ ἐντελεχείᾳ.

CHAPTER III

ARGUMENT

Elaboration of the reasoning of the last chapter, showing that the school of Parmenides, in addition to its faulty deductions, rests on a false basis, for it ignores the conceptual distinctions between the different ways of existing and the different senses in which we can say of A that it 'is' B (186 a 4–10).

Melissus is hopelessly inconsequent. His syllogisms are faulty. He confuses the time *at which a given state of things 'begins' with the* place *from which one thing, or territory, 'begins' and at which another ends. Nor does he see the difference between saying that a thing cannot 'begin to be' unless it does so at some definite point of time, and saying that it cannot 'begin to change' unless it does so at some definite point within itself; or between saying that it is 'all one' and that it is incapable of movement or even of change* (a 10–22).

[*Aristotle next examines the possible meanings of Parmenides' doctrine 'Only the One Being* is.' *Parmenides ignored the various different senses of 'is'* (a 22–25); *and, even if 'is' has only one sense, the inference that there can be only one thing which* is *was false* (a 25–32). *If 'is'*

Τόν τε δὴ τρόπον τοῦτον ἐπιοῦσιν ἀδύνατον φαίνεται τὰ ὄντα ἓν εἶναι· καὶ ἐξ ὧν ἐπιδεικνύουσι,

sense in which a whole can be regarded as being the sum of the parts into which it can be divided. This latter case brought our philosophers to a stand ; and they had to admit multiplicity in unity, unwillingly, as though it were a paradox. And so, of course, it would be, if 'one' and 'many' were used in the sense in which they contradict each other, but not if, for instance, we are regarding the 'one' both in its actual unity and its potential multiplicity.

CHAPTER III

ARGUMENT (*continued*)

means that the subject of his proposition is identical with Being, *and this subject has attributes, it can be shown that (on his principles) the One Being is a nonentity* (a 32–b 12).

The One Being cannot have any magnitude (b 12–14).

A term which 'is' in the substantive sense (e.g. *Man*) *contains in its definition other terms* (e.g. *Biped, Animal*). *The Eleatics cannot explain the status of these defining terms either as accidents or as indivisible constituents without admitting the existence of plurality* (b 14–35)

Plato is criticized for 'giving in' to Parmenides' arguments, without attacking his fundamental premisses (187 a 1–10).

'Being,' then, cannot be 'one' in the sense in which the Eleatics assert this (a 10).

(*H. Maier*, Syllogistik des Aristoteles (*Tübingen*, 1896), *ii. pp.* 277 *ff., classifies the fallacies committed by Aristotle's predecessors through confusing the various senses of 'is' in propositions. Maier's results are summarized by Mr. Ross*, Metaphysics (1924), *i. p. lxxxviii.)—C.*]

Following up this line, we shall see that it is impossible for 'all things to be one,' and we shall have

186 a λύειν οὐ χαλεπόν. ἀμφότεροι γὰρ ἐριστικῶς συλλογίζονται καὶ Μέλισσος καὶ Παρμενίδης—καὶ γὰρ ψευδῆ λαμβάνουσι καὶ ἀσυλλόγιστοί εἰσιν αὐτῶν οἱ λόγοι—μᾶλλον δ' ὁ Μελίσσου φορτικὸς καὶ οὐκ ἔχων ἀπορίαν [ἀλλ' ἑνὸς ἀτόπου δοθέντος τἆλλα συμβαίνει· τοῦτο δ' οὐθὲν χαλεπόν].[1]

Ὅτι μὲν οὖν παραλογίζεται Μέλισσος, δῆλον· οἴεται γὰρ εἰληφέναι, εἰ τὸ γενόμενον ἔχει ἀρχὴν ἅπαν, ὅτι καὶ τὸ μὴ γενόμενον οὐκ ἔχει. εἶτα καὶ τοῦτο ἄτοπον, τὸ παντὸς οἴεσθαι εἶναι ἀρχὴν τοῦ πράγματος καὶ μὴ τοῦ χρόνου, καὶ γενέσεως μὴ τῆς ἁπλῆς ἀλλὰ καὶ ἀλλοιώσεως, ὥσπερ οὐκ ἀθρόας γινομένης μεταβολῆς.

Ἔπειτα διὰ τί ἀκίνητον, εἰ ἕν; ὥσπερ γὰρ καὶ τὸ μέρος ἓν ὄν—τοδὶ τὸ ὕδωρ—κινεῖται ἐν ἑαυτῷ, διὰ τί οὐ καὶ πᾶν; ἔπειτα ἀλλοίωσις διὰ τί οὐκ

[1] [See note on 185 a 12.—C.]

---

[a] He argues in the form 'if all *A's* are *not-B's*, then all *not-A's* are *B's*,' which is a false syllogism; for the negation of the antecedent does not carry the negation of the consequent, though the negation of the consequent does carry that of the antecedent. Thus the correct inference from 'all *A's* are *not-B's*' would be 'all *B's* are *not-A's*.'

[b] [Melissus, *Frag.* 1, 'That which was, was always and always will be. For if it had come into being, before it came into being it must have been nothing; if, then, it was nothing, nothing could ever come out of nothing.' *Frag.* 2, 'Since then, it did not come into being, and since it is and always was and always will be, it has no beginning or end, but is unlimited.' Aristotle unfairly accuses Melissus of arguing from 'no beginning in time' to 'no beginning (limit) in space' (*cf. Soph. El.* 167 b 13), and from 'no absolute coming into being out of nothing' to 'no beginning of change of quality.' At 253 b 25, freezing is given as an instance of ἀθρόα ἀλλοίωσις.—C.]

[c] [*Or possibly*, 'for just as a part (of the world), *e.g.* this mass of water, though it is one, can *move on its own axis*

no difficulty in refuting the arguments by which the assertion that they are is supported. For both Parmenides and Melissus argue sophistically, inasmuch as they make unsound assumptions and argue unsoundly from them. But Melissus offends the more grossly of the two, so as really to raise no valuable point for discussion.

The false reasoning of Melissus is palpable;[a] for, assuming that 'all that comes into existence has a beginning,' he deduces from it 'all that does not come into existence has no beginning.' And, moreover, the assumption itself 'whatever comes into existence has a beginning' is untenable, in so far as 'began some-when' is taken (as Melissus takes it) to be equivalent to 'begins some-where' (so that if the Universe 'had no beginning' it 'can have no limit,' and is 'unbounded'): and again in so far as no distinction is made between the thing itself having to begin-to-be at 'some particular point of time,' and a modification of the thing having to start from 'some particular point within the thing itself,'—as if there could not be a simultaneous modification over the whole field affected.[b]

And again why should unity involve rigidity? For if a definite body of water, regarded as a unit without internal distinctions of quality, may have currents of motion within itself, why not the Universe?[c] And, in any case, why not modifications other than those

(without change of place), why should not the universe do the same?' So Simplicius, 112. 16. *Cf.* 240 a 30, ἐπὶ τῆς σφαίρας καὶ ὅλως τῶν ἐν αὑτοῖς κινουμένων. The argument is, at any rate, valid as applied to Parmenides' limited spherical Being, which could rotate without losing its unity in any sense. The arguments in this paragraph seem not to be directed only against Melissus.—C.]

186 a ἂν εἴη; ἀλλὰ μὴν οὐδὲ τῷ εἴδει οἷόν τε ἓν εἶναι (πλὴν τῷ ἐξ οὗ· οὕτως δὲ ἓν καὶ τῶν φυσικῶν τινες λέγουσιν, ἐκείνως δ' οὔ)· ἄνθρωπος γὰρ ἵππου ἕτερον τῷ εἴδει, καὶ τἀναντία ἀλλήλων.

Καὶ πρὸς Παρμενίδην δὲ ὁ αὐτὸς τρόπος τῶν λόγων, καὶ εἴ τινες ἄλλοι εἰσὶν ἴδιοι· καὶ ἡ λύσις τῇ μὲν ὅτι ψευδής, τῇ δὲ ὅτι οὐ συμπεραίνεται—ψευδὴς μὲν ᾗ ἁπλῶς λαμβάνει τὸ ὂν λέγεσθαι, λεγομένου πολλαχῶς, ἀσυμπέραντος δὲ ὅτι, εἰ μόνα τὰ λευκὰ ληφθείη, σημαίνοντος ἓν τοῦ λευκοῦ, οὐθὲν ἧττον πολλὰ τὰ λευκὰ καὶ οὐχ ἕν. οὔτε γὰρ τῇ συνεχείᾳ ἓν ἔσται τὸ λευκὸν οὔτε τῷ λόγῳ· ἄλλο γὰρ ἔσται τὸ εἶναι λευκῷ καὶ τὸ δεδεγμένῳ, καὶ οὐκ ἔσται παρὰ τὸ λευκὸν οὐθὲν χωριστόν· οὐ γὰρ ᾗ χωριστὸν ἀλλὰ τῷ εἶναι ἕτερον τὸ λευκὸν καὶ ᾧ ὑπάρχει. ἀλλὰ τοῦτο Παρμενίδης οὔπω ἑώρα.

Ἀνάγκη δὴ λαβεῖν μὴ μόνον ἓν σημαίνειν τὸ ὄν, καθ' οὗ ἂν κατηγορηθῇ, ἀλλὰ καὶ ὅπερ ὂν καὶ

---

[a] [Both Parmenides and Melissus denied the reality of any change.—C.]

[b] [*i.e.* imagine a philosopher who asserts, ' Only what *is white* exists,' instead of Parmenides' proposition : ' Only what *is* (τὸ ὄν) exists.' Even so, there might still be any number of white things in existence. Similarly Parmenides' assertion is consistent with the existence of any number of things that *are*. Also, ' what is white ' need not be a single continuous thing (ἕν, συνεχές, Parm. 8. 6).—C.]

[c] [Even if all that exists were one white thing, there would still be a conceptual difference between whiteness and being a subject which has ' white ' as attribute, without involving the separate existence of any second thing (for the attribute does not exist separately). *Cf.* Maier, *Syllogistik*, ii. 282, 284.—C.]

of local movement?[a] But of course the Universe cannot really be homogeneous, like a mass of water (except in the sense of its ultimate constituent being uniform; in which sense, though not in the other, some of the Physicists also have maintained its unity); for a man is patently different in kind from a horse, and opposites are different in kind from each other.

The same kind of argument will apply to Parmenides, as well as such other arguments as particularly apply to his treatment; for here too the refutation turns on the falsity of his assumption and the unsoundness of his deductions. His assumption is false inasmuch as he treats 'being' as having only one meaning, whereas in reality it has several. And his inferences are false, because, even if we accepted such a proposition as 'nothing that is not white exists,' and if 'white' had only one meaning, still the white things would be many and not one.[b] Obviously not one in the sense of a homogeneous *continuum*. Nor in the sense of a conceptual identity, for there remains a conceptual distinction between the subject in which the whiteness is seated, and the qualification of 'being white,' and that distinction does not involve the separate existence of anything alongside of 'that which is white';[c] because the plurality is established not by there being something separate, but by there being a conceptual distinction between white and the subject in which it inheres. But Parmenides had not yet arrived at this principle.

Parmenides, then, must assume not only that the word 'is,' whatever it may be predicated of, has only one meaning, but also that it means 'is identical-

186 a ὅπερ ἕν. τὸ γὰρ συμβεβηκὸς καθ' ὑποκειμένου τινὸς λέγεται· ὥστε ᾧ συμβέβηκε τὸ ὂν οὐκ ἔσται
186 b (ἕτερον γὰρ τοῦ ὄντος)· ἔσται τι ἄρα οὐκ ὄν. οὐ δὴ ἔσται ἄλλῳ ὑπάρχον τὸ ὅπερ ὄν. οὐ γὰρ ἔσται ὄντι αὐτῷ εἶναι, εἰ μὴ πολλὰ τὸ ὂν σημαίνει οὕτως ὥστε εἶναί τι ἕκαστον· ἀλλ' ὑπόκειται τὸ ὂν σημαίνειν ἕν. εἰ οὖν τὸ ὅπερ ὂν μηδενὶ συμβέβηκεν, ἀλλ' ἐκείνῳ, τί μᾶλλον τὸ ὅπερ ὂν σημαίνει τὸ ὂν ἢ μὴ ὄν; εἰ γὰρ ἔσται τὸ ὅπερ ὂν ταὐτὸ καὶ λευκόν, τὸ λευκῷ δ' εἶναι μὴ ἔστιν ὅπερ ὄν (οὐδὲ γὰρ συμβεβηκέναι αὐτῷ οἷόν τε τὸ ὄν· οὐθὲν γὰρ ὂν ὃ οὐχ ὅπερ ὄν), οὐκ ἄρα ὂν τὸ λευκόν—οὐχ οὕτω δὲ ὥσπερ τι μὴ ὄν, ἀλλ' ὅλως μὴ ὄν. τὸ ἄρα ὅπερ ὂν οὐκ ὄν· ἀληθὲς γὰρ εἰπεῖν ὅτι λευκόν, τοῦτο δὲ οὐκ ὂν ἐσήμαινεν· ὥστ' εἰ

with-Being,' and that 'is one' means 'is identical-with-Unity.' ('Being' will then no longer be regarded as an attribute); for an attribute is ascribed to some subject (other than itself); consequently, the subject to which 'being' (supposing it to be an attribute) is ascribed will have no being at all; for it will be other than 'being' (the attribute ascribed to it) and so will be something which (simply) is not. Accordingly 'identical-with-Being' (the sole meaning we have given to the word 'is') cannot be an attribute of some subject other than itself. For in that case its subject cannot be a thing which is, unless 'being' denotes a plurality of things in the sense that each is some thing that is. But it has been assumed that 'being' denotes only one thing. If, then, what is identical-with-Being is not an attribute of anything else, but (a subject, so that other things are attributes) of it, is there any reason to say that 'identical-with-Being' denotes that which *is* (a real entity) any more than that which *is not* (a nonentity)? (I can show that there is no reason.) For suppose the thing which is identical-with-Being also has the attribute 'white,' and that 'being white' is not identical-with-Being—(the only sense in which it can 'be' at all), for it cannot even have 'being' as an attribute, because (*ex hypothesi*) nothing except what is identical-with-Being has any being at all—then it follows that white *is not*, and that not merely in the sense that it *is not this or that*, but in the sense of an absolute nonentity. Accordingly, that which is identical-with-Being (our subject) will be a nonentity; for (we assumed that) it is true to say of it that it is white, and this means that it is a nonentity. So that even if (to escape this diffi-

186 b καὶ τὸ λευκὸν σημαίνει ὅπερ ὄν, πλείω ἄρα σημαίνει τὸ ὄν.

[Οὐ τοίνυν οὐδὲ μέγεθος ἕξει τὸ ὄν, εἴπερ ὅπερ ὂν τὸ ὄν· ἑκατέρῳ γὰρ ἕτερον τὸ εἶναι τῶν μορίων.][1]

Ὅτι δὲ διαιρεῖται τὸ ὅπερ ὂν εἰς ὅπερ ὄν τι ἄλλο, καὶ τῷ λόγῳ φανερόν· οἷον ὁ ἄνθρωπος εἰ ἔστιν ὅπερ ὄν τι, ἀνάγκη καὶ τὸ ζῷον ὅπερ ὄν τι εἶναι καὶ τὸ δίπουν. εἰ γὰρ μὴ ὅπερ ὄν τι, συμβεβηκότα ἔσται· ἢ οὖν τῷ ἀνθρώπῳ ἢ ἄλλῳ τινὶ ὑποκειμένῳ. ἀλλ' ἀδύνατον· συμβεβηκός τε γὰρ λέγεται τοῦτο, ἢ ὃ ἐνδέχεται ὑπάρχειν καὶ μὴ ὑπάρχειν, ἢ οὗ ἐν τῷ λόγῳ ὑπάρχει τὸ ᾧ συμβέβηκεν[2] (οἷον τὸ μὲν καθῆσθαι ὡς χωριζόμενον, ἐν δὲ τῷ σιμῷ ὑπάρχει ὁ λόγος ὁ τῆς ῥινὸς ᾗ φαμὲν συμβεβηκέναι τὸ σιμόν). ἔτι ὅσα ἐν τῷ ὁριστικῷ λόγῳ ἔνεστιν ἢ ἐξ ὧν ἐστιν, ἐν τῷ λόγῳ τῷ τούτων οὐκ ἐνυπάρχει ὁ λόγος ὁ τοῦ ὅλου (οἷον ἐν τῷ δίποδι ὁ τοῦ ἀνθρώπου, ἢ ἐν τῷ λευκῷ ὁ τοῦ λευκοῦ ἀνθρώπου). εἰ τοίνυν ταῦτα τοῦτον ἔχει τὸν τρόπον καὶ τῷ ἀνθρώπῳ συμβέβηκε τὸ δίπουν, ἀνάγκη χωριστὸν εἶναι αὐτό (ὥστε ἐνδέχοιτο ἂν μὴ δίπουν εἶναι τὸν ἄνθρωπον), ἢ ἐν τῷ λόγῳ τῷ τοῦ δίποδος ἐνέσται ὁ τοῦ ἀνθρώπου

[1] The words bracketed add nothing to what has been said in the previous chapter and they interrupt the context here. I have treated them as an interpolation.

[2] I reject, as an unmeaning interpolation, the words ἢ ἐν ᾧ ὁ λόγος ὑπάρχει ᾧ συμβέβηκεν (after 20 συμβέβηκεν). [The words are omitted in FI and ignored by Simplicius.—C.]

[a] [I have substituted for W.'s rendering of this argument (a 32, ἀνάγκη δὴ—b 12, σημαίνει τὸ ὄν), my own version, improved in one or two details by Prof. Ross.—C.]

[b] [*i.e.*, if 'being' is to be confined to the One Being, which is identical-with-Being and identical-with-Unity.—C.]

culty) we say that the term 'white' denotes what is identical-with-Being, then 'being' has more than one meaning.[a]

[Nor can 'the existent' have any magnitude, if it is to exclude plurality;[b] for one part of it will have an existence distinguishable from that of another.[c]]

But the analysing of a substantive entity into other substantive entities (so far from being anything startling) is clearly illustrated by definitions. For instance, if 'man' signifies a substantive existence, so must 'animal' and 'biped' also. Otherwise they would be accidental attributes, whether of 'man' or of some other subject. But that is impossible, for an accidental attribute must either be separable, so as sometimes to apply and sometimes not (*e.g.* that the man is 'sitting down'), or else must include its subject in its own definition (as 'snub' includes in its definition the definition of 'nose,' of which we say that snubness is an attribute). Now, the terms of the definition (which are the constituent principles of the thing defined) do not themselves severally contain, in *their* definitions, any mention of the whole thing which they combine to define. For instance, 'man' does not enter into the definition of 'biped,' nor does that of 'white man' enter into the definition of 'white.' Therefore, if 'biped' were an attribute of man at all, it would have to be a separable one, so that the man might, on occasion, not be a biped; the only alternative (as we have said) being that 'man' should be included in the

[c] [*Cf.* Plato's criticism (*Soph.* 244 E) of Parmenides' One Being as a sphere extended in space. (The circumference of a sphere must be distinguishable from its centre, etc.)—C.]

186 b λόγος. ἀλλ᾿ ἀδύνατον· ἐκεῖνο γὰρ ἐν τῷ ἐκείνου λόγῳ ἔνεστιν. εἰ δ᾿ ἄλλῳ συμβέβηκε τὸ δίπουν καὶ τὸ ζῷον καὶ μὴ ἔστιν ἑκάτερον ὅπερ ὄν τι, καὶ ὁ ἄνθρωπος ἂν εἴη τῶν συμβεβηκότων ἑτέρῳ. ἀλλὰ τὸ ὅπερ ὄν τι ἔστω μηδενὶ συμβεβηκός, καὶ καθ᾿ οὗ ἄμφω, καὶ ἑκάτερον καὶ τὸ ἐκ τούτων λεγέσθω· ἐξ ἀδιαιρέτων ἄρα τὸ πᾶν.

137 a Ἔνιοι δ᾿ ἐνέδοσαν τοῖς λόγοις ἀμφοτέροις, τῷ μὲν ὅτι πάντα ἓν εἰ τὸ ὂν ἓν σημαίνει, ὅτι ἔστι τὸ μὴ ὄν, τῷ δὲ ἐκ τῆς διχοτομίας, ἄτομα ποιήσαντες μεγέθη. φανερὸν δὲ καὶ ὅτι οὐκ ἀληθὲς ὡς, εἰ

---

[a] The whole section, b 23–35, ἔτι ὅσα . . . ἄρα τὸ πᾶν is of difficult and doubtful interpretation. My rendering derives some countenance from the note of Simplicius (133. 13 *sq.*) that a very obvious way of establishing the plurality of existing things would have been to define 'genus' (ἔτι προχειρότατον ἦν δεῖξαι πολλὰ τὰ ὄντα διὰ τοῦ ἐκθέσθαι τὸν τοῦ γένους ὁρισμόν· κατὰ πλειόνων γὰρ τὸ γένος), and further from the option given us by Philoponus (79. 13, *sq.*) to take ἀδιαιρέτων, in the last line, as equivalent to οὐσιῶν. [Aristotle here takes the proposition whose predicate defines its subject, *e.g.* 'Man is a biped animal.' (In his view 'biped' and 'animal' are *essential* attributes—καθ᾿ αὑτά, *Anal. Post.* 73 a 34—as distinct from *separable accidents*, *e.g.* 'sitting,' and *inseparable accidents*, *e.g.* 'snub,' which contains in its definition its subject 'nose'; for whatever is snub must be a nose.) What is the status of these defining predicates, according to our opponent? 'Biped' is not an *inseparable accident*, for it does not contain 'man' in its definition. Nor is it a *separable accident*, for man must be biped. And our opponent does not recognize accidental predication at all, but holds that whatever is must be a substantive existent. So 'biped' and 'animal' must be substantive existents, and so must 'man.' If so, either 'man,' a substantive existent, must be divisible into other substantive existents (*ll.* 14-16), and we shall have a plurality;

definition of 'biped'; and this is not so, for it is the other way about, 'biped' being included in the definition of 'man.' And if both 'biped' and 'animal' were attributes of some other subject than man and were not themselves, severally, subjects at all, then man himself would belong to the class of 'things attributed to a subject.' We must, then, absolutely lay it down, that the substantivally existent is not an attribute of something else; and also that what is true in this respect of the elements of a definition, severally and collectively, is also true of the thing which they define. The universe then is composed of a plurality of distinct individual entities.[a]

Note that some thinkers have given in to both the Eleatic arguments—to the argument that, if 'being' has only one meaning, all things are one, by conceding that 'what is not' exists;[b] and also to the argument from dichotomy by supposing the existence of indivisible magnitudes.[c] Now it is obvious that, from the premisses 'being has only one meaning' and

or every such substantive existent must be indivisible, and we shall have a universe consisting of a plurality of indivisibles (*ll.* 34–35). See Maier, *Syllogistik*, ii. 286. —C.]

[b] [Plato, in the *Sophist*, 'gave in' to Parmenides to the extent that he 'thought that all things that are would be one (viz. Being itself), if one did not refute Parmenides' saying: Never will this be proved, that things that are-not are, and he supposed it necessary to prove that what-is-not is' (*Met.* 1088 a 35). *Cf.* Ross, *Met.* I. xc.; Maier, *Syllogistik*, ii. 289.—C.]

[c] [Alexander (Simplic. 138. 3) saw here a reference to Zeno's arguments against the view that reality consists of a plurality of indivisible monads, and to Xenocrates, the Platonist, who maintained the existence of indivisible lines. Zeno, *frag.* 3; Diels, *Vors.* 19 A 22.—C.]

ἓν σημαίνει τὸ ὂν καὶ μὴ οἷόν τε ἅμα τὴν ἀντίφασιν, οὐκ ἔσται οὐθὲν μὴ ὄν· οὐθὲν γὰρ κωλύει (μὴ ἁπλῶς εἶναι ἀλλὰ) μὴ ὄν τι εἶναι τὸ μὴ ὄν. τὸ δὲ δὴ φάναι παρ' αὐτὸ τὸ ὂν ὡς εἰ μή τι ἔσται ἄλλο, ἓν πάντα ἔσεσθαι, ἄτοπον. τίς γὰρ μανθάνει αὐτὸ τὸ ὂν εἰ μὴ τὸ ὅπερ ὄν τι εἶναι; εἰ δὲ τοῦτο, οὐδὲν ὅμως κωλύει πολλὰ εἶναι τὰ ὄντα, ὥσπερ εἴρηται.

Ὅτι μὲν οὖν οὕτως ἓν εἶναι τὸ ὂν ἀδύνατον, δῆλον.

---

[a] [Aristotle seems to criticize Plato (somewhat unfairly) for not attacking Parmenides' fundamental premisses and inferences more drastically in the *Sophist*. W. rendered the foregoing paragraph a 1-10 more freely as follows: 'Note that some thinkers have not shrunk from the conclusion that, if " existent " has only one meaning, there are things that do not exist; and also that, if magnitude were infinitely divisible, there could be no motion—they themselves believing in atomic magnitudes. And indeed it is sufficiently obvious that we might take " existent " to have only one meaning and might still escape the Heracleitean paradox; for it is quite possible that a thing might " have a being "

CHAPTER IV

ARGUMENT

Of the Physicists (who do recognize plurality and change) there are two schools. One finds unity in a universal underlying material (water, air, fire, or some intermediate form), and diversity in some such condition as degree of compression (or, in Plato's case, the other way round, variations in a ' more or less ' of materiality, and unity in the participation in the Idea). The other holds distinct substances to have been already present in the primal ' mixture ' or confusum (187 a 12-26).

Examination and refutation of the doctrine of Anax-

'contradictories cannot co-exist,' it is not a true inference that there is nothing which 'is not'; for 'what is not' may very well (not 'exist' absolutely, but) be 'what is *not this or that*.' But to assert that, if there is to be nothing over and above 'just what is,' it will follow that all things are one, is absurd. For who would take this expression 'just what is' to mean anything but 'something that substantively exists'? But if it means that, there is nothing against the things that exist being a plurality, as we have seen.[a]

It is clear, then, that the existent cannot be all 'one' in this sense.

in a general sense, without having the "substantive being" to which we had agreed to confine the term "existence." But, in any case, even if "it is" could be said of nothing save the primarily existent, the assertion that all things are one would be senseless; for who would take the "primarily" existent to mean anything but the "substantively" existent? And in that case why should not there be a multiplicity of "primarily existent" things, as we see to be the actual case?'—C.]

CHAPTER IV

ARGUMENT (*continued*)

agoras that all things were confounded together till 'Intelligence' disentangled and arranged them (a 26-188 a 5).

But, though the Anaxagorean theory of mixtures that cannot be unmixed (so that nothing can exist apart by itself in its purity) breaks down, yet there really are such things as qualities and characteristics that cannot exist apart by themselves. So that, if Anaxagoras had included 'particles' of them *in his primal 'mixture,' he would have had the best of reasons for saying that they could not be 'sifted out' in their purity, any more than 'magnitude' could, since there is nothing so small as to have no size at all* (a 5-18).

187 a 12 Ὡς δ᾿ οἱ φυσικοὶ λέγουσι, δύο τρόποι εἰσίν. οἱ μὲν γὰρ ἓν ποιήσαντες τὸ ὂν σῶμα τὸ ὑποκείμενον —ἢ τῶν τριῶν τι ἢ ἄλλο ὅ ἐστι πυρὸς μὲν πυκνότερον ἀέρος δὲ λεπτότερον—τἆλλα γεννῶσι πυκνότητι καὶ μανότητι πολλὰ ποιοῦντες. ταῦτα δ᾿ ἐστὶν ἐναντία, καθόλου δ᾿ ὑπεροχὴ καὶ ἔλλειψις, ὥσπερ τὸ μέγα φησὶ Πλάτων καὶ τὸ μικρόν, πλὴν ὅτι ὁ μὲν ταῦτα ποιεῖ ὕλην τὸ δὲ ἓν τὸ εἶδος, οἱ δὲ τὸ μὲν ἓν τὸ ὑποκείμενον ὕλην τὰ δ᾿ ἐναντία διαφορὰς καὶ εἴδη. οἱ δ᾿ ἐκ τοῦ ἑνὸς ἐνούσας τὰς ἐναντιότητας ἐκκρίνεσθαι, ὥσπερ Ἀναξίμανδρός φησι, καὶ ὅσοι δ᾿ ἓν καὶ πολλά φασιν εἶναι,[1] ὥσπερ Ἐμπεδοκλῆς καὶ Ἀναξαγόρας· ἐκ τοῦ μίγματος

[1] [After εἶναι E has τὰ ὄντα.—C.]

---

[a] [Whether this phrase could describe any factor in Anaximander's system (as most of the ancient commentators supposed) is still disputed. See Ross on *Met.* 988 a 30.—C.]

[b] Alexander (quoted and supported by Simplicius) tells us of Plato's hearers mentioning the ἀόριστος δυάς, ἣν μέγα καὶ μικρὸν ἔλεγεν (the indeterminate dyad, which he [Plato] called great and small). Though Plato has nowhere in his written works developed the 'great and small' nomenclature, yet its connexion with the 'indeterminate dyad' (*cf. Philebus*) makes it fairly certain that it stands for the purely dimensional 'not-anything' that occupied in Plato's system the place taken by the conceptually abstracted *hyle* or 'matter' in Aristotle. (*Cf.* note on 201 b 20.)

We turn now to the Physicists. There are two schools of them. Those of the one school reduce existence to unity by positing a single underlying substance—whether one of the familiar three, or a something that is denser than fire and rarer than air [a]—and arrive at a plurality by conceiving all else to be generated from it by condensation and rarefaction. Now dense and rare are opposites and may be brought under the more general conception of excess and defect. So Plato,[b] too, has his 'great and small,' only he makes matter consist in this diversifying antithesis, and finds unity in the Idea; whereas the others find unity in the underlying matter and distinctions and forms in the opposites, or, like Anaximander, extract the contrasts themselves out of the indeterminate prime substance.[c] The other school, to which Empedocles and Anaxagoras belong, start from the first with both unity and multiplicity;

[c] The association of Anaximander with the first school, in the translation, cannot be justified by the text. But I follow Philoponus in thinking that a rearrangement is necessary, and that the text cannot be accepted as sound in its present form. He does not undertake to restore it. [The division between the two groups is marked by οἱ δέ (*l.* 20) answering οἱ μέν (*l.* 12). *Cf. Met.* 1069 b 20, where Anaximander is grouped with Empedocles and Anaxagoras. He would be assigned to this second group because his Unlimited contained (ἐνούσας) a plurality of opposites which were 'separated out' (ἐκκρίνεσθαι, as φησί seems to imply, was perhaps his own word). So, though his Unlimited stuff was not a mechanical mixture of distinct things, a plurality was originally *in* it and emerged out of it. Hence Aristotle regards this Unlimited as analogous to the primitive 'mixtures' of Empedocles and Anaxagoras, from which 'they also separate out' their plurality (καὶ οὗτοι ἐκκρίνουσι τἆλλα). Theophrastus (Diels, *Vors. Anaximandros*, 9, 9*a*) also couples Anaximander with Anaxagoras.—C.]

187 a γὰρ καὶ οὗτοι ἐκκρίνουσι τἆλλα. διαφέρουσι δ' ἀλλήλων τῷ τὸν μὲν περίοδον ποιεῖν τούτων, τὸν δ' ἅπαξ, καὶ τὸν μὲν ἄπειρα τά τε ὁμοιομερῆ καὶ τἀναντία, τὸν δὲ τὰ καλούμενα στοιχεῖα μόνον.

Ἔοικε δὲ Ἀναξαγόρας ἄπειρα οὕτως οἰηθῆναι διὰ τὸ ὑπολαμβάνειν τὴν κοινὴν δόξαν τῶν φυσικῶν εἶναι ἀληθῆ, ὡς οὐ γινομένου οὐδενὸς ἐκ τοῦ μὴ ὄντος· διὰ τοῦτο γὰρ οὕτω λέγουσιν 'ἦν ὁμοῦ τὰ πάντα,' καὶ τὸ γίνεσθαι τοιόνδε καθέστηκεν ἀλλοιοῦσθαι· οἱ δὲ σύγκρισιν καὶ διάκρισιν.

Ἔτι δ' ἐκ τοῦ γίνεσθαι ἐξ ἀλλήλων τἀναντία· ἐνυπῆρχεν ἄρα. εἰ γὰρ πᾶν μὲν τὸ γινόμενον ἀνάγκη γίνεσθαι ἢ ἐξ ὄντων ἢ ἐκ μὴ ὄντων, τούτων δὲ τὸ μὲν ἐκ μὴ ὄντων γίνεσθαι ἀδύνατον (περὶ γὰρ ταύτης ὁμογνωμονοῦσι τῆς δόξης ἅπαντες οἱ περὶ φύσεως), τὸ λοιπὸν ἤδη συμβαίνειν ἐξ ἀνάγκης ἐνόμισαν ἐξ ὄντων μὲν καὶ ἐνυπαρχόντων γίνεσθαι,
187 b διὰ μικρότητα δὲ τῶν ὄγκων ἐξ ἀναισθήτων ἡμῖν. διό φασι πᾶν ἐν παντὶ μεμῖχθαι, διότι πᾶν ἐκ

---

[a] [Anaxagoras 'has an unlimited number—his substances composed of homogeneous parts and also his opposites.' He specifies four pairs (hot, cold ; wet, dry ; dense, rare ; bright, dark, *frags.* 4, 12). It is not clear if this was a complete list, nor whether Aristotle here means that he had an indefinite number of opposites. See note on 184 b 20. —C.]

[b] [That Empedocles' elements were not each unlimited in quantity Aristotle states at 203 a 18, and Empedocles' own language (*frag.* 17) supports him.—C.]

[c] [*Cf. De gen. et corr.* 314 a 6 ff., which seems to imply that Anaxagoras somewhere used the word ἀλλοιοῦσθαι to describe the process by which the many arise out of the one mixture. This use is, from Aristotle's standpoint, incorrect ; ἀλλοίωσις would be more properly applied to the 'modification' of the single element (water, air) of the monists. The

for they assume an undistinguished *confusum*, from which the constituents of things are sifted out. But they differ in this, that Empedocles supposes the course of Nature to return upon itself, coming round again periodically to its starting-point; while Anaxagoras makes it move on continuously without repeating itself. Moreover, he assumes an unlimited number of distinguishable substances, from the first, as well as an unlimited number of uniform particles in each substance;[a] whereas Empedocles has only his four so-called elements.[b]

Anaxagoras appears to have based his conviction that the primal substances are unlimited in number on his uncompromising acceptance of the dogma, common to all the Physicists, that 'nothing can come out of what does not exist.' This made him declare that originally 'all things existed together' and explain that genesis was nothing more than the modification induced by setting them in order; whereas the same dogma made the others attribute genesis to transforming combination and resolution.[c]

Further, Anaxagoras argued from the genesis of unlikes from each other that they were already in each other; for since whatever comes to be must arise either out of what exists or out of what does not exist, and since the latter was universally held to be impossible, it remained that all things arose out of what existed, and so must be there already, only in particles so minute as to escape our senses. So he and his school argued that particles of every-

other pluralists (οἱ δέ, Empedocles, *frag.* 8, the Atomists) saw that it must be a process of 'combination' and 'separation' of unchanging elements. Anaxagoras, however, also clearly asserted this in Fragment 17. See Diels, *Vors.* 46 A 52; Zeller, I.[6] 1209.—C.]

187 b παντὸς ἑώρων γινόμενον· φαίνεσθαι δὲ διαφέροντα καὶ προσαγορεύεσθαι ἕτερα ἀλλήλων ἐκ τοῦ μάλισθ' ὑπερέχοντος διὰ πλῆθος ἐν τῇ μίξει τῶν ἀπείρων· εἰλικρινῶς μὲν γὰρ ὅλον λευκὸν ἢ μέλαν ἢ γλυκὺ ἢ σάρκα ἢ ὀστοῦν οὐκ εἶναι, ὅτου δὲ πλεῖστον ἕκαστον ἔχει, τοῦτο δοκεῖν εἶναι τὴν φύσιν τοῦ πράγματος.

Εἰ δὴ τὸ μὲν ἄπειρον ᾗ ἄπειρον ἄγνωστον, τὸ μὲν κατὰ πλῆθος ἢ κατὰ μέγεθος ἄπειρον ἄγνωστον πόσον τι, τὸ δὲ κατ' εἶδος ἄπειρον ἄγνωστον ποιόν τι. τῶν δ' ἀρχῶν ἀπείρων οὐσῶν καὶ κατὰ πλῆθος καὶ κατ' εἶδος, ἀδύνατον εἰδέναι τὰ ἐκ τούτων· οὕτω γὰρ εἰδέναι τὸ σύνθετον ὑπολαμβάνομεν, ὅταν εἰδῶμεν ἐκ τίνων καὶ πόσων ἐστίν.

Ἔτι δ' εἰ ἀνάγκη, οὗ τὸ μόριον ἐνδέχεται ὁπηλικονοῦν εἶναι κατὰ μέγεθος καὶ μικρότητα, καὶ αὐτὸ ἐνδέχεσθαι (λέγω δὲ τῶν τοιούτων τι μορίων, εἰς ὃ ἐνυπάρχον διαιρεῖται τὸ ὅλον), εἰ δὲ ἀδύνατον ζῷον ἢ φυτὸν ὁπηλικονοῦν εἶναι κατὰ μέγεθος καὶ μικρότητα, φανερὸν ὅτι οὐδὲ τῶν μορίων ὁτιοῦν· ἔσται γὰρ καὶ τὸ ὅλον ὁμοίως. σὰρξ δὲ καὶ ὀστοῦν καὶ τὰ τοιαῦτα μόρια ζῴου, καὶ οἱ καρποὶ τῶν

---

[a] [Aristotle is paraphrasing Anaxagoras, *frag.* 11 (and elsewhere), ἐν παντὶ παντὸς μοῖρα ἔνεστι; *frag.* 12, μοῖραι δὲ πολλαὶ πολλῶν εἰσι. παντάπασι δὲ οὐδὲν ἀποκρίνεται οὐδὲ διακρίνεται ἕτερον ἀπὸ τοῦ ἑτέρου πλὴν νοῦ . . . ἕτερον δὲ οὐδέν ἐστιν ὅμοιον οὐδενί, ἀλλ' ὅτων πλεῖστα ἔνι ταῦτα ἐνδηλότατα ἓν ἕκαστόν ἐστι καὶ ἦν. He now begins a series of objections. —C.]

[b] [Anaxagoras, *frag.* 3, οὔτε γὰρ τοῦ σμικροῦ ἐστι τό γε ἐλάχιστον, ἀλλ' ἔλασσον ἀεί. τὸ γὰρ ἐὸν οὐκ ἔστι τὸ μὴ (τομῇ Zeller) οὐκ εἶναι. ἀλλὰ καὶ τοῦ μεγάλου ἀεί ἐστι μεῖζον.—C.]

[c] Not the sections of a mathematical line, for instance, which do not exist in advance before they have been con-

thing must exist in everything else, since they saw all kinds of things emerging from each other. And they held that things have a different appearance and receive different names according to the prevalence of one constituent or another in the mixture; and that accordingly, no such thing exists as pure black or white or sweet or flesh or bone, but the nature of a thing is judged by what it has most of in it.[a]

But (1) if a thing has no limit under some certain aspect, then we cannot define it in respect of that said aspect as to which it is unlimited. Thus we cannot say how great a thing is, or how many of it there are, if it has no limit as to size or number; nor can we say what kind of thing it is if there is no limit to its diversity. If, then, the constituents of a thing are unlimited both as to magnitude and as to kind, we cannot know what that thing is; for we reckon to know about a thing that is put together when we know the quality and quantity of its components.

Again, (2) a thing, the parts of which can be of any magnitude, great or small,[b] can itself be of any magnitude—I am speaking of parts into which, as already existing in it, the whole can be resolved.[c] If this is a necessary consequence, and if it is impossible that an animal or plant should exceed all limit in greatness or smallness, the same must be true of any part of it; for the whole and the parts must bear a definite proportion to each other.[d] Now flesh and bone and so forth are parts of animals,

stituted by actual section. There are no atomic lengths out of which a given line is made.

[d] [*Literally*, 'for otherwise it will be possible for the whole similarly (to be of any size).'—C.]

187 b φυτῶν. δῆλον τοίνυν ὅτι ἀδύνατον σάρκα ἢ ὀστοῦν ἢ ἄλλο τι ὁπηλικονοῦν εἶναι τὸ μέγεθος ἐπὶ τὸ μεῖζον ἢ ἐπὶ τὸ ἔλαττον.

Ἔτι εἰ πάντα μὲν ἐνυπάρχει τὰ τοιαῦτα ἐν ἀλλήλοις καὶ μὴ γίνεται ἀλλ' ἐκκρίνεται ἐνόντα, λέγεται δὲ ἀπὸ τοῦ πλείονος, γίνεται δὲ ἐξ ὁτουοῦν ὁτιοῦν—οἷον ἐκ σαρκὸς ὕδωρ ἐκκρινόμενον καὶ σὰρξ ἐξ ὕδατος—ἅπαν δὲ σῶμα πεπερασμένον ἀναιρεῖται ὑπὸ σώματος πεπερασμένου, φανερὸν ὅτι οὐκ ἐνδέχεται ἐν ἑκάστῳ ἕκαστον ὑπάρχειν. ἀφαιρεθείσης γὰρ ἐκ τοῦ ὕδατος σαρκός, καὶ πάλιν ἄλλης γενομένης ἐκ τοῦ λοιποῦ ἀποκρίσει, εἰ καὶ ἀεὶ ἐλάττων ἔσται ἡ ἐκκρινομένη, ἀλλ' ὅμως οὐχ ὑπερβαλεῖ μέγεθός τι τῇ σμικρότητι. ὥστ' εἰ μὲν στήσεται ἡ ἔκκρισις, οὐχ ἅπαν ἐν παντὶ ἐνέσται (ἐν γὰρ τῷ λοιπῷ ὕδατι οὐκ ἐνυπάρξει σάρξ)· εἰ δὲ μὴ στήσεται ἀλλ' ἀεὶ ἕξει ἀφαίρεσιν, ἐν πεπερασμένῳ μεγέθει ἴσα πεπερασμένα ἐνέσται ἄπειρα τὸ πλῆθος· τοῦτο δ' ἀδύνατον.

Πρὸς δὲ τούτοις, εἰ ἅπαν μὲν σῶμα ἀφαιρεθέντος τινὸς ἔλαττον ἀνάγκη γίνεσθαι, τῆς δὲ σαρκὸς ὥρισται τὸ ποσὸν καὶ μεγέθει καὶ μικρότητι,
188 a φανερὸν ὅτι ἐκ τῆς ἐλαχίστης σαρκὸς οὐθὲν ἐκκριθήσεται σῶμα· ἔσται γὰρ ἔλαττον τῆς ἐλαχίστης.

---

[a] [For ἀναιρεῖν (exhaust, use up) *cf.* 238 a 27.—C.]

[b] [ἴσα so understood by Alexander and Themistius, *ap.* Simplic. 169. 17, σάρκες ἴσαι τὸ μέγεθος. But is Aristotle alluding to Anaxagoras, *frag.* 3, καὶ ἴσον ἐστὶ (τὸ μεῖζον) τῷ σμικρῷ πλῆθος and *frag.* 6, ἴσαι μοῖραί εἰσι τοῦ τε μεγάλου καὶ τοῦ σμικροῦ πλῆθος? 'In a limited magnitude there will always be *as many* limited particles (as ever), in fact an unlimited number.'—C.]

[c] [Aristotle assumes this as proved above (187 b 13 ff.).

and fruits parts of plants. It is clear then that neither flesh nor bone nor anything of the kind can be great or small beyond limit.

(3) Further, if on the one hand (as Anaxagoras holds) all such things are already there in each other and do not come into existence but are merely sifted out from where they are and take their names from their dominant constituents, and anything can be sifted out of anything (water out of flesh or flesh out of water), but, on the other hand, any limited body must use up[a] any other limited body in such a process, then clearly it is impossible for every thing to exist in every thing. For if flesh has been sifted out of a given body of water, and then more flesh is sifted out of the remaining water, even if the successive extracts constantly diminish in quantity they cannot diminish below the minimum particle. Consequently, if, when they reach this minimum, the process is of necessity arrested, then it will no longer be true that everything is contained in everything, for there will be no flesh in the water that is left over. If, on the other hand, you can always go on sifting out one minimum particle at a time, it follows that there are an unlimited number of bodies of equal size[b] contained in a limited body, which is impossible.

(4) Besides, since the subtraction of anything from a given body must reduce the size of that body, and since a mass of flesh cannot be indefinitely great or small,[c] it is clear that from the minimum of flesh no other body can be extracted, for that would reduce it below its minimum.

Anaxagoras would deny that there can be a minimum piece of flesh—an ἐλάχιστον. See p. 44, note *b*.—C.]

188 a Ἔτι δ' ἐν τοῖς ἀπείροις σώμασιν ἐνυπάρχοι ἂν ἤδη σὰρξ ἄπειρος καὶ αἷμα καὶ ἐγκέφαλος, κεχωρισμένα μέντοι ἀπ' ἀλλήλων, οὐθὲν δ' ἧττον ὄντα, καὶ ἄπειρον ἕκαστον· τοῦτο δ' ἄλογον.

Τὸ δὲ μηδέποτε διακριθήσεσθαι οὐκ εἰδότως μὲν λέγεται, ὀρθῶς δὲ λέγεται· τὰ γὰρ πάθη ἀχώριστα. εἰ οὖν ἐμέμικτο τὰ χρώματα καὶ αἱ ἕξεις, ἐὰν διακριθῶσιν, ἔσται τι λευκὸν ἢ ὑγιαῖνον οὐχ ἕτερόν τι ὂν οὐδὲ καθ' ὑποκειμένου. ὥστε ἄτοπος τὰ ἀδύνατα ζητῶν ὁ νοῦς, εἴπερ βούλεται μὲν διακρῖναι, τοῦτο δὲ ποιῆσαι ἀδύνατον καὶ κατὰ τὸ ποσὸν καὶ κατὰ τὸ ποιόν—κατὰ μὲν τὸ ποσὸν ὅτι οὐκ ἔστιν ἐλάχιστον μέγεθος, κατὰ δὲ τὸ ποιὸν ὅτι ἀχώριστα τὰ πάθη.

Οὐκ ὀρθῶς δὲ οὐδὲ τὴν γένεσιν λαμβάνει τῶν ὁμοιοειδῶν. ἔστι μὲν γὰρ ὡς ὁ πηλὸς εἰς πηλοὺς διαιρεῖται, ἔστι δ' ὡς οὔ· καὶ οὐχ ὁ αὐτὸς τρόπος, ὡς πλίνθοι ἐξ οἰκίας καὶ οἰκία ἐκ πλίνθων, οὕτω δὲ καὶ ὕδωρ καὶ ἀὴρ ἐξ ἀλλήλων καὶ εἰσὶ καὶ γίνονται.

---

[a] [*Literally*, 'separate of course from one another' (μέντοι in this sense, Plato, *Rep*. 339 B; *Charm*. 159 C), *i.e.* the bits of any one substance being separate from other bits of the *same* substance. καὶ ἄπειρον ἕκαστον 'and each (substance existing) in unlimited amount.'—C.]

[b] ἔτι . . . ἄλογον. Anaxagoras apparently assumed molecules of tissues and of all distinguishable *components* of organic things, but not actual microscopic *organisms*. Thus, bone or fruit, being integral parts of animal or plant, were made up of molecules like themselves, and must be as many times as large as a molecule as they had, in each case, molecules in them. And the organism, plant or animal,

(5) Again, in each of the unlimited number of primal substances an unlimited amount of flesh and blood and brains would exist, not indeed gathered together in recognizable aggregates,[a] but still existing. So that each of these substances would exist without limit within each of the others, which would also exist without limit within it. And this is impossible.[b]

And yet this notion of something being there that can never become completely distinct is very sound, though Anaxagoras had not got the right hold of it. For modifications [c] cannot exist apart, by themselves. So if such things as colours and states were included in the primal Anaxagorean *confusum* and could have been wholly disengaged from it, there would have resulted a ' white ' and a ' healthy ' that was nothing else but itself—not even to the extent of having any subject ! So it was just as well that ' Intelligence ' did not desire to sever such things out ; for it could not be done, either with respect to quantity, since there is no minimum magnitude,[d] nor with respect to quality, for modifications cannot exist apart by themselves.

Nor was his conception of genesis from similar particles adequate ; for although in one way a piece of mud is divided into smaller pieces of the same stuff—mud, in another way it is not (but is divided into earth and water—dissimilar parts). Also, water and air do not consist of one another or come out of one another in the same way that bricks come out of a house that is broken up, or a house consists of bricks.

was itself made up of these differing component parts. Ultimately then a definite proportion must subsist between the size of a molecule and the size of any given organism.

[c] [πάθη, 'affections' or 'qualities' which inhere in substances. —C.]

[d] [*Cf. De gen. et corr.* A 2.—C.]

188 a Βέλτιον δὲ ἐλάττω καὶ πεπερασμένα λαβεῖν, ὅπερ ποιεῖ Ἐμπεδοκλῆς.

CHAPTER V

ARGUMENT

Surveying the ground we have covered, we may note that, while most of the thinkers were intent on considering the material of which things are composed, and put their speculations on that subject in the foreground, they all assume less obtrusively some antithesis between opposing or contrasted states and movements, which determine the changes that take place in the material. And these pairs of opposites have every right to be considered ' principles ' (188 a 19-b 26).

Πάντες δὴ τἀναντία ἀρχὰς ποιοῦσιν, οἵ τε λέγοντες ὅτι ἓν τὸ πᾶν καὶ μὴ κινούμενον (καὶ γὰρ Παρμενίδης θερμὸν καὶ ψυχρὸν ἀρχὰς ποιεῖ, ταῦτα δὲ προσαγορεύει πῦρ καὶ γῆν) καὶ οἱ μανὸν καὶ πυκνόν, καὶ Δημόκριτος τὸ πλῆρες καὶ κενόν, ὧν τὸ μὲν ὡς ὂν τὸ δ᾿ ὡς οὐκ ὂν εἶναί φησιν· ἔτι θέσει, σχήματι, τάξει· ταῦτα δὲ γένη ἐναντίων· θέσεως ἄνω κάτω, πρόσθεν ὄπισθεν, σχήματος γωνία, εὐθύ, περιφερές.

Ὅτι μὲν οὖν τἀναντία πως πάντες ποιοῦσι τὰς ἀρχάς, δῆλον. καὶ τοῦτο εὐλόγως· δεῖ γὰρ τὰς

---

[a] [πεπερασμένα may mean that *each* of Empedocles' elements is *limited in quantity*, *cf*. 187 a 26.—C.]

[b] [Parmenides, *frags*. 8. 53 ; 9. *Met*. 986 b 18.—C.]

[c] [Anaximenes first used these terms ; but Aristotle (187 a 15) and Theophrastus (Simplic. 149. 32) attribute the conception to the other Ionian monists. Diels, *Vors*. 3 A 5.—C.]

[d] [*Cf*. *Met*. 985 b 4.—C.]

Empedocles, then, was, so far, sounder in assuming a small limited [a] number of prime substances.

CHAPTER V

ARGUMENT (*continued*)

This fundamental agreement amongst all the superficial contradictions in the several systems is the more significant because it seems to have been arrived at unintentionally, and even unconsciously, by men who were looking in other directions. At any rate there it is, and it is obviously what we want ; for if we assume, as a primary fact, some such ' contrasted possibilities ' in our material, we at once escape from the deadlock of the ' nothing-out-of-what-is-not already-there ' dogma, and can bridge the chasm between the ' already ' and the ' not yet ' there (b 26-189 a 10).

This brings us to observe that all these thinkers assume as principles some ' couple ' of antithetical qualities or forces, and this whether they declare the sum of things to be one and rigid (for even Parmenides erects ' hot ' and ' cold,' which he calls ' fire ' and ' earth,' into principles) [b] ; or whether they speak of ' rare ' and ' dense ' [c] ; or whether with Democritus, they speak of ' solidity ' and ' vacancy ' (the one regarded as the ' existent ' and the other as the ' non-existent '), and further distinguish the atoms by position, shape, and order,[d] all of which are expressed in antithetical couples : position as ' above and below,' ' before and behind,' and shape as ' angular, straight, or curved.'

Clearly, then, they all assume certain numbers of antithetical couples as principles ; and not without

188 a ἀρχὰς μήτε ἐξ ἀλλήλων εἶναι μήτε ἐξ ἄλλων, καὶ ἐκ τούτων πάντα· τοῖς δ' ἐναντίοις τοῖς πρώτοις ὑπάρχει ταῦτα—διὰ μὲν τὸ πρῶτα εἶναι μὴ ἐξ ἄλλων, διὰ δὲ τὸ ἐναντία μὴ ἐξ ἀλλήλων. ἀλλὰ δεῖ τοῦτο καὶ ἐπὶ τοῦ λόγου σκέψασθαι πῶς συμβαίνει.

Ληπτέον δὴ πρῶτον ὅτι πάντων τῶν ὄντων οὐθὲν οὔτε ποιεῖν πέφυκεν οὔτε πάσχειν τὸ τυχὸν ὑπὸ τοῦ τυχόντος, οὐδὲ γίνεται ὁτιοῦν ἐξ ὁτουοῦν, ἂν μή τις λαμβάνῃ κατὰ συμβεβηκός. πῶς γὰρ ἂν γένοιτο τὸ λευκὸν ἐκ μουσικοῦ, πλὴν εἰ μὴ συμβεβηκὸς εἴη τῷ μὴ λευκῷ ἢ τῷ μέλανι τὸ μουσικόν; ἀλλὰ λευκὸν μὲν γίνεται ἐξ οὐ λευκοῦ, 188 b καὶ τούτου οὐκ ἐκ παντὸς ἀλλ' ἐκ μέλανος ἢ τῶν μεταξύ, καὶ μουσικὸν οὐκ ἐκ μουσικοῦ, πλὴν οὐκ ἐκ παντὸς ἀλλ' ἐξ ἀμούσου ἢ εἴ τι αὐτῶν ἐστι μεταξύ. οὐδὲ δὴ φθείρεται εἰς τὸ τυχὸν πρῶτον, οἷον τὸ λευκὸν οὐκ εἰς τὸ μουσικόν (πλὴν εἰ μή ποτε κατὰ συμβεβηκός) ἀλλ' εἰς τὸ μὴ λευκόν, καὶ οὐκ εἰς τὸ τυχὸν ἀλλ' εἰς τὸ μέλαν ἢ τὸ μεταξύ·

reason, for 'principles,' being themselves primary, must not be derived either from each other or from anything else, and all other things must arise out of them. And the terms of a primary antithesis fulfil this condition ; for, because they are primary, they cannot be derived from anything else, and because they are antithetical, they cannot rise out of each other. But we must go more closely into the question what this means and how it works out.

Note, then, to begin with, that things cannot act upon each other and turn into one another at random, unless it be by incidental concomitance ; for how (to take an example) could culture, as such, 'become' pallor ? A man of culture, who had been of a swarthy complexion, might indeed become pale, and so the person we had described as 'cultured' might come to be a person we describe as 'pale,' but only incidentally to the fact that the cultured person who became pale had other qualities concomitantly with his culture ; for he cannot 'become' pale on the mere strength of already having some quality other than pallor to start from, unless that quality is on the specific line of antithesis to pallor—swarthiness or some intermediate shade. So too a man cannot 'become' cultured merely because he already has some characteristic which is not culture, unless it is the specific characteristic of complete or partial want of culture. And all this holds of the loss of qualifications just as much as of their acquisition. A man does not become cultured when he ceases to be pale (unless it be by incidental concomitance), for what 'ceasing to be pale' means is not only becoming something not pale, but something on the line antithetical to pallor that leads to swarthiness. So too,

188 b ὡς δ' αὔτως καὶ τὸ μουσικὸν εἰς τὸ μὴ μουσικόν, καὶ τοῦτο οὐκ εἰς τὸ τυχὸν ἀλλ' εἰς τὸ ἄμουσον ἢ εἴ τι αὐτῶν ἐστι μεταξύ.

Ὁμοίως δὲ τοῦτο καὶ ἐπὶ τῶν ἄλλων· ἐπεὶ καὶ τὰ μὴ ἁπλᾶ τῶν ὄντων ἀλλὰ σύνθετα κατὰ τὸν αὐτὸν ἔχει λόγον, ἀλλὰ διὰ τὸ μὴ τὰς ἀντικειμένας διαθέσεις ὠνομάσθαι λανθάνειν τοῦτο συμβαίνει. ἀνάγκη γὰρ πᾶν τὸ ἡρμοσμένον ἐξ ἀναρμόστου γίνεσθαι καὶ τὸ ἀνάρμοστον ἐξ ἡρμοσμένου, καὶ φθείρεσθαι τὸ ἡρμοσμένον εἰς ἀναρμοστίαν, καὶ ταύτην οὐ τὴν τυχοῦσαν ἀλλὰ τὴν ἀντικειμένην. διαφέρει δ' οὐθὲν ἐπὶ ἁρμονίας εἰπεῖν ἢ τάξεως ἢ συνθέσεως· φανερὸν γὰρ ὅτι ὁ αὐτὸς λόγος. ἀλλὰ μὴν καὶ οἰκία καὶ ἀνδριὰς καὶ ὁτιοῦν ἄλλο γίνεται ὁμοίως· ἥ τε γὰρ οἰκία γίνεται ἐκ τοῦ μὴ συγκεῖσθαι ἀλλὰ διῃρῆσθαι ταδὶ ὡδί, καὶ ὁ ἀνδριὰς καὶ τῶν ἐσχηματισμένων τι ἐξ ἀσχημοσύνης· καὶ ἕκαστον τούτων τὰ μὲν τάξις, τὰ δὲ σύνθεσίς τίς ἐστιν.

Εἰ τοίνυν τοῦτ' ἔστιν ἀληθές, ἅπαν ἂν γίγνοιτο τὸ γιγνόμενον καὶ φθείροιτο τὸ φθειρόμενον ἢ ἐξ ἐναντίων ἢ εἰς ἐναντία καὶ τὰ τούτων μεταξύ. τὰ δὲ μεταξὺ ἐκ τῶν ἐναντίων ἐστίν (οἷον χρώματα

[a] [*Or* 'in tune.' Aristotle refers specially to the 'attunement' or adjustment (ἁρμονία) of a musical instrument.—C.]

if culture lapses from itself, it cannot lapse into any chance thing you may name other than itself, but only into a more or less complete 'want of culture.'

And it is the same with everything else; for structural combinations no less than isolated qualities obey the same law; only that we fail to note it because we have not specific names for the several 'absences of structure' corresponding to different structural forms. But, all the same, anything that is articulated [a] must rise out of something from which that particular articulation is absent; and if, in its turn, it falls out of articulation it must go back again to the absence of the specific articulation it had. It makes no difference whether we speak of 'harmony' or of arrangement or of combination in this connexion; for the principle is clearly the same. And so too with a house or a statue or any such product. For what the house replaces by being made is the unordered relation of the materials to each other; and what passes away in the making of the statue, or any other shapely work, is the unshapeliness of the material; for all such things are constituted either by the formative disposition or the combining of the material or materials.

If then all this is so, it would seem that whenever anything comes into existence or passes out of it, the movement is along the determined line between the terms of some contrast; or (if we start from some intermediate state) the movement is towards one of the extremes. And since the intermediates are compounded in various degrees out of the opposite terms of the contrasted couple (colours, for instance,

188 b ἐκ λευκοῦ καὶ μέλανος)· ὥστε πάντ' ἂν εἴη τὰ φύσει γινόμενα ἢ ἐναντία ἢ ἐξ ἐναντίων.

Μέχρι μὲν οὖν ἐπὶ τοσοῦτον σχεδὸν συνηκολουθήκασι καὶ τῶν ἄλλων οἱ πλεῖστοι, καθάπερ εἴπομεν πρότερον· πάντες γὰρ τὰ στοιχεῖα καὶ τὰς ὑπ' αὐτῶν καλουμένας ἀρχάς, καίπερ ἄνευ λόγου τιθέντες, ὅμως τἀναντία λέγουσιν, ὥσπερ ὑπ' αὐτῆς τῆς ἀληθείας ἀναγκασθέντες. διαφέρουσι δ' ἀλλήλων τῷ τοὺς μὲν πρότερα τοὺς δ' ὕστερα λαμβάνειν, καὶ τοὺς μὲν γνωριμώτερα κατὰ τὸν λόγον τοὺς δὲ κατὰ τὴν αἴσθησιν. οἱ μὲν γὰρ θερμὸν καὶ ψυχρόν, οἱ δ' ὑγρὸν καὶ ξηρόν, ἕτεροι δὲ περιττὸν καὶ ἄρτιον, οἱ δὲ νεῖκος καὶ φιλίαν αἰτίας τίθενται τῆς γενέσεως· ταῦτα δ' ἀλλήλων διαφέρει κατὰ τὸν εἰρημένον τρόπον. ὥστε ταὐτὰ λέγειν πως καὶ ἕτερα ἀλλήλων—ἕτερα μὲν ὥσπερ
189 a καὶ δοκεῖ τοῖς πλείστοις, ταὐτὰ δὲ ᾗ ἀνάλογον· λαμβάνουσι γὰρ ἐκ τῆς αὐτῆς συστοιχίας· τὰ μὲν γὰρ περιέχει, τὰ δὲ περιέχεται τῶν ἐναντίων.

Ταύτῃ τε δὴ ὡσαύτως λέγουσι καὶ ἑτέρως, καὶ χεῖρον καὶ βέλτιον, καὶ οἱ μὲν γνωριμώτερα κατὰ τὸν λόγον (ὥσπερ εἴρηται πρότερον) οἱ δὲ κατὰ

---

[a] The modern reader must familiarize himself with the conception, deeply entrenched in the Greek and the scholastic mind, that the members of any defined group of things may be regarded as having a directly linear relation to each other, according as they manifest the characteristic of the group in the perfection of its standard representative or only in some less pronounced degree.

[b] [The Pre-Socratics did not use the word στοιχεῖον, but used ἀρχή of their primitive stuff.—C.]

[c] [The Pythagoreans, who regard as the 'elements of number' the 'Odd' (the principle of Limit) and the 'Even'

out of white and black), it follows that all things that come into existence in the course of nature are either opposites themselves or are compounded of opposites.[a]

Up to this point, as already indicated, we may claim the pretty general consent of all the systems ; for all thinkers posit their elements or 'principles,' as they call them ;[b] and, though they give no reasoned account of these 'principles,' nevertheless we find—as though truth itself drove them to it in spite of themselves—that they are really talking about contrasted couples. But they differ in the order they follow, some starting from what is most luminous to the intelligence, others from what is more directly cognizable by the senses. Thus some posit 'hot and cold,' or 'wet and dry,' as causes of becoming, and others 'odd and even,'[c] or 'amity and conflict,'[d] which illustrates what I am saying. Thus they agree in one respect, while differing in others. Their differences are obvious and are universally recognized ; but what is not seen so generally is that they are all analogous in so far as they all rest upon the same fundamental conception of antithesis,[e] though some express it in a wider and some in a narrower formula.

To this extent, then, they agree and differ, and do worse or better one than the other ; some, as already said, beginning with what is more accessible to intelligence and others with what is more accessible to

(Unlimited), *Met.* 986 a 15. These are more abstract and remote principles than 'hot and cold,' etc.—C.]

[d] [Empedocles, whose Φιλότης and Νεῖκος were opposed moving causes, *cf. Met.* 985 a 30 *οὐ μίαν ποιήσας τὴν τῆς κινήσεως ἀρχὴν ἀλλ' ἑτέρας καὶ ἐναντίας.*—C.]

[e] [*Cf.* *συστοιχία* used of the Pythagoreans' list of ten opposite principles, headed by Limit—Unlimited, Odd—Even, *Met.* 986 a 22.—C.]

189 a τὴν αἴσθησιν—τὸ μὲν γὰρ καθόλου κατὰ τὸν λόγον γνώριμον, τὸ δὲ καθ' ἕκαστον κατὰ τὴν αἴσθησιν· ὁ μὲν γὰρ λόγος τοῦ καθόλου, ἡ δ' αἴσθησις τοῦ κατὰ μέρος—οἷον τὸ μὲν μέγα καὶ τὸ μικρὸν κατὰ τὸν λόγον, τὸ δὲ μανὸν καὶ τὸ πυκνὸν κατὰ τὴν αἴσθησιν.

Ὅτι μὲν οὖν ἐναντίας δεῖ τὰς ἀρχὰς εἶναι φανερόν.

CHAPTER VI

ARGUMENT

'Antithesis,' involving duality, being established as essential to natural changes, there must be at least two 'principles' between which the contrast exists; and the hypothesis of an unlimited number of principles being ruled out, we must look for a definite number, not less than two, but not unnecessarily large; for if a smaller number suffices, the hypothesis of a larger number, even if not demonstrably false, is at least superfluous (189 a 11-20).

Must we then go beyond two? Perhaps so, for it is

Ἐχόμενον δ' ἂν εἴη λέγειν πότερον δύο ἢ τρεῖς ἢ πλείους εἰσίν.

Μίαν μὲν γὰρ οὐχ οἷόν τε, ὅτι οὐχ ἓν τὸ ἐναντίον, ἀπείρους δ', ὅτι οὐκ ἐπιστητὸν τὸ ὂν ἔσται. μία τε ἐναντίωσις ἐν παντὶ γένει ἑνί, ἡ δ' οὐσία ἕν

[a] [The Platonic antithesis. See note on 187 a 17.—C.]

[b] [*Cat.* 6 a 15. Contraries (ἐναντία) are defined, by a spatial metaphor, as 'things which, within the same class (γένος), are separated by the greatest possible distance.' *Cf. Met.* Δ 10.—C.] See Ross, *Aristotle, Physics*, Vol. I., Comment. pp. 314 ff.

sense ; for the general is approached by the intelligence and the particular by the senses, since the mind grasps the universal principle and the senses the partial application. Thus, 'great and small'[a] are mental conceptions, while 'thick and thin' are sense impressions.

But in any case it is clear that the 'principles' must form a contrasted couple.

CHAPTER VI

ARGUMENT (*continued*)

difficult to conceive of the 'opposites' acting directly upon each other, or constituting anything in combination. A 'material' element for them to act upon seems essential ; but there is no advantage in positing more than one such principle (a 20-b 27).

We can see, therefore, that the ultimate principles of Nature can be reduced either to three or to two, but we must examine the question very carefully before committing ourselves to either side of this alternative (b 27-29).

The ground might now seem to be clear for discussing whether the ultimate principles of Nature are two, or three, or some greater but limited number.

A single principle will not do, for we have established antithesis as an ultimate constituent in Nature, and antithesis involves duality. Nor can the factors of Nature be unlimited, or Nature could not be made the object of knowledge. Now, as far as antithesis goes, two principles would be enough, for every defined class comes under one general antithesis,[b] and the whole sum of 'things that exist

189 a τι γένος. καὶ ὅτι ἐνδέχεται ἐκ πεπερασμένων· βέλτιον δ' ἐκ πεπερασμένων, ὥσπερ Ἐμπεδοκλῆς, ἢ ἐξ ἀπείρων· πάντα γὰρ ἀποδιδόναι οἴεται, ὅσαπερ Ἀναξαγόρας ἐκ τῶν ἀπείρων. ἔτι δέ ἐστιν ἄλλα ἄλλων πρότερα ἐναντία, καὶ γίνεται ἕτερα ἐξ ἄλλων (οἷον γλυκὺ καὶ πικρὸν καὶ λευκὸν καὶ μέλαν)· τὰς δ' ἀρχὰς ἀεὶ δεῖ μένειν. ὅτι μὲν οὖν οὔτε μία οὔτε ἄπειροι, δῆλον ἐκ τούτων.

Ἐπεὶ δὲ πεπερασμέναι, τὸ μὴ ποιεῖν δύο μόνον ἔχει τινὰ λόγον. ἀπορήσειε γὰρ ἄν τις πῶς ἢ ἡ πυκνότης τὴν μανότητα ποιεῖν τι πέφυκεν ἢ αὕτη τὴν πυκνότητα· ὁμοίως δὲ καὶ ἄλλη ὁποιαοῦν ἐναντιότης· οὐ γὰρ ἡ φιλία τὸ νεῖκος συνάγει καὶ ποιεῖ τι ἐξ αὐτοῦ, οὐδὲ τὸ νεῖκος ἐξ ἐκείνης, ἀλλ' ἄμφω ἕτερόν τι τρίτον. ἔνιοι δὲ καὶ πλείω

---

[a] The student must resist the temptation to take οὐσία and γένος in this passage in the sense that first suggests itself, of ' substance ' and ' category ' respectively. I follow Simplicius (198. 28) in taking οὐσία to stand here and elsewhere in this chapter for ' all the concrete things that constitute Nature and are capable of change,' τὰ φυσικὰ πάντα πράγματα, as distinct from the immaterial existences touched upon, or dealt with, in the *De caelo* and the *Metaphysica*, and led up to in the *Physics*. As to their all being embraced under one antithesis *cf.* Introduction to Bk. I. p. 6.

[b] ' Sweet and bitter ' are regarded as the two extremes between which all savours stand, and between which all changes of savour move. ' White and black ' are the corresponding extremes of colour. But these ' couples ' must be regarded as special cases of some more general contrast, *e.g.* ' hot and cold ' (Parmenides). [*De gen. et corr.* B 2, explains that secondary couples of contrasted qualities (*e.g.* sweet-bitter) are derived from the primary (tangible) couples, hot-cold, wet-dry, whose combinations constitute the nature of the ' simple bodies ' (earth, air,

in Nature,' as such, forms a defined class.[a] Since, then, one antithesis will suffice, it is better not to go beyond it ; for the more limited, if adequate, is always preferable, as we saw in the case of Empedocles, who claims to get everything out of his four substances that Anaxagoras claims to get out of his unlimited number. Or we may put it that some antitheses are more general than others, and some are derived from others. Such subordinate antitheses are exemplified by 'sweet and bitter' or 'black and white.' But the fundamental principles must be in evidence in *every* case.[b] Obviously, then, since antithesis is fundamental and implies duality, the ultimate principles, though limited in number, must be more than one.

But, granting them to be limited, there are reasons for supposing them to be more than two. For (1) a man might well be at a loss to conceive how 'density' itself, for instance, could possibly make 'rarity's' self into anything, or 'rarity' density's And so too with any other antithesis ; for, I suppose, we are not asked to think of 'amity' drawing 'hostility' closer together and thereby constructing something out of it, nor *vice versa*, rather than both of them acting upon a third something.[c] (2) And

water, fire). The indefinitely numerous secondary couples cannot be *permanent* principles.—C.]

[c] [*Cf. De gen. et corr.* A 6-10, on 'action and passion,' and 329 b 20 ff. δεῖ δε ποιητικὰ καὶ παθητικὰ εἶναι ἀλλήλων τὰ στοιχεῖα. *Cf.* Joachim, *ad loc.* Aristotle's primary contraries (hot-cold, wet-dry) enable the 'simple bodies' to *act on* one another and to compose substances of a higher order. But Anaximenes' 'density' cannot *act on* 'rarity' or *make* it anything, nor could Empedocles' 'amity' *produce* anything out of 'hostility' by its characteristic power of 'drawing together' unlike elements into a unity.—C.]

189 a λαμβάνουσιν ἐξ ὧν κατασκευάζουσιν τὴν τῶν ὄντων φύσιν. πρὸς δὲ τούτοις ἔτι κἂν τόδε τις ἀπορήσειεν, εἰ μή τις ἑτέραν ὑποτίθησι τοῖς ἐναντίοις φύσιν· οὐθενὸς γὰρ ὁρῶμεν τῶν ὄντων οὐσίαν τἀναντία· τὴν δ' ἀρχὴν οὐ καθ' ὑποκειμένου δεῖ λέγεσθαί τινος (ἔσται γὰρ ἀρχὴ τῆς ἀρχῆς· τὸ γὰρ ὑποκείμενον ἀρχή, καὶ πρότερον δοκεῖ τοῦ κατηγορουμένου εἶναι). ἔτι οὐκ εἶναί φαμεν οὐσίαν ἐναντίαν οὐσίᾳ. πῶς οὖν ἐκ μὴ οὐσιῶν οὐσία ἂν εἴη; ἢ πῶς ἂν πρότερον μὴ οὐσία οὐσίας εἴη;

Διόπερ εἴ τις τόν τε πρότερον ἀληθῆ νομίσειεν
189 b εἶναι λόγον καὶ τοῦτον, ἀναγκαῖον, εἰ μέλλει διασώσειν ἀμφοτέρους αὐτούς, ὑποτιθέναι τι τρίτον· ὥσπερ φασὶν οἱ μίαν τινὰ φύσιν εἶναι λέγοντες τὸ πᾶν, οἷον ὕδωρ ἢ πῦρ ἢ τὸ μεταξὺ τούτων. δοκεῖ δὲ τὸ μεταξὺ μᾶλλον· πῦρ γὰρ ἤδη καὶ γῆ καὶ ἀὴρ καὶ ὕδωρ μετ' ἐναντιοτήτων συμπεπλεγμένα ἐστίν· διὸ καὶ οὐκ ἀλόγως ποιοῦσιν οἱ τὸ ὑποκείμενον

[a] [*e.g.* Democritus's infinite atoms, Empedocles' four elements (Simplic.).—C.]

[b] Aristotle states this as an argument that might well suggest itself to a man. He does not commit himself to its validity. As a matter of fact, he inclines to regard both the contrasted terms of the antithesis and the material on which they act as logical abstractions, neither of which has actual existence except as associated with the other in a *compositum*. *Cf.* Introduction, pp. liv-lv.

[c] ['Natural existence,' *i.e.* a 'substance' in the primary and fullest sense. A substance has no contrary, but, while remaining one and the same, admits contrary qualities (*Cat.* 3 b 24, 4 a 10).—C.] See also p. 60, note *a*.

[d] [*Cf.* 187 a 14, note.—C.]

indeed some thinkers have assumed more than one such principle out of which to construct the nature of things.[a] (3) Besides, one might encounter another difficulty in supposing there to be nothing in Nature underlying the antithetical couple; for we never see the antithetical principles, by themselves, constituting the substantive existence of anything we know in Nature, and a true principle cannot be the mere attribute of something else, for the subject to which it was attributed would be a principle anterior to it, inasmuch as it would be presupposed in the very fact of having this attribute predicated of it.[b] (4) Nor can we make 'natural existence' the term of an antithesis,[c] for things move from one term of an antithesis to the other, and since nothing 'exists in Nature' that is contrasted with 'natural existence,' how could any such existent move out of what is not there for it to move out of? Or how could such a non-existent be a presupposition of the existent?

Thus, if our former insistence on the two terms of some antithesis being principles is sound, and if we are now convinced that these antithetical principles need something to work on, and if we are to preserve both these conclusions, must we not necessarily posit a third principle as the subject on which the antithetical principles act? And it is this third principle that those physicists are really thinking about who say that the universe has but one single constituent, water or fire or some intermediate substance.[d] Of these the hypothesis of an intermediate substance seems best, for fire, earth, air, and water, are themselves implicated with contrasted properties; so that there is reason in distinguishing the universal

189 b ἕτερον τούτων ποιοῦντες, τῶν δ' ἄλλων οἱ ἀέρα· καὶ γὰρ ὁ ἀὴρ ἥκιστα ἔχει τῶν ἄλλων διαφορὰς αἰσθητάς, ἐχόμενον δὲ τὸ ὕδωρ. ἀλλὰ πάντες γε τὸ ἓν τοῦτο τοῖς ἐναντίοις σχηματίζουσιν, οἷον πυκνότητι καὶ μανότητι καὶ τῷ μᾶλλον καὶ ἧττον. ταῦτα δ' ἐστὶν ὅλως ὑπεροχὴ δηλονότι καὶ ἔλλειψις (ὥσπερ εἴρηται πρότερον)· καὶ ἔοικε παλαιὰ εἶναι καὶ αὕτη ἡ δόξα, ὅτι τὸ ἓν καὶ ὑπεροχὴ καὶ ἔλλειψις ἀρχαὶ τῶν ὄντων εἰσί (πλὴν οὐ τὸν αὐτὸν τρόπον, ἀλλ' οἱ μὲν ἀρχαῖοι τὰ δύο μὲν ποιεῖν τὸ δὲ ἓν πάσχειν, τῶν δ' ὕστερόν τινες τοὐναντίον τὸ μὲν ἓν ποιεῖν τὰ δὲ δύο πάσχειν φασὶ μᾶλλον).

Τὸ μὲν οὖν τρία φάναι τὰ στοιχεῖα εἶναι ἔκ τε τούτων καὶ ἐκ τοιούτων ἄλλων ἐπισκοποῦσι δόξειεν ἂν ἔχειν τινὰ λόγον (ὥσπερ εἴπομεν), τὸ δὲ πλείω τριῶν οὐκέτι. πρὸς γὰρ τὸ πάσχειν ἱκανὸν τὸ ἕν· εἰ δὲ τεττάρων ὄντων δύο ἔσονται ἐναντιώσεις, δεήσει χωρὶς ἑκατέρας ὑπάρχειν ἑτέραν τινὰ μεταξὺ φύσιν· εἰ δ' ἐξ ἀλλήλων δύνανται γεννᾶν δύο οὖσαι, περίεργος ἂν ἡ ἑτέρα τῶν

---

[a] [*Or*, 'and of the rest (who make it some one element) those who make it air (are the most reasonable).'—C.]

[b] [187 a 16.—C.]

[c] [See 187 a 18.—C.]

[d] Whereas Aristotle (as against Empedocles) took it as a direct datum of observation that the 'four elements' themselves pass into each other. Therefore there cannot be any two or more ultimate material principles, acted upon by distinct antithetical 'couples' irreducible to a common principle. [The argument is: 'If there were four *active contrasted* principles, then (1) if each couple is in-

material element from all of them; though some have identified it with air,[a] inasmuch as the special characteristics of air are less obtrusive on the senses than those of the others, water coming next in this respect. But the Physicists, each and all, mould this universal subject matter by such antithetical principles as density and rarity, or 'more and less,' all which couples may be reduced to 'excess and defect' as already pointed out.[b] In reality, this doctrine, that the 'one' and 'excess and defect' are the principles of all that is, turns out to have been in possession of men's minds from of old; but not always in the same way, for the earlier thinkers made the 'one' the receptive subject and the contrasted 'two' the active agents, whereas more recently the 'two' have sometimes been regarded as the subject passively acted upon and the 'one' as the agent.[c]

So from these and other considerations it appears that there is much to be said for assuming three principles, but not for going beyond three. For one passive principle is enough. And if there were supposed to be four altogether, and they were to constitute two couples, each couple would demand a subject of its own to act upon, apart from the other couple;[d] or, if one were derivative from the other, it would not count among the 'principles,' for it

dependent of the other, each will need a separate passive principle (whereas one is enough); (2) if one couple is derivable from the other, there will really be only one primary couple; the second will be superfluous. This interpretation of the second alternative is Porphyry's, in Simplicius (206. 3), who objects that γεννᾶσθαι (for γεννᾶν) would be required, and himself understands εἰ δὲ ἐξ ἀλλήλων πάντα γεννῶσι.—C.]

189 b ἐναντιώσεων εἴη. ἅμα δὲ καὶ ἀδύνατον πλείους εἶναι ἐναντιώσεις τὰς πρώτας. ἡ γὰρ οὐσία ἕν τι γένος ἐστὶ τοῦ ὄντος, ὥστε τῷ πρότερον καὶ ὕστερον διοίσουσιν ἀλλήλων αἱ ἀρχαὶ μόνον, ἀλλ' οὐ τῷ γένει· ἀεὶ γὰρ ἐν ἑνὶ γένει μία ἐναντίωσίς ἐστιν, πᾶσαί τε αἱ ἐναντιώσεις ἀνάγεσθαι δοκοῦσιν εἰς μίαν.

Ὅτι μὲν οὖν οὔτε ἓν τὸ στοιχεῖον οὔτε πλείω δυοῖν ἢ τριῶν, φανερόν· τούτων δὲ πότερον (καθάπερ εἴπομεν) ἀπορίαν ἔχει πολλήν.

CHAPTER VII

INTRODUCTORY NOTE

The theses of this critically important chapter present no very great difficulties in themselves; but the hints in which they are set forth leave so much to be developed and connected together by the reader, as to suggest that here, as so often elsewhere in Aristotle's writings, we have notes of lectures rather an elaborated argument.

Moreover, this chapter is at the same time an example of the opposite characteristics of repetition and over-elaborated redundancy, which often raises the suspicion that here, as in many of Aristotle's works, our present texts show traces of the fusion of two different recensions.

Torstrik, whose attempt to disentangle two recensions from the present text of the *De Anima* is well known, seems to have had his suspicions first roused by a study of the *Physics*.[a] The actual existence of two (and in some Books three) separate recensions of the *Ethics*, with fragmentary evidence in the same direction with regard to other works of Aristotle, affords a basis for Torstrik's

[a] *Aristotelis De Anima Libri Tres.* Recensuit Adolphus Torstrik. Berolini, 1862, p. ii.

would not be primary. Nor could there be two primary antitheses; for 'natural existence' groups 'all that is in Nature' under a single aspect, so that the supposed pairs of principles could only differ from each other in priority, and could not belong to different series; for there is never more than one general antithesis common to one group, and it seems that all antitheses can be brought under one general antithesis.

It is clear, then, that there must be more than one element or principle, and that there cannot be more than two or three. But, within these limits, the decision as between two and three presents great difficulties.

CHAPTER VII

INTRODUCTORY NOTE (*continued*)

hypothesis; and what we know of the recovery and editing of Aristotle's manuscripts in the time of Sulla and Cicero gives it additional verisimilitude.

So much for the text. The translation presents special difficulties; for a great part of the chapter is devoted to the careful examination of certain Greek words and forms of expression, with their symmetries, balancings, and ambiguities, to which there are no English equivalents; though some general conception of their nature may be gathered from considering the difficulties in which the English writer may often find himself when he wishes to distinguish between the transitive and neuter significance of such verbs as, for example, 'change' and 'move.'

Thus, if A 'changes' B (transitive), B itself 'changes' (neuter), and, in this sense, it is B only that 'changes,' and not A at all. Or if A is 'moving' B (transitive), it

INTRODUCTORY NOTE (*continued*)

is A only, and not B, that is ' moving ' in this sense, but yet both are ' moving ' in the intransitive sense.

Similarly in Greek the one verb *gignesthai* may mean either ' become ' (A becomes B), or ' come into existence ' (B comes into existence). Again, in some cases, but not in all, to say that A ' becomes ' (*gignetai*) B implies, as its reciprocal, that B has taken the place of (*gignetai ek*) A, and is there instead of it. This want of symmetry and its rationale must be inquired into. But yet again, this very phrase (*gignetai ek*) is itself ambiguous, for it may mean either that B has ' taken the place ' of A (blackness taken the place of whiteness, for example, so that the whiteness is no longer there), or that B is ' made of ' A (the statue made of bronze). [In the case of black replacing white, the subject of *gignetai* is commonly the thing which persists through the change : ' a man *becomes* black *from* being white.' When B is ' made of ' A, A persists and is called τὸ ἐξ οὗ ἐνυπάρχοντος—the factor *out of which* the generation occurs, but which *remains a constituent* of

ARGUMENT

Taking the illustration of the bronze and the Hermes as our starting-point, we see that in every change there is a factor (the bronze) that persists, a factor (the Hermes-form) that has ' come into being ' in the process of change, and a factor (the amorphousness of the unfounded bronze) that has disappeared and been superseded. Here, then, we have the ' material ' on which the change has been wrought, and the terms of antithesis between which the change itself takes place, the ' material ' being the persistent factor, and the ' terms of the antithesis ' the factors that come and go.

In other cases the persistence and identity of the ' material ' is not so obvious ; for a number of the characteristics of an object (its colour, shape, consistency, size, etc.) may all be changing at once, each along its own antithetical line, so that the acorn, for example, cannot well be said to ' persist '

INTRODUCTORY NOTE (*continued*)

the thing generated (unlike the 'shortage,' which is superseded by the 'form').—C.]

Obvious truth had been obscured and missed, and serious fallacies fallen into, by Greek thinkers for want of a careful analysis of these linguistic confusions and distinctions. The best I have been able to do, without intolerable violence to English idiom, is sometimes to give a make-shift paraphrase (which loses much of the relevance and all of the sharpness of outline of the original), and sometimes to be content with simply pointing out the identity in Greek of two expressions that have no resemblance to each other in English.

In the 'Argument,' however, I have endeavoured to disentangle the main theses of the chapter from the verbal analyses in which they are involved, and at the same time to give them the expansions which (when read in the light of other passages in Aristotle's works) they seem at once to warrant and to demand.

It is to the 'Argument,' therefore, rather than to the translation, that the English reader must look for guidance.

ARGUMENT (*continued*)

in the oak; but here, too, there is a 'something to start with' as the subject of a continuous change; and the change, however complex, can always be resolved into a combination of antithetical movements.[a]

But what about this 'something to start with' itself? The bronze or the seed, from which we chose to start, stands at the end, as well as at the beginning, of a certain development. What is its *material, and what did it start from itself? We can trace it back and back as far as to the four elemental substances, but no further. These substances,*

[a] The difference and underlying identity of these two kinds of change is developed in the *De gen. et cor.* Its relation to the divergent theory expounded in *Physics* v. will be examined in its place. See p. 116, note *b*.

ARGUMENT (*continued*)

however, cannot themselves be accepted as the ultimate material, for they have antithetical characteristics and can be transmuted into each other by antithetical changes. The cold-moist water can become the hot-moist vapour (or air) by antithetical change ; and the moisture itself, which has so far persisted, is the term of an antithesis, and may disappear in its turn, if the hot-moist vapour passes into the hot-dry fire.

And yet analogy and conceptual analysis forbid us to resolve the objective-sensible world altogether into antitheses acting upon each other (cf. Gen. Introd. p. xxv). *We have already seen not only that dry cannot directly act upon moist, without a subject to act upon and be incorporated in, but also that moist, as such, cannot ' become ' or ' be replaced by ' hot, as such, under any circumstances (any more than solid, as such, could become, or be replaced by, blue, as such), for moist and hot, respectively, are terms each of its own antithesis, along the line of which alone it can change.*

So although, as we trace the material back, it yields more and more to analysis into antitheses, and never ceases to evade us when we pursue it, yet we cannot quite get rid of it. Nay, it is itself the one and only ' persistent ' and

189 b 30 Ὧδ' οὖν ἡμεῖς λέγωμεν πρῶτον περὶ πάσης γενέσεως ἐπελθόντες· ἔστι γὰρ κατὰ φύσιν τὰ κοινὰ πρῶτον εἰπόντας οὕτω τὰ περὶ ἕκαστον ἴδια θεωρεῖν.

Φαμὲν γὰρ γίνεσθαι ἐξ ἄλλου ἄλλο καὶ ἐξ ἑτέρου ἕτερον ἢ τὰ ἁπλᾶ λέγοντες ἢ τὰ συγκείμενα. λέγω δὲ τοῦτο ὡδί. ἔστι γὰρ γίνεσθαι ἄνθρωπον

---

[a] It is convenient to call it ' matter ' in English, though there is only one word in Greek (*hyle*) both for the immediate and for the ultimate material of which anything is made.

ARGUMENT (*continued*)

'continuous' and 'non-contrasted' principle of the persistent objective-sensible world. It is always there, though we can never isolate it.

The 'principles,' then, are a coupled or antithetical 'two' in a way; but, when all is said and done, they are also 'three' in another way, which other way cannot be ignored.

Well, but suppose we accept the 'material' principle less grudgingly, and lay down as our fundamental basis the 'ultimate material' [a] *of all things, and then consider the collectivity of the perceptible attributes that (on the analogy of the statue) we call the 'form' of whatever it is that we are examining. Are not these 'two' principles, 'matter' and 'form,' enough? Statically perhaps they are, but not dynamically; for (as we shall see more fully in the next chapter) the bronze, for instance, can only 'become' the statue in virtue of 'not being' the statue already. It is a condition of its 'acquiring' the form, that it should 'not yet have' it. 'Shortage,' therefore, or 'privation,' is an essential factor in 'becoming,' even though we should try to reduce the 'antithesis' to a single term and make that one term do all the work by its presence or absence.*

So again, however hard we plead for 'two, in one way,' we have to add 'but three in another way.'

In advancing now to the formulation of a positive theory, let us begin with the general conception of 'change' (that is to say, of things 'coming into existence' altogether, or 'becoming this or that' in particular which they were not before). For the natural order of exposition, as we have seen, is to start from the general principle and proceed to the special applications.

Note, then, that in speaking of one thing becoming another, or one thing coming out of, or in the place of, another, we may use either (1) simple or (2) complex terms. I mean that we can say either (1)

189 b 35 μουσικόν, ἔστι δὲ τὸ μὴ μουσικὸν γίνεσθαι μουσικόν,
190 a ἢ τὸν μὴ μουσικὸν ἄνθρωπον ἄνθρωπον μουσικόν. ἁπλοῦν μὲν οὖν λέγω τὸ γιγνόμενον τὸν ἄνθρωπον καὶ τὸ μὴ μουσικόν, καὶ ὃ γίγνεται ἁπλοῦν τὸ μουσικόν· συγκείμενον δὲ καὶ ὃ γίγνεται καὶ τὸ γιγνόμενον, ὅταν τὸν μὴ μουσικὸν ἄνθρωπον φῶμεν γίγνεσθαι μουσικὸν ἄνθρωπον.

Τούτων δὲ τὸ μὲν οὐ μόνον λέγεται τόδε γίγνεσθαι ἀλλὰ καὶ ἐκ τοῦδε (οἷον ἐκ μὴ μουσικοῦ μουσικός)· τὸ δ' οὐ λέγεται ἐπὶ πάντων· οὐ γὰρ ἐξ ἀνθρώπου ἐγένετο μουσικός, ἀλλ' ἄνθρωπος ἐγένετο μουσικός.

Τῶν δὲ γινομένων ὡς τὰ ἁπλᾶ λέγομεν γίγνεσθαι, τὸ μὲν ὑπομένον γίγνεται τὸ δ' οὐχ ὑπομένον· ὁ μὲν γὰρ ἄνθρωπος ὑπομένει μουσικὸς γινόμενος ἄνθρωπος καὶ ἔστι, τὸ δὲ μὴ μουσικὸν καὶ τὸ ἄμουσον οὔτε ἁπλῶς οὔτε συντιθέμενον ὑπομένει.

Διωρισμένων δὲ τούτων, ἐξ ἁπάντων τῶν γιγνομένων τοῦτο ἔστι λαβεῖν, ἐάν τις ἐπιβλέψῃ (ὥσπερ λέγομεν), ὅτι δεῖ τι ἀεὶ ὑποκεῖσθαι τὸ γινόμενον, καὶ τοῦτο εἰ καὶ ἀριθμῷ ἐστιν ἕν, ἀλλ' εἴδει γε οὐχ ἕν· τὸ γὰρ εἴδει λέγω καὶ λόγῳ ταὐτόν· οὐ γὰρ ταὐτὸν τὸ ἀνθρώπῳ καὶ τὸ ἀμούσῳ εἶναι. καὶ τὸ

[a] [This conclusion is Aristotle's main point. Down to a 31 he is thinking mainly of changes involving the gain of a positive attribute. In what sense does this something 'come to be out of what is not'? There is always a subject *x* which is there before and after the change; and even before it gains the new attribute, it is conceptually complex —*x* which is not-*A* (subject + 'shortage').—C.]

that a ' man ' becomes cultured, or that the ' uncultured ' in him is replaced by culture, or (2) that the ' uncultured man ' becomes a ' cultivated man.' In this case (1) the ' man ' (who acquires culture), and his state of ' unculture ' (which is replaced by culture) and the ' culture ' itself (which was not, but has ' come to be ') are all what I call ' simple ' terms ; whereas (2) both the ' uncultivated man,' who became something he was not, and the ' cultivated man ' that he became, are what I call ' composite ' terms.

And note that in some of these cases we can say, not only that a thing ' becomes so-and-so,' but also that it does so ' *from being* so-and-so ' ; *e.g.* a man becomes cultivated from being uncultivated. But we cannot use this expression in all cases ; for he does not become cultivated ' *from being* a man ' ; on the contrary, he becomes a cultivated *man*.

And of the two simple terms, ' man ' and ' uncultivated ' (both of which we said ' became ' something), one (the ' man ') persists when he has become a cultivated man ; but the other (the ' uncultivated ' or ' non-cultivated ' in him) does not persist either in the simple ' cultivated ' which we say it has ' become ' or in the composite ' cultivated man.'

Observing these distinctions, we may reach a principle of universal application (if we ' observingly distil it out,' as they say), namely that in all cases of becoming there must always be a subject—the thing which becomes or changes, and this subject, though constituting a unit, may be analysed into two concepts and expressed in two terms with different definitions ; for the definition of ' man ' is distinct from the definition of ' uncultivated.' [a]

190 a μὲν ὑπομένει, τὸ δ᾽ οὐχ ὑπομένει—τὸ μὲν μὴ ἀντικείμενον ὑπομένει (ὁ γὰρ ἄνθρωπος ὑπομένει), τὸ μουσικὸν δὲ καὶ τὸ ἄμουσον οὐχ ὑπομένει, οὐδὲ τὸ ἐξ ἀμφοῖν συγκείμενον (οἷον ὁ ἄμουσος ἄνθρωπος).

Τὸ δ᾽ ἔκ τινος γίγνεσθαί τι (καὶ μὴ τόδε γίγνεσθαί τι) μᾶλλον μὲν λέγεται ἐπὶ τῶν μὴ ὑπομενόντων (οἷον ἐξ ἀμούσου μουσικὸν γίνεσθαι, ἐξ ἀνθρώπου δὲ οὔ)· οὐ μὴν ἀλλὰ καὶ ἐπὶ τῶν ὑπομενόντων ἐνίοτε λέγεται ὡσαύτως· ἐκ γὰρ χαλκοῦ ἀνδριάντα γίγνεσθαί φαμεν, οὐ τὸν χαλκὸν ἀνδριάντα. τὸ μέντοι ἐκ τοῦ ἀντικειμένου καὶ μὴ ὑπομένοντος ἀμφοτέρως λέγεται—καὶ ἐκ τοῦδε τόδε, καὶ τόδε τόδε· καὶ γὰρ ἐξ ἀμούσου καὶ ὁ ἄμουσος γίνεται μουσικός. διὸ καὶ ἐπὶ τοῦ συγκειμένου ὡσαύτως· καὶ γὰρ ἐξ ἀμούσου ἀνθρώπου καὶ ὁ ἄμουσος ἄνθρωπος γίγνεσθαι λέγεται μουσικός.

Πολλαχῶς δὲ λεγομένου τοῦ γίγνεσθαι, καὶ τῶν μὲν οὐ γίγνεσθαι ἀλλὰ τόδε τι γίγνεσθαι, ἁπλῶς δὲ γίγνεσθαι τῶν οὐσιῶν μόνων, κατὰ μὲν τἆλλα φανερὸν ὅτι ἀνάγκη ὑποκεῖσθαί τι τὸ γιγνόμενον—καὶ γὰρ ποσὸν καὶ ποιὸν καὶ πρὸς ἕτερον καὶ ποτὲ

And the one persists while the other disappears—the one that persists being the one that is *not* embraced in an antithesis ; for it is the ' man ' that persists, and neither the simple ' cultured ' or ' uncultivated ' nor the composite ' uncultivated man.'

When we speak of something ' becoming *from* or *out of* ' whatever it may be (rather than of its ' becoming so-and-so '), we generally mean by the thing *from* or *out of* which the becoming takes place, the non-persistent term or aspect : thus, we speak of becoming cultivated from being uncultivated, not from being a man. Still the expression ' out of ' is used sometimes of the factor which persists : we say a statue is made *out of* bronze, not that the bronze *becomes* a statue (in the sense of ceasing to be bronze). When, however, the thing *from* or *out of* which the becoming occurs is the contrasted, non-persistent term, both expressions are used : we can say of a thing, *e.g.* ' the uncultivated ' (man), either that he ' becomes this ' (cultivated) or that he ' becomes this (cultivated) *from being* that (uncultivated).' Hence it is the same with the composite terms : we say that the ' uncultivated man ' becomes cultivated and also that he becomes so ' *from being* an uncultivated man.'

But there is (in Greek) a further ambiguity ; for the same word (*gignesthai*) is employed either of a thing ' coming to be ' in the absolute sense of ' coming into existence,' or in the sense of ' coming to be this or that ' which it was not before ; and it is only of a concrete thing, as such, that we can speak of its ' coming to be ' in the full sense of coming into existence. Now in all other cases of change, whether of quantity or quality or relation or time or place, it

190 a καὶ ποῦ γίνεται ὑποκειμένου τινός, διὰ τὸ μόνην τὴν οὐσίαν μηθενὸς κατ' ἄλλου λέγεσθαι ὑπο-
190 b κειμένου, τὰ δ' ἄλλα πάντα κατὰ τῆς οὐσίας—ὅτι δὲ καὶ αἱ οὐσίαι καὶ ὅσα ἄλλα ἁπλῶς ὄντα ἐξ ὑποκειμένου τινὸς γίνεται, ἐπισκοποῦντι γένοιτ' ἂν φανερόν· ἀεὶ γὰρ ἔστι τι ὃ ὑποκεῖται, ἐξ οὗ γίνεται τὸ γιγνόμενον, οἷον τὰ φυτὰ καὶ τὰ ζῷα ἐκ σπέρματος. γίγνεται δὲ τὰ γιγνόμενα ἁπλῶς τὰ μὲν μετασχηματίσει (οἷον ἀνδριὰς ἐκ χαλκοῦ), τὰ δὲ προσθέσει (οἷον τὰ αὐξανόμενα), τὰ δ' ἀφαιρέσει (οἷον ἐκ τοῦ λίθου ὁ Ἑρμῆς), τὰ δὲ συνθέσει (οἷον οἰκία), τὰ δ' ἀλλοιώσει (οἷον τὰ τρεπόμενα κατὰ τὴν ὕλην). πάντα δὲ τὰ οὕτω γινόμενα φανερὸν ὅτι ἐξ ὑποκειμένων γίνεται.

Ὥστε δῆλον ἐκ τῶν εἰρημένων ὅτι τὸ γινόμενον ἅπαν ἀεὶ σύνθετόν ἐστι, καὶ ἔστι μέν τι γινόμενον, ἔστι δέ τι ὃ τοῦτο γίνεται, καὶ τοῦτο διττόν· ἢ γὰρ τὸ ὑποκείμενον ἢ τὸ ἀντικείμενον. λέγω δὲ ἀντικεῖσθαι μὲν τὸ ἄμουσον, ὑποκεῖσθαι δὲ τὸν

[a] [τὸ γινόμενον means the total thing—either subject + 'shortage' (before the change) or subject + form (after the change). There must be a subject + 'shortage' in *all* 'becoming'—both change and 'absolute becoming.'—C.]

is obvious that there must be some underlying subject which undergoes the change, since it is only a concrete something that can have that 'substantive existence,' the characteristic of which is that it can itself be predicated of no other subject, but is itself the subject of which all the other categories are predicated ; but on further consideration it will be equally obvious that a substance also, or anything, whether natural or artificial, that exists independently, proceeds from something that may be regarded as the subject of that change which results in its coming into being ; for in every case there is something already there, out of which the resultant thing comes ; for instance the sperm of a plant or animal. The processes by which things 'come into existence' in this absolute sense may be divided into (1) change of shape, as with the statue made of bronze, or (2) additions, as in things that grow, or (3) subtractions, as when a block of marble is chipped into a Hermes, or (4) combination, as in building a house, or (5) such modifications as affect the properties of the material itself. Clearly, then, all the processes that result in anything 'coming to exist' in this absolute sense start with some subject that is already there to undergo the process.

From all this it is clear that anything that 'becomes'[a] is always complex : there is (1) something that begins to exist (the new element of form), and (2) something that 'comes to be this' (comes to have this form) ; and this second thing may be regarded under two aspects—as the subject which persists, or as the contrasted qualification (which the new form will replace). For instance, in the uncultivated man who becomes cultivated, 'uncultivated' is the contrasted qualification, 'man,' the subject ; or, when

ἄνθρωπον· καὶ τὴν μὲν ἀσχημοσύνην καὶ τὴν ἀμορφίαν καὶ τὴν ἀταξίαν τὸ ἀντικείμενον, τὸν δὲ χαλκὸν ἢ τὸν λίθον ἢ τὸν χρυσὸν τὸ ὑποκείμενον. φανερὸν οὖν ὡς, εἴπερ εἰσὶν αἰτίαι καὶ ἀρχαὶ τῶν φύσει ὄντων, ἐξ ὧν πρώτων εἰσὶ καὶ γεγόνασι μὴ κατὰ συμβεβηκὸς ἀλλ' ἕκαστον ὃ λέγεται κατὰ τὴν οὐσίαν, ὅτι γίγνεται πᾶν ἔκ τε τοῦ ὑποκειμένου καὶ τῆς μορφῆς· σύγκειται γὰρ ὁ μουσικὸς ἄνθρωπος ἐξ ἀνθρώπου καὶ μουσικοῦ τρόπον τινά· διαλύσεις γὰρ τοὺς λόγους εἰς τοὺς λόγους τοὺς ἐκείνων. δῆλον οὖν ὡς γίνοιτ' ἂν τὰ γιγνόμενα ἐκ τούτων.

Ἔστι δὲ τὸ μὲν ὑποκείμενον ἀριθμῷ μὲν ἕν, εἴδει δὲ δύο—ὁ μὲν γὰρ ἄνθρωπος καὶ ὁ χρυσὸς καὶ ὅλως ἡ ὕλη ἀριθμητή· τόδε γάρ τι μᾶλλον, καὶ οὐ κατὰ συμβεβηκὸς ἐξ αὐτοῦ γίνεται τὸ γιγνόμενον, ἡ δὲ στέρησις καὶ ἡ ἐναντίωσις συμβεβηκός—ἓν δὲ τὸ εἶδος, οἷον ἡ τάξις ἢ ἡ μουσικὴ ἢ τῶν ἄλλων τι τῶν οὕτω κατηγορουμένων. διὸ ἔστι μὲν ὡς δύο λεκτέον εἶναι τὰς ἀρχάς, ἔστι δ' ὡς τρεῖς· καὶ ἔστι μὲν ὡς τἀναντία—οἷον εἴ τις λέγοι τὸ μουσικὸν καὶ τὸ ἄμουσον, ἢ τὸ θερμὸν καὶ τὸ ψυχρόν, ἢ τὸ ἡρμοσμένον καὶ τὸ ἀνάρ-

the statue is made, the contrasted qualification is the unshapeliness, formlessness, want of purposeful arrangement; the subject is the bronze or marble or gold. If, then, we grant that the things of Nature have ultimate determinants and principles which constitute them, and also that we can speak of them 'coming to be' not in an incidental but in an essential sense—so as to come to be the things they are and which their names imply, not having been so before—then it is obvious that they are composed, in every case, of the underlying subject and the 'form' which their defining properties give to it; for the cultivated man is in a way 'compact' of the subject 'man' and the qualification 'cultivated,' for the definition of such a *compositum* may always be resolved into the definitions of these two components. Clearly then these are the elements, or factors, out of which things that 'come to be' arise.

Now the subject is numerically one thing, but has two conceptually distinct aspects, for the man, or the gold, or the material factor in general is a thing that can be counted, since it may almost be regarded as a concrete individual thing and is not an incidental factor in the generation of what comes into being; whereas the negation of the emergent qualification or the presence of its opposite is incidental. On the other hand, the form—*e.g.* the 'order' or the 'culture' or any other such predicable qualification—is also one thing. So there is a sense in which the ultimate principles of the sum of changing things are two, but a sense in which they are three; for the actual change itself takes place between the terms of an antithesis, such as cultivated and uncultivated, hot and cold, articulated and unarticulated, and so forth; but from

190 b μοστον—ἔστι δ᾽ ὡς οὔ· ὑπ᾽ ἀλλήλων γὰρ πάσχειν τἀναντία ἀδύνατον. λύεται δὲ καὶ τοῦτο διὰ τὸ ἄλλο εἶναι τὸ ὑποκείμενον· τοῦτο γὰρ οὐκ ἐναντίον. ὥστε οὔτε πλείους τῶν ἐναντίων αἱ ἀρχαὶ τρόπον τινά, ἀλλὰ δύο ὡς εἰπεῖν τῷ ἀριθμῷ, οὔτ᾽ αὖ 191 a παντελῶς δύο, διὰ τὸ ἕτερον ὑπάρχειν τὸ εἶναι αὐτοῖς, ἀλλὰ τρεῖς· ἕτερον γὰρ τὸ ἀνθρώπῳ καὶ τὸ ἀμούσῳ εἶναι, καὶ τὸ ἀσχηματίστῳ καὶ χαλκῷ.

Πόσαι μὲν οὖν αἱ ἀρχαὶ τῶν περὶ γένεσιν φυσικῶν, καὶ πῶς πόσαι, εἴρηται· καὶ δῆλόν ἐστιν ὅτι δεῖ ὑποκεῖσθαί τι τοῖς ἐναντίοις καὶ τἀναντία δύο εἶναι.

Τρόπον δέ τινα ἄλλον οὐκ ἀναγκαῖον· ἱκανὸν γὰρ ἔσται τὸ ἕτερον τῶν ἐναντίων ποιεῖν τῇ ἀπουσίᾳ καὶ παρουσίᾳ τὴν μεταβολήν. ἡ δ᾽ ὑποκειμένη φύσις ἐπιστητὴ κατ᾽ ἀναλογίαν· ὡς γὰρ πρὸς ἀνδριάντα χαλκὸς ἢ πρὸς κλίνην ξύλον ἢ πρὸς ἄλλο[1] τι τῶν ἐχόντων μορφὴν ἡ ὕλη καὶ τὸ ἄμορφον ἔχει πρὶν λαβεῖν τὴν μορφήν, οὕτως αὕτη πρὸς οὐσίαν ἔχει καὶ τὸ τόδε τι καὶ τὸ ὄν. μία μὲν οὖν ἀρχὴ αὕτη (οὐχ οὕτω μία οὖσα οὐδὲ οὕτως ἓν ὡς

[1] [ἄλλο Simplic. 226. 7: ἄλλων F: τῶν ἄλλων cett.—C.]

---

[a] [*Literally*, 'and in a sense (we may describe these principles as) "the contraries" . . . in a sense, not'; *i.e.* according as the ambiguous phrases τὸ μουσικόν, τὸ ἄμουσον, etc., mean (1) the subject with the (positive or negative) qualification—this is not really a 'contrary'—or (2) the (negative) qualification only—the contrary proper, which cannot be acted upon by its contrary (*cf. Met.* 1075 a 27 ff.).—C.]

[b] [*Or* (reading ὂν for ἓν with E), 'not having the same sort of unity nor the same sort of existence as a concrete individual.'—C.]

another point of view these two principles are inadequate, for they cannot possibly act or be acted upon directly by each other.[a] This difficulty, however, disappears if we admit, as a third principle, a non-antithetical 'subject.' So in a sense there are no principles except the terms of opposition, and it may be said that they are two in number and no more; but there is a sense also in which we cannot quite admit this and must go on to three, because of the conceptual distinction that exists in them; for instance in the uncultivated man, between his being a man and his being uncultivated, or in the unshaped bronze, between its being unshaped and its being bronze.

It is now clear, then, how many are the principles of things in the changing world of Nature, and in what sense; namely that there is something that underlies all opposites, and that opposition involves two terms.

But, if we take it another way, we may escape the duality of the opposition by considering one of its terms taken singly as competent, by its absence or presence, to accomplish the whole change. Then there will only be the 'ultimately underlying' factor in Nature in addition to this formal principle to reckon with. And of this 'underlying' factor we can form a conception by analogy; for it will bear the same relation to concrete things in general, or to any specific concrete thing, which the bronze bears to the statue before it has been founded, or the wood to the couch, or the crude material of any object that has determined form and quality to that object itself. This ultimate material will count as one principle (not, of course, one[b] in the sense of a concrete 'indi-

191 a τὸ τόδε τι), μία δὲ ἡ[1] ὁ λόγος, ἔτι δὲ τὸ ἐναντίον τούτῳ ἡ στέρησις.

Ταῦτα δὲ πῶς δύο καὶ πῶς πλείω, εἴρηται ἐν τοῖς ἄνω. πρῶτον μὲν οὖν ἐλέχθη ὅτι ἀρχαὶ τἀναντία μόνον, ὕστερον δ' ὅτι ἀνάγκη καὶ ἄλλο τι ὑποκεῖσθαι καὶ εἶναι τρία· ἐκ δὲ τῶν νῦν φανερὸν τίς ἡ διαφορὰ τῶν ἐναντίων, καὶ πῶς ἔχουσιν αἱ ἀρχαὶ πρὸς ἀλλήλας, καὶ τί τὸ ὑποκείμενον. πότερον δὲ οὐσία τὸ εἶδος ἢ τὸ ὑποκείμενον, οὔπω δῆλον· ἀλλ' ὅτι αἱ ἀρχαὶ τρεῖς καὶ πῶς τρεῖς, καὶ τίς ὁ τρόπος αὐτῶν, δῆλον.

Πόσαι μὲν οὖν καὶ τίνες εἰσὶν αἱ ἀρχαί, ἐκ τούτων θεωρείσθωσαν.

[1] [ἡ codd. (Torstrik, *Philologus*, xii. (1857) 520). Simplic. 233. 4, interprets '*and one the* (*principle*, ἀρχή), "*the definition.*"' Prof. Ross tells me he prefers this reading and interpretation to the editors' corrections ᾗ, ἤ, <τὸ εἶδος> ἤ, and compares ἡ τὸ AB, '*the* (*line*) "*the AB.*"'—C.]

CHAPTER VIII

ARGUMENT

Absolute non-existence must be distinguished from incidental non-existence. (*What Aristotle, in this chapter, calls 'shortage'* (steresis) *is defined elsewhere as the 'negation of something within a defined class,'* i.e. '*within an antithesis,*' Met. 1011 b 19 ἡ δὲ στέρησις ἀπόφασίς ἐστιν ἀπό τινος ὡρισμένου γένους.) *Non-existence as such is absolute; but the negation of the form, as incidental to the material, is an essential factor in the 'becoming' of anything. And the non-existence of the 'shortage'* (*itself*

vidual '); and the collectivity of determining qualities implied by the thing's definition is also one principle; and further there is the opposite of this, namely the ' being without ' or ' shortage ' of it.

How the principles, then, can be taken as two, and how that enumeration appears to need supplementing, has now been shown. First it appeared as though the ' terms of an antithesis ' constituted all the principles necessary; but then we saw that something must underlie them, constituting a third. And now we see that the two terms of the opposition itself stand on a different footing from each other; and we see how all the principles are related to each other, and what we are to understand by the ' underlying subject.' It remains to consider whether the modifying ' form ' or the modified ' matter ' is to be regarded as the more ' essential ' factor of a thing; but that there are three principles altogether, and in what sense they are three, has already been demonstrated.

Let this, then, suffice as to the number of the principles and as to what they are.

CHAPTER VIII

ARGUMENT (*continued*)

a negation) is incidentally necessary to the presence of the negated form in the thing that emerges from the change.

The impossibility of anything coming into or passing out of existence can only be maintained with respect to the absolutely, not with respect to the incidentally, non-existent. This is the root of the fallacy of Parmenides.

(At this critical point of Aristotle's exposition the text, as we have it, is elliptical almost to the point of unintelligibility, unless supplemented from other sources.)

191 a 23 Ὅτι δὲ μοναχῶς οὕτω λύεται καὶ ἡ τῶν ἀρχαίων ἀπορία, λέγωμεν μετὰ ταῦτα. ζητοῦντες γὰρ οἱ κατὰ φιλοσοφίαν πρῶτοι τὴν ἀλήθειαν καὶ τὴν φύσιν τὴν τῶν ὄντων ἐξετράπησαν οἷον ὁδόν τινα ἄλλην ἀπωσθέντες ὑπὸ ἀπειρίας, καί φασιν οὔτε γίνεσθαι τῶν ὄντων οὐδὲν οὔτε φθείρεσθαι, διὰ τὸ ἀναγκαῖον μὲν εἶναι γίγνεσθαι τὸ γιγνόμενον ἢ ἐξ ὄντος ἢ ἐκ μὴ ὄντος, ἐκ δὲ τούτων ἀμφοτέρων ἀδύνατον εἶναι· οὔτε γὰρ τὸ ὂν γίνεσθαι (εἶναι γὰρ ἤδη), ἔκ τε μὴ ὄντος οὐδὲν ἂν γενέσθαι· ὑποκεῖσθαι γάρ τι δεῖ. καὶ οὕτω δὴ τὸ ἐφεξῆς συμβαῖνον αὔξοντες οὐδ᾽ εἶναι πολλά φασιν ἀλλὰ μόνον αὐτὸ τὸ ὄν.

Ἐκεῖνοι μὲν οὖν ταύτην ἔλαβον τὴν δόξαν διὰ τὰ εἰρημένα· ἡμεῖς δὲ λέγομεν ὅτι τὸ ἐξ ὄντος ἢ ἐκ μὴ ὄντος γίνεσθαι, ἢ τὸ μὴ ὂν ἢ τὸ ὂν ποιεῖν τι ἢ πάσχειν, ἢ ὁτιοῦν τόδε γίνεσθαι, ἕνα μὲν τρόπον
191 b οὐδὲν διαφέρει ἢ τὸ τὸν ἰατρὸν ποιεῖν τι ἢ πάσχειν, ἢ τὸ ἐξ ἰατροῦ εἶναί τι ἢ γίγνεσθαι· ὥστε, ἐπειδὴ τοῦτο διχῶς λέγεται, δῆλον ὅτι καὶ τὸ ἐξ ὄντος καὶ τὸ ὂν ἢ ποιεῖν ἢ πάσχειν. οἰκοδομεῖ μὲν οὖν ὁ

---

[a] [*Cf.* 187 a 27 τὴν κοινὴν δόξαν τῶν φυσικῶν.—C.]

[b] [*Literally*, 'they denied even the existence of a plurality of things, and allowed existence only to "Being itself"'—the inference drawn by Parmenides, *frag*. 8, though avoided later by the pluralist physicists.—C.]

[c] [ἕνα μὲν τρόπον. This first explanation goes down to b 27.—C.]

It remains to show that the conclusion we have reached not only solves the problem of genesis, but furnishes the only escape from the blind alley into which the first speculations on the subject led their authors. For when first they began to reason on the truth of things and the nature of all that exists, pioneers as they were, they fell upon a false track for want of a clue, and maintained that nothing at all could either come into existence or pass out of it ; [a] for they argued that, if a thing comes into existence, it must proceed either out of the existent or out of the non-existent, both of which were impossible ; for how could anything 'come out of' the existent, since it is already there ? and obviously it could not come out of the non-existent, for what it comes out of must be there for it to come out of, and the non-existent is not there at all. And so, developing the logical consequences of this, they went on to say that the actually and veritably 'existent' is not many, but only one.[b]

Such then is their dogma ; but, as for us, we maintain that when we speak of anything 'coming to be,' whether out of existence or non-existence, or of the non-existent or the existent acting or being acted on in any way, or of anything at all 'becoming this or that,' one explanation is as follows : [c] It is much the same as saying that a 'physician' does or experiences something, or that he has 'become' (and now is) something that he has 'turned into,' instead of remaining a physician. For all these expressions are ambiguous, and this ambiguity is clearly analogous to the ambiguity concealed under our language when we speak of what 'the existent has turned into,' or of the existent 'doing' this or 'experiencing' that. For if the physician builds a

ἰατρὸς οὐχ ᾗ ἰατρὸς ἀλλ' ᾗ οἰκοδόμος, καὶ λευκὸς γίνεται οὐχ ᾗ ἰατρὸς ἀλλ' ᾗ μέλας· ἰατρεύει δὲ καὶ ἀνίατρος γίνεται ᾗ ἰατρός. ἐπεὶ δὲ μάλιστα λέγομεν κυρίως τὸν ἰατρὸν ποιεῖν τι ἢ πάσχειν ἢ γίγνεσθαι ἐξ ἰατροῦ, ἐὰν ᾗ ἰατρὸς ταῦτα πάσχῃ ἢ ποιῇ ἢ γίνηται, δῆλον ὅτι καὶ τὸ ἐκ μὴ ὄντος γίγνεσθαι τοῦτο σημαίνει τὸ ᾗ μὴ ὄν. ὅπερ ἐκεῖνοι μὲν οὐ διελόντες ἀπέστησαν, καὶ διὰ ταύτην τὴν ἄγνοιαν τοσοῦτον προσηγνόησαν ὥστε μηθὲν οἴεσθαι γίγνεσθαι μηδὲ εἶναι τῶν ἄλλων, ἀλλ' ἀνελεῖν πᾶσαν τὴν γένεσιν.

Ἡμεῖς δὲ καὶ αὐτοί φαμεν γίγνεσθαι μὲν οὐδὲν ἁπλῶς ἐκ μὴ ὄντος, πώς[1] μέντοι γίγνεσθαι ἐκ μὴ ὄντος, οἷον κατὰ συμβεβηκός· ἐκ γὰρ τῆς στερήσεως—ὅ ἐστι καθ' αὑτὸ μὴ ὄν—οὐκ ἐνυπάρχοντος γίγνεταί τι· (θαυμάζεται δὲ τοῦτο καὶ ἀδύνατον οὕτω δοκεῖ γίγνεσθαί τι ἐκ μὴ ὄντος)· ὡσαύτως δὲ οὐδ' ἐξ ὄντος οὐδὲ τὸ ὂν γίγνεσθαι, πλὴν κατὰ συμ-

[1] [πὼς scripsi; *cf.* 1030 a 22 τὸ τί ἐστιν ἁπλῶς μὲν τῇ οὐσίᾳ (ὑπάρχει), πὼς δὲ τοῖς ἄλλοις: ὅπως E: ὅμως cett.—C.]

---

[a] [*More literally*, 'and since, when we speak of the physician doing or experiencing something or ceasing to be a physician and becoming something else, we are using the words in their most proper sense if all this happens to him *qua* physician, plainly in the same way 'coming out of the non-existent' properly means 'coming out of the non-existent *as such*.'—C.]

[b] [*Literally*, 'for out of (*i.e.* in place of) the shortage—a thing which in itself " is not "—there comes to be something, the shortage not being a factor that is retained in the result' (as the matter is retained). See p. 68, Introd. Note.—C.]

house, it is not *qua* physician but *qua* builder that he does so ; or if he becomes light in complexion, it is *qua* dark in complexion not *qua* physician that he changes ; whereas if he exercises the healing art, or drops or loses that art, so as to become a non-physician, it is *qua* physician that he does so. And so, just as strange conceptions might be formed as to what a physician could or could not do or suffer or become, if we were always thinking of him in his primary and direct capacity as a physician, but applied our conclusions to him in *all* his actual or possible capacities, so, obviously, if we always argue from the non-existent *qua* non-existent, but apply our conclusions to the incidentally non-existent as well, we shall fall into analogous errors.[a] And it was just because the earlier thinkers failed to grasp this analytical distinction, that they piled misconception upon misconception to the pitch of actually concluding that there was no such thing as genesis and that nothing at all ever came to be, or was, except the one and only 'existent.'

Now we, too (who recognize both 'form' and 'lack of form,' or 'shortage,' as factors in becoming), assert that nothing can 'come to be,' in the absolute sense, out of the non-existent, but we declare nevertheless that all things which come to be owe their existence to the incidental non-existence of something ; for they owe it to the 'shortage' from which they started 'being no longer there.'[b] And if it seems an amazing paradox to maintain that anything derives in this way from the non-existent, yet it is really quite true. Moreover, it is equally true that it is only in this same incidental sense that anything can derive from the existent either, or 'what is'

βεβηκός· οὕτω δὲ καὶ τοῦτο γίγνεσθαι τὸν αὐτὸν τρόπον οἷον εἰ ἐκ ζῴου ζῷον γίγνοιτο, καὶ ἐκ τινὸς ζῴου τὶ ζῷον (οἷον εἰ κύων ἐξ ἵππου γίγνοιτο). γίγνοιτο μὲν γὰρ ἂν οὐ μόνον ἐκ τινὸς ζῴου ὁ κύων ἀλλὰ καὶ ἐκ ζῴου, ἀλλ' οὐχ ᾗ ζῷον (ὑπάρχει γὰρ ἤδη τοῦτο)· εἰ δέ τι μέλλει γίγνεσθαι ζῷον μὴ κατὰ συμβεβηκός, οὐκ ἐκ ζῴου ἔσται· καὶ εἴ τι ὄν, οὐκ ἐξ ὄντος,—οὐδ' ἐκ μὴ ὄντος· τὸ γὰρ ἐκ μὴ ὄντος εἴρηται ἡμῖν τί σημαίνει, ὅτι ᾗ μὴ ὄν. ἔτι δὲ καὶ τὸ εἶναι ἅπαν ἢ μὴ εἶναι οὐκ ἀναιροῦμεν.

Εἷς μὲν δὴ τρόπος οὗτος, ἄλλος δ' ὅτι ἐνδέχεται ταὐτὰ λέγειν κατὰ τὴν δύναμιν καὶ τὴν ἐνέργειαν· τοῦτο δ' ἐν ἄλλοις διώρισται δι' ἀκριβείας μᾶλλον.

Ὥσθ' (ὅπερ ἐλέγομεν) αἱ ἀπορίαι λύονται δι' ἃς ἀναγκαζόμενοι ἀναιροῦσι τῶν εἰρημένων ἔνια· διὰ γὰρ τοῦτο τοσοῦτον καὶ οἱ πρότερον ἐξετράπησαν τῆς ὁδοῦ τῆς ἐπὶ τὴν γένεσιν καὶ φθορὰν καὶ ὅλως μεταβολήν· αὕτη γὰρ ἂν ὀφθεῖσα ἡ φύσις ἔλυσεν αὐτῶν πᾶσαν τὴν ἄγνοιαν.

---

[a] [ὑπάρχει γὰρ ἤδη τοῦτο seems to refer to an implied clause καὶ ζῷον ἂν γίγνοιτο, ἀλλ' οὐχ ᾗ ζῷον.—C.]

[b] [*i.e.* εἴ τι (μέλλει γίγνεσθαι) ὂν (μὴ κατὰ συμβεβηκὸς) οὐκ ἐξ ὄντος (ἔσται).—C.]

[c] [*Cf. Met.* Z, 7-9, and Θ. *De gen. et corr.* A 3, refers to our passage and admits that it leaves obscure the problem how there can be an 'unqualified coming-to-be' of a substance.—C.]

[d] [If αὕτη is read (and no variant is recorded), αὕτη ἡ φύσις means 'this nature (entity),' viz. matter qualified by shortage. See 191 b 35-192 a 6.—C.]

can come into being. In this sense, however, this does occur in the same way as (for instance) if 'an animal' should turn into 'an animal,' or a particular animal—say, a horse—should turn into another particular animal—say, a dog. The dog would come into being, not only 'out of' a particular animal, but out of 'an animal' (and it would become 'an animal'),[a] but only incidentally, not *qua* animal, since it was already an animal and could not 'turn into' what it already was. If anything, then, is to 'turn into' an animal, otherwise than incidentally, it must be non-animal at the start and must come to be animal in the process. Similarly, if a thing is to become or turn into an existent otherwise than incidentally, it cannot start from what exists;[b]—though neither can it start from the non-existent, for we have explained that this means 'from the non-existent as such.' At the same time we do not do away with the principle that everything must either be or not be.

This then is one way of formulating the solution of the problem. But there is also an alternative formula based on the distinction between existing as a potentiality and existing as an actuality. But this is developed more fully elsewhere.[c]

It is thus (as I have said) that the difficulties are solved which led to the denial of some of the obvious facts that we have now discussed; for it was this fundamental misconception that threw the earlier thinkers so far off the track concerning 'coming into existence' and 'passing out of it,' and the nature of change in general. Had they detected the existence of the entity we have described,[d] it would have released them from all the confusion.

CHAPTER IX

ARGUMENT

The Platonists approached the true solution, though (in spite of their phraseology) they never really formulated the essential triad ; for they confounded ' matter ' and ' shortage ' together (the positive ' seat ' and the privative ' emptiness '), under one concept of ' the non-existent ' or ' not-anything ' (191 b 35-192 a 12).

The metaphors in which they delight are quite intelligible, if we take them separately and apply them to the appropriate

Ἡμμένοι μὲν οὖν καὶ ἕτεροί τινές εἰσιν αὐτῆς, ἀλλ' οὐχ ἱκανῶς. πρῶτον μὲν γὰρ ὁμολογοῦσιν ἁπλῶς γίνεσθαι ἐκ μὴ ὄντος, ᾗ Παρμενίδην ὀρθῶς λέγειν· εἶτα φαίνεται αὐτοῖς, εἴπερ ἐστὶν ἀριθμῷ μία, καὶ δυνάμει μία μόνον εἶναι. τοῦτο δὲ διαφέρει πλεῖστον. ἡμεῖς μὲν γὰρ ὕλην καὶ στέρησιν ἕτερόν φαμεν εἶναι, καὶ τούτων τὸ μὲν οὐκ ὂν εἶναι κατὰ συμβεβηκός—τὴν ὕλην—τὴν δὲ στέρησιν καθ' αὑτήν, καὶ τὴν μὲν ἐγγὺς καὶ οὐσίαν πως—τὴν ὕλην—τὴν δὲ στέρησιν οὐδαμῶς· οἱ δὲ τὸ μὴ ὂν τὸ μέγα καὶ τὸ μικρὸν ὁμοίως, ἢ τὸ

---

[a] [αὐτῆς refers to αὕτη ἡ φύσις (*l.* 33), *i.e.* matter qualified by shortage.—C.]

[b] [*Cf.* 190 b 23. The subject (ὑποκείμενον) is numerically one thing (a piece of bronze), but conceptually two, as bronze qualified by the *shortage* of the form it will receive. This shortage was overlooked.—C.]

[c] [Simplicius understood ἐγγὺς to mean ἐγγὺς οὐσίας: 'comes near to being substance, and in a sense is substance.' *Cf.* 190 b 25 τόδε γάρ τι μᾶλλον.—C.]

CHAPTER IX

ARGUMENT (*continued*)

'matter' or 'shortage' as the case may be; but they become self-contradictory, if we apply them all to the undifferentiated 'negation' (a 13-25).

[*Matter, properly distinguished from the shortage, is eternal. The nature of Form is a question that belongs to metaphysics. We are concerned only with natural and perishable forms* (a 25-b 4).—*C.*]

Certain other thinkers have approached the view we have now expounded,[a] but without effectively reaching it. For they began by accepting as true the conception of Parmenides that 'coming to be' could mean nothing but emerging from the non-existent. Then, under the one concept of the 'non-existent,' they united both the 'matter' and the 'shortage' of our system, failing to distinguish between 'numerical unity' of subject (which we admit) and 'unanalysable unity' of aspect or potentiality (which we deny).[b] But this distinction which they ignored makes all the difference. For we distinguish between 'matter' and 'shortage' (or absence of form), and assert that the one, namely matter as such, represents the incidental non-existence of attributes, whereas the other, namely shortage as such, is the direct negation or non-existence of the form of which it is the shortage. So that matter, though never existing in isolation, may be pretty well taken as constituting the 'concrete being' of which it is the basis, but shortage not in the least so.[c] These philosophers, on the other hand, conceive of the non-existent as

192 a συναμφότερον ἢ τὸ χωρὶς ἑκάτερον. ὥστε παντελῶς ἕτερος ὁ τρόπος οὗτος τῆς τριάδος κἀκεῖνος· μέχρι μὲν γὰρ δεῦρο προῆλθον, ὅτι δεῖ τινα ὑποκεῖσθαι φύσιν, ταύτην μέντοι μίαν ποιοῦσιν (καὶ γὰρ εἴ τις δυάδα ποιεῖ, λέγων μέγα καὶ μικρὸν αὐτήν, οὐθὲν ἧττον ταὐτὸ ποιεῖ· τὴν γὰρ ἑτέραν παρεῖδεν).

Ἡ μὲν γὰρ ὑπομένουσα συναιτία τῇ μορφῇ τῶν γινομένων ἐστίν, ὥσπερ μήτηρ· ἡ δ' ἑτέρα μοῖρα τῆς ἐναντιώσεως πολλάκις ἂν φαντασθείη τῷ πρὸς τὸ κακοποιὸν αὐτῆς ἀτενίζοντι τὴν διάνοιαν οὐδ' εἶναι τὸ παράπαν· ὄντος γάρ τινος θείου καὶ ἀγαθοῦ καὶ ἐφετοῦ, τὸ μὲν ἐναντίον αὐτῷ φαμεν εἶναι, τὸ δὲ ὃ πέφυκεν ἐφίεσθαι καὶ ὀρέγεσθαι αὐτοῦ κατὰ τὴν ἑαυτοῦ φύσιν. τοῖς δὲ συμβαίνει τὸ ἐναντίον ὀρέγεσθαι τῆς ἑαυτοῦ φθορᾶς. καίτοι οὔτε αὐτὸ ἑαυτοῦ οἷόν τε ἐφίεσθαι τὸ εἶδος (διὰ τὸ μὴ εἶναι ἐνδεές) οὔτε τὸ ἐναντίον· φθαρτικὰ γὰρ

---

[a] [See 187 a 17 note.—C.]

[b] [*Literally*, 'none the less he makes (" great " and " small ") the same (single) thing.'—C.]

[c] [Plato, *Timaeus*, 50 c. 'We must conceive three kinds: (1) that which comes to be, (2) that in which it comes to be, (3) that from which it is copied when it is born into existence. And we may liken (2) the recipient to a mother, (3) the model to a father, and (1) that which is between them to a child.' The matter to be moulded, he adds, must be entirely formless before the forms are impressed on it.—C.]

[d] [*Cf. Met.* Δ 22, on meanings of 'privation.'—C.]

[e] [*Or*, 'since there is something divine, good, and desirable,' viz. either the God of Aristotle's system, as Final Cause, or the principle of Form generally.—C.]

the 'great' and the 'small' alike,[a] whether severally or conjointly. So they too have a triad of the 'great,' the 'small,' and the 'Idea' (or form), but this triad is really quite different from ours of 'matter,' 'shortage,' and 'form'; for, although they go so far with us as to recognize the necessity of some underlying subject, yet in truth the 'great and small' of which it consists can only be equated with our 'matter,' and is not a dyad at all. It is true indeed that you may now and again find it spoken of as the 'dyad' of great and small, but that makes no difference,[b] for the other member of our antithesis (shortage, namely) is systematically ignored.

Now we, who distinguish between matter and shortage, can very well see why matter, which co-operates with form in the genesis of things, may be conceived as their matrix or womb.[c] And we can also see how a man who concentrates his mind on the negative and defect-involving character of shortage[d] may come to think of it as purely non-existent. Then, if we were to think of 'existence' as something august and good and desirable,[e] we might think of shortage as the evil *contradiction* of this good, but of matter as a something the very nature of which is to *desire* and yearn towards the actually existent. But the school of thought we are examining, inasmuch as it identifies matter and shortage, falls into the position of representing the opposite of existence as yearning for its own destruction. But how can either form or shortage really desire form? Not form itself, because it has no lack of it; and not shortage, which is the antithesis of form, because the terms of an antithesis, being

192 a ἀλλήλων τὰ ἐναντία. ἀλλὰ τοῦτ᾿ ἔστιν ἡ ὕλη—ὥσπερ ἂν εἰ θῆλυ ἄρρενος καὶ αἰσχρὸν καλοῦ· πλὴν οὐ καθ᾿ αὑτὸ αἰσχρὸν ἀλλὰ κατὰ συμβεβηκός, οὐδὲ θῆλυ ἀλλὰ κατὰ συμβεβηκός.

Φθείρεται δὲ καὶ γίνεται ἔστι μὲν ὥς, ἔστι δ᾿ ὡς οὔ· ὡς μὲν γὰρ τὸ ἐν ᾧ, καθ᾿ αὑτὸ φθείρεται (τὸ γὰρ φθειρόμενον ἐν τούτῳ ἐστίν, ἡ στέρησις), ὡς δὲ κατὰ τὴν δύναμιν, οὐ καθ᾿ αὑτό, ἀλλ᾿ ἄφθαρτον καὶ ἀγένητον ἀνάγκη αὐτὴν εἶναι. εἴτε γὰρ ἐγίγνετο, ὑποκεῖσθαί τι δεῖ πρῶτον, τὸ ἐξ οὗ ἐνυπάρχοντος· τοῦτο δ᾿ ἐστὶν αὐτῆς ἡ φύσις, ὥστ᾿ ἔσται πρὶν γενέσθαι—λέγω γὰρ ὕλην τὸ πρῶτον ὑποκείμενον ἑκάστῳ, ἐξ οὗ γίνεταί τι ἐνυπάρχοντος μὴ κατὰ συμβεβηκός—εἴτε φθείρεται, εἰς τοῦτο ἀφίξεται ἔσχατον, ὥστε ἐφθαρμένη ἔσται πρὶν φθαρῆναι.

---

[a] [*Or*, ' But this (factor which desires form) is Matter. . . .' From this point onwards Aristotle is rather stating his own view than criticizing Plato's.—C.]

[b] [Or Aristotle's own matter + ' shortage,' to which the following statements are applicable.—C.]

[c] [*Or* ' for the thing that perishes—the shortage—is in it,' and so the matter, considered merely as what contains the shortage, *ceases to be* that, ' as such.'—C.]

[d] φθείρεται δὲ . . . ἀνάγκη αὐτὴν εἶναι. The translation of this very perplexing passage must be regarded as no more than a suggestion. It puzzled Simplicius so much that he was driven to supposing that Aristotle misapplied (κατεχρήσατο) the phrase τὸ ἐν ᾧ, and really meant the opposite, τὸ ἐν αὐτῇ by it.

[e] [ἐξ οὗ ἐνυπάρχοντος, some ' matter ' *from* which it would arise but which would *persist as a factor in it* after it had come into being.—C.]

[f] [*Or*, ' and if matter perishes, this (matter itself) will be what it is ultimately reduced to; so that it will have to have perished already before perishing.' In sum : if matter

mutually destructive, cannot desire each other. So that[a] if (to borrow their own metaphors) we are to regard matter as the female desiring the male or the foul desiring the fair, the desire must be attributed not to the foulness itself, as such, but to a subject that is foul or female incidentally.

As for the undifferentiated matter-shortage of the school we have been discussing,[b] it may be regarded as either perishable or imperishable ; for if we think of it as the bare *seat-of-shortage*, it perishes, as such, for ' shortage ' is exactly what does perish in it,[c] on its receiving the form ; but if we are considering it as the potentiality of receiving forms, it cannot perish, as such, but must necessarily be exempt both from destruction and genesis.[d] For if it ' came to be,' there must have been some subject already there for it to proceed from,[e] and just ' being there as the subject ' is precisely what constitutes the nature of ' matter ' itself, so that it must have been before it came to be. For what I mean by matter is precisely the ultimate underlying subject, common to all the things of Nature, presupposed as their substantive, not incidental, constituent. And again, the destruction of a thing means the disappearance of everything that constitutes it except just that very underlying subject which its existence presupposes, and if this perished, then the thing that presupposes it would have perished with it by anticipation before it came into existence.[f]

were to come into existence or perish, it would have to come out of itself or perish into itself—matter being precisely the starting-point of coming-to-be and the terminus of perishing. So matter would have to be, before it could come-to-be, and to have perished, before it could have anything to perish into.—C.]

192 a Περὶ δὲ τῆς κατὰ τὸ εἶδος ἀρχῆς, πότερον μία ἢ πολλαί, καὶ τίς ἢ τίνες εἰσί, δι' ἀκριβείας τῆς πρώτης φιλοσοφίας ἔργον ἐστὶ διορίσαι, ὥστε
192 b εἰς ἐκεῖνον τὸν καιρὸν ἀποκείσθω. περὶ δὲ τῶν φυσικῶν καὶ φθαρτῶν εἰδῶν ἐν τοῖς ὕστερον δεικνυμένοις ἐροῦμεν.

Ὅτι μὲν οὖν εἰσὶν ἀρχαί, καὶ τίνες, καὶ πόσαι τὸν ἀριθμόν, διωρίσθω ἡμῖν οὕτως· πάλιν δὲ ἄλλην ἀρχὴν ἀρξάμενοι λέγωμεν.

---

[a] [See *Met.* Z Θ; and Λ where the existence of eternal and immutable form without matter is specially discussed. —C.]

So much for 'matter'; but the detailed examination and determination of 'form' as a principle, and the question of its unity or plurality, and its nature (singular or plural), is the business of First Philosophy; so let it be deferred till we come to that.[a] It is with natural and perishable forms only that we shall deal in the sequel of this treatise.

Let this suffice for the demonstration of the existence of principles in Nature and the determination of what they are and how many in number. In the next book we must make a fresh start with fresh questions.

EXCURSUS

ON SQUARING THE CIRCLE

(Bk. I. Ch. ii. 185 a 16)

Popular conceptions on the subject of squaring the circle are so vague and so entirely erroneous that to do justice to Aristotle's handling of the question will be impossible without some introductory remarks. In the first place, what is squaring the circle? It is reducing the area of a circle to an equal area bounded by straight lines, for any plane rectilinear figure can be quite easily converted into a square of the same size.

The problem then is to find a plane rectilinear figure equal in area to a given circle. That it is inherently possible that there should be such a rectilinear area is obvious (for the size of the area does not depend on the shape) and a very convincing proof of this was already known to Aristotle as we shall see.[a] No contradiction of this is in any way intended by the statement that the circle cannot be squared. The question is not whether there is or can be a rectilinear area which will equal the area swept out by a given radius rotating round one of its extreme points (for that is the most enlightening definition of a circle), but whether such an area can be diagrammatically constructed. It is perfectly easy to construct it, but not within the limits prescribed for constructions by Euclid, and evidently by the tradition he followed. For Euclid undertook to show how far he could carry elementary Geometry on the condition of having no implements allowed to him but compasses and a ruler, and of never

[a] Bryson's proof, see note *a*, p. 19.

making use of any diagrammatic figure which he had not been able to construct under these conditions. Even these instruments, the compasses and the ruler, he only allows himself to use under certain restrictions, but with these we need not concern ourselves. He claims no logical superiority for constructions in which these two instruments alone have been used, above other constructions which require other instruments; but he is simply trying how far he can carry the science forward on this very narrow basis. There are many simple operations which cannot be so performed (for instance, with ruler and compasses only you cannot, in general, trisect an angle), and among them is the construction of a rectilinear figure equal in area to a given circle.

Archimedes (†212 B.C.) did not restrict his reasoning to figures constructed under the Euclidean tradition, and he tells us that the right-angle triangle of which the sides containing the right angle are respectively equal to the circumference and radius of the circle, is equal in area to the circle. He proves this by demonstrating rigorously that any area greater than that of the triangle will be greater than that of the circle, and any area less than that of the triangle will be less than that of the circle.[a] There is no suggestion of construction here, his reasoning does not depend on being able to construct the triangle and it is obvious that such a triangle exists. Therefore, though he makes no attempt here to solve the problem as originally conceived, he completely disposes of the misconception that, because a square equal to a given circle cannot be constructed by means of compasses and ruler only, there can be no such square.

Having thus reduced the problem of measuring the area of a circle to that of measuring its circumference, he goes on to show how successive approximations can be made to the numerical expression of π (the ratio of the circumference to the radius).

Again there is a vague idea in some minds that the reason why the circle cannot be squared is that π is in-

[a] Archimedes, On the Measurement of the Circle.

commensurable, *i.e.* cannot be exactly expressed numerically; and that you cannot construct a figure involving incommensurable ratios. But if that were the difficulty it would be impossible to construct a circle itself from a given radius.

There is no inherent difficulty in constructing a line that is incommensurable with a given line. The simplest case of all is that of a side and the diagonal of a square, and it is generally supposed to be historically true that it was in this connexion that the problem of incommensurables was first confronted. Drawing the diagonal is a simple act of construction, and it is only by the construction that we can assign a perfectly definite length to the diagonal relatively to that of the side, the length of the diagonal being incommensurable with that of the side. So too by drawing a circle to a radius we obtain the circumference, the length of the circumference being incommensurable with that of the radius.

Aristotle has been blamed for his contemptuous treatment of Antiphon,[a] because it is assumed that he ought to have seen in him an incipient attempt to apply the principle of approximations which was developed by Archimedes and ultimately by Newton and Leibniz in the calculus. But this is hardly reasonable, for, according to the account given by Simplicius, Antiphon does not really supply so much as a hint of this method. He simply assumes at the end what he might as well have assumed at the beginning, that the circumference is already a polygon with very minute sides; in other words he believes in atomic (*i.e.* indivisible) lines; which belief lies at the root of endless confusion of thought.[b]

Aristotle is perfectly clear and explicit on this question, yet his teaching has been misunderstood or neglected even by such men as Galileo[c] and Leibniz.[d]

[a] See Bk. I. Ch. ii. p. 18, l. 18.
[b] See Gen. Introd. pp. lxxx f. and lxxxvii ff.
[c] Galileo, *Discorsi e Dimostrazioni Matematiche, intorno a due nuove scienze* (Leyden, 1538); *Opere* (Edizione Nazionale), vol. viii. pp. 80 *sqq.* and 91 *sqq.*
[d] *Œuvres inédits de Leibniz*, p. 106 Couturat.

Galileo has exposed Aristotle's worst mistakes, but he has fallen into others of his own by neglect of Aristotle's principles in these matters. And the adoption of Leibniz's system of notation in preference to that of Newton, while amply justified on practical grounds, still tends to impede the progress of young students to perfect clarity of mind to this day.

BOOK II

INTRODUCTION

[From the review, partly critical, partly constructive, of his predecessors' imperfect apprehensions, Aristotle, according to his practice, turns to his own positive doctrine.

The scope of Physics—the science of Nature—must first be determined by defining the meaning of ' nature.' If we consider the whole field of natural (as opposed to artificial or manufactured) products—the ' simple bodies ' (earth, water, air, fire) and the living organisms whose bodies are composed of these elements—we find that they all possess one peculiar characteristic, a tendency to move and change in various ways, due to some impulse internal to themselves. This inner source of movement (or of rest) is ' nature.' Whatever exhibits it ' has a nature ' of its own and ' exists by nature,' and consequently comes within the scope of Physics.

It is the business of the physical philosopher to account completely for the existence of such objects and for the changes they undergo—in other words, to know them through their ' causes.' The main object of this Book is to determine how many kinds of ' cause ' must be taken into account.

Aristotle introduces this problem by setting his own definition of the ' nature ' of a thing in comparison with the two chief types of popular and philosophic opinion upon the question, wherein lies the nature or ultimate essence of things. The pre-Socratic philosophers, speaking generally, found this in the matter of which bodies consist ; and they carried the analysis of matter down to some ultimate stuff or stuffs which they called the 'nature of things.' The speculation inspired by Socrates saw the nature of a thing in the 'form' or essential character, given in the definition of 'what it is to be' a thing of that sort. This is the truer view; and consequently this form or essence must be identical with the internal source of movement by which we have defined natural objects. The form is also the final goal towards which the development of the thing moves and in which that development comes to rest.

Such forms as these have their being in the changing world of Nature; they are immersed in matter, and cannot be studied in abstraction from matter. Thus Physics is distinguished, on the one hand, from mathematics, which isolates in thought the forms it studies from the power of movement, and on the other hand from First Philosophy (metaphysics), which is concerned with pure immaterial forms—God, the intelligences of the heavenly spheres, the human reason.

The rest of the Book is occupied with the character of the 'causes' of natural things and of their motions and changes. The term 'cause' has a wider sense than in the English use. It covers all the 'conditions necessary but not separately sufficient to account for the existence of a thing' (see Ross, *Aristotle*, p. 73). Greek philosophy had always been intent upon the task of discovering, not laws of succession in phenomena, but what things in themselves *are*. Hence among the 'causes' of a thing—what we need to know in order to give a full account of it—the two internal constituents, the matter and the form, stand first. As already hinted in the first chapter, the form of the thing is also the moving or 'efficient' cause of its coming into being and behaving as it does; and the form is also the 'end,' wherein its nature is fully realized. Thus there are four 'causes'—material, formal, efficient, and final—to be known, if the existence and behaviour of any natural object is to be fully accounted for.

Popular thought further recognizes the agency of Luck or Chance; and the inveterate belief that events just 'happen' by the blind working of forces pushing from behind, without any intelligent direction, has been countenanced or tacitly assumed by men of science. To such a view of the world Aristotle, as the successor of Plato and Socrates, is fundamentally opposed. He does not, however, deny validity to the conceptions of Chance and Luck; he sets himself to analyse them and to find for them a meaning and a sphere consistent with the belief that the workings of Nature cannot be completely understood as the outcome of chance or of mere necessity, but reveal the presence of some kind of purpose aiming at a goal desired. The Book closes with an attempt to establish this teleology.—C.].

B

CHAPTER I

ARGUMENT

[*The scope of Physics is now determined by defining 'nature.' 'Natural things' or 'things which have a nature,' as distinct from artificial products as such, contain an internal tendency to move* (i.e. *be moved or changed*) *in certain ways. Thus, the elements and inanimate compounds of them tend to rise or fall in space and undergo change; living things move about, change, and grow. The 'nature'* is *this innate tendency to movement and change. Its existence is an obvious fact of experience* (192 b 8–193 a 10).

192 b 8 Τῶν γὰρ ὄντων τὰ μὲν ἔστι φύσει, τὰ δὲ δι' ἄλλας αἰτίας—φύσει μὲν τά τε ζῷα καὶ τὰ μέρη αὐτῶν καὶ τὰ φυτὰ καὶ τὰ ἁπλᾶ τῶν σωμάτων (οἷον γῆ καὶ πῦρ καὶ ἀὴρ καὶ ὕδωρ)· ταῦτα γὰρ εἶναι καὶ τὰ τοιαῦτα φύσει φαμέν. πάντα δὲ τὰ ῥηθέντα φαίνεται διαφέροντα πρὸς τὰ μὴ φύσει συνεστῶτα. τὰ μὲν γὰρ φύσει ὄντα πάντα φαίνεται ἔχοντα ἐν ἑαυτοῖς ἀρχὴν κινήσεως καὶ στάσεως—τὰ μὲν κατὰ τόπον, τὰ δὲ κατ' αὔξησιν καὶ φθίσιν, τὰ δὲ κατ' ἀλλοίωσιν—κλίνη δὲ καὶ ἱμάτιον καὶ εἴ τι τοιοῦτον ἄλλο γένος ἐστίν, ᾗ μὲν τετύχηκε τῆς κατηγορίας ἑκάστης καὶ καθ' ὅσον ἐστὶν ἀπὸ τέχνης, οὐδεμίαν ὁρμὴν ἔχει μεταβολῆς

BOOK II

CHAPTER I

ARGUMENT *(continued)*

By some the 'nature' of a concrete thing is identified with the matter it consists of, and the 'nature of things' in general, with an ultimate eternal stuff. Others find the nature of a thing in its 'form.' Taking this latter view himself, Aristotle restates his definition of 'nature' in terms that imply his identification of the internal principle of movement with the form, existing in the concrete thing; and he supports this by several arguments (a 10–b 21).—C.]

Some things exist, or come into existence, by nature; and some otherwise. Animals and their organs, plants, and the elementary substances—earth, fire, air, water—these and their likes we say exist by nature. For all these seem distinguishable from those that are not constituted by nature; and the common feature that characterizes them all seems to be that they have within themselves a principle of movement (or change) and rest—in some cases local only, in others quantitive, as in growth and shrinkage, and in others again qualitive, in the way of modification. But a bedstead or a garment or the like, in the capacity which is signified by its name and in so far as it is craft-work, has within itself no such inherent trend towards change, though

192b ἔμφυτον, ᾗ δὲ συμβέβηκεν αὐτοῖς εἶναι λιθίνοις ἢ γηΐνοις ἢ μικτοῖς, ἐκ τούτων ἔχει καὶ κατὰ τοσοῦτον, ὡς οὔσης τῆς φύσεως ἀρχῆς τινος καὶ αἰτίας τοῦ κινεῖσθαι καὶ ἠρεμεῖν ἐν ᾧ ὑπάρχει πρώτως καθ' αὑτὸ καὶ μὴ κατὰ συμβεβηκός. λέγω δὲ τὸ μὴ κατὰ συμβεβηκός, ὅτι γένοιτ' ἂν αὐτὸς αὑτῷ τις αἴτιος ὑγιείας ὢν ἰατρός· ἀλλ' ὅμως οὐ καθὸ ὑγιάζεται τὴν ἰατρικὴν ἔχει, ἀλλὰ συμβέβηκε τὸν αὐτὸν ἰατρὸν εἶναι καὶ ὑγιαζόμενον· διὸ καὶ χωρίζεταί ποτ' ἀπ' ἀλλήλων. ὁμοίως δὲ καὶ τῶν ἄλλων ἕκαστον τῶν ποιουμένων· οὐδὲν γὰρ αὐτῶν ἔχει τὴν ἀρχὴν ἐν ἑαυτῷ τῆς ποιήσεως, ἀλλὰ τὰ μὲν ἐν ἄλλοις καὶ ἔξωθεν (οἷον οἰκία καὶ τῶν ἄλλων τῶν χειροκμήτων ἕκαστον), τὰ δ' ἐν αὑτοῖς μὲν ἀλλ' οὐ καθ' αὑτά, ὅσα κατὰ συμβεβηκὸς αἴτια γένοιτ' ἂν αὑτοῖς.

Φύσις μὲν οὖν ἐστι τὸ ῥηθέν· φύσιν δὲ ἔχει ὅσα τοιαύτην ἔχει ἀρχήν. καὶ ἔστι πάντα ταῦτα οὐσία· ὑποκείμενον γάρ τι, καὶ ἐν ὑποκειμένῳ ἐστὶν ἡ φύσις ἀεί.

Κατὰ φύσιν δὲ ταῦτά τε καὶ ὅσα τούτοις ὑπάρχει

---

[a] [I have repunctuated the text, taking ἐκ τούτων not with μικτοῖς but with ἔχει: 'it has one derived from these (natural substances) and in so far (as it is composed of them).—C.]

[b] [English cannot reproduce the ambiguity of ἡ φύσις, which (like the French *la nature*) may mean 'the nature' (of any natural thing) or 'Nature' collectively, the sum total of such natures.—C.]

[c] [Strictly, 'of *being moved* (κινεῖσθαι) and being at rest' (not of *initiating* motion from within). The heavenly bodies, which have no principle of rest, are here ignored.—C.]

[d] Have a substantive existence, ***i.e.*** are concrete entities. *Cf.* Gen. Introd. pp. lii-liii.

owing to the fact of its being composed of earth or stone or some mixture of substances, it incidentally has within itself the principles of change which inhere primarily in these materials.[a] For nature[b] is the principle and cause of motion and rest[c] to those things, and those things only, in which she inheres primarily, as distinct from incidentally. What I mean by ' as distinct from incidentally ' is like this : If a man were a physician and prescribed successfully for himself, the patient would cure himself ; but it would not be *qua* patient that he possessed the healing art, though in this particular case it happened that the physician's personality coincided with that of the patient, which is not always the case. And so it is with all manufactured or ' made ' things : none of them has within itself the principle of its own making. Generally this principle resides in some external agent, as in the case of the house and its builder, and so with all hand-made things. In other cases, such as that of the physician-patient, though the patient does indeed contain in himself the principle of action, yet he does so only incidentally, for it is not *qua* subject acted on that he has in himself the causative principle of the action.

This, then, being what we mean by ' nature,' anything that has in itself such a principle as we have described may be said to ' possess a nature ' of its own inherently. And all such things have a substantive existence[d] ; for each of them is a substratum or ' subject ' presupposed by any other category, and it is only in such substrata that nature ever has her seat.

Further, not only nature itself and all things that ' have a nature,' but also the behaviour of these things

192 b καθ' αὑτά, οἷον τῷ πυρὶ φέρεσθαι ἄνω· τοῦτο γὰρ
193 a φύσις μὲν οὐκ ἔστιν, οὐδ' ἔχει φύσιν, φύσει δὲ καὶ κατὰ φύσιν ἐστίν.

Τί μὲν οὖν ἐστιν ἡ φύσις, εἴρηται, καὶ τί τὸ φύσει καὶ κατὰ φύσιν. ὡς δ' ἔστιν ἡ φύσις, πειρᾶσθαι δεικνύναι γελοῖον· φανερὸν γὰρ ὅτι τοιαῦτα τῶν ὄντων ἐστὶ πολλά. τὸ δὲ δεικνύναι τὰ φανερὰ διὰ τῶν ἀφανῶν οὐ δυναμένου κρίνειν ἐστὶ τὸ δι' αὑτὸ καὶ μὴ δι' αὑτὸ γνώριμον. ὅτι δ' ἐνδέχεται τοῦτο πάσχειν, οὐκ ἄδηλον· συλλογίσαιτο γὰρ ἄν τις ἐκ γενετῆς ὢν τυφλὸς περὶ χρωμάτων· ὥστε ἀνάγκη τοῖς τοιούτοις περὶ τῶν ὀνομάτων εἶναι τὸν λόγον, νοεῖν δὲ μηδέν.

Δοκεῖ δ' ἡ φύσις καὶ ἡ οὐσία τῶν φύσει ὄντων ἐνίοις εἶναι τὸ πρῶτον ἐνυπάρχον ἑκάστῳ ἀρρύθμιστον καθ' ἑαυτό, οἷον κλίνης φύσις τὸ ξύλον, ἀνδριάντος δ' ὁ χαλκός. (σημεῖον δέ φησιν Ἀντιφῶν ὅτι, εἴ τις κατορύξειε κλίνην καὶ λάβοι δύναμιν ἡ σηπεδὼν ὥστε ἀνεῖναι βλαστόν, οὐκ ἂν γενέσθαι κλίνην ἀλλὰ ξύλον, ὡς τὸ μὲν κατὰ συμβεβηκὸς ὑπάρχον—τὴν κατὰ νόμον διάθεσιν καὶ τὴν τέχνην—τὴν δ' οὐσίαν οὖσαν ἐκείνην ἣ καὶ διαμένει ταῦτα πάσχουσα συνεχῶς). εἰ δὲ καὶ τούτων ἕκαστον πρὸς ἕτερόν τι ταὐτὸ τοῦτο

---

[a] [*Or* 'a thing's proximate constituent, in itself unformed,' *e.g.* the wood of a bedstead, which is 'proximate' to the form, as contrasted with the remoter elements into which the wood might be analysed—'simple bodies' or 'ultimate matter.' *Cf. Met.* 1015 a 7.—C.]

[b] [Antiphon, *frag.* 15 (Diels, *Vors.*^{4} ii. 295) εἴ τις κατορύξειε κλίνην καὶ ἡ σηπεδὼν τοῦ ξύλου ἔμβιος γένοιτο, οὐκ ἂν γένοιτο κλίνη ἀλλὰ ξύλον. For σηπεδών (putrefaction) as cause of

in virtue of their inherent characteristics is spoken of as 'natural.' For instance, for fire actually to rise, as distinct from having the tendency to rise, neither *is* nature nor *has* a nature; but it comes about 'by nature' and is 'natural.'

Such, then, are the definitions of 'nature,' of what exists 'by nature,' and of what is 'natural.' Any attempt to prove that nature, in this sense, is a reality would be childish; for it is patent that many things corresponding to our definitions do actually exist; and to set about proving the obvious from the unobvious betrays confusion of mind as to what is self-evident and what is not. Such confusion, however, is not unknown, though it is like a man born blind arguing about colours, and amounts to reasoning about names without having any corresponding concept in the mind.

Now some hold that the nature and substantive existence of natural products resides in their material [a] on the analogy of the wood of a bedstead or the bronze of a statue. (Antiphon [b] took it as an indication of this that if a man buried a bedstead and the sap in it took force and threw out a shoot it would be tree and not bedstead that came up, since the artificial arrangement of the material by the craftsman is merely an incident that has occurred to it, whereas its essential and natural quality is to be found in that which persists continuously throughout such experiences.) And in like manner, it is thought,[c] if the materials themselves bear to yet other substances the same relation which the manufactured

generation *cf.* Plato, *Phaedo* 96 B (Burnet *ad loc.*), Hippocr. Περὶ σαρκῶν 3.—C.]

[c] [εἶναι (*l.* 20) depends on δοκεῖ (*l.* 10).—C.]

193 a πέπονθεν—οἷον ὁ μὲν χαλκὸς καὶ ὁ χρυσὸς πρὸς ὕδωρ, τὰ δ' ὀστᾶ καὶ ξύλα πρὸς γῆν, ὁμοίως δὲ καὶ τῶν ἄλλων ὁτιοῦν—ἐκεῖνα τὴν φύσιν εἶναι καὶ τὴν οὐσίαν αὐτῶν. διόπερ οἱ μὲν πῦρ, οἱ δὲ γῆν, οἱ δ' ἀέρα φασίν, οἱ δὲ ὕδωρ, οἱ δ' ἔνια τούτων, οἱ δὲ πάντα ταῦτα τὴν φύσιν εἶναι τὴν τῶν ὄντων. ὃ γάρ τις αὐτῶν ὑπέλαβε τοιοῦτον, εἴτε ἓν εἴτε πλείω, τοῦτο καὶ τοσαῦτά φησιν εἶναι τὴν ἅπασαν οὐσίαν, τὰ δὲ ἄλλα πάντα πάθη τούτων καὶ ἕξεις καὶ διαθέσεις· καὶ τούτων μὲν ὁτιοῦν εἶναι ἀίδιον —οὐ γὰρ εἶναι μεταβολὴν αὐτοῖς ἐξ αὐτῶν—τὰ δ' ἄλλα γίγνεσθαι καὶ φθείρεσθαι ἀπειράκις.

Ἕνα μὲν οὖν τρόπον οὕτως ἡ φύσις λέγεται, ἡ πρώτη ἑκάστῳ ὑποκειμένη ὕλη τῶν ἐχόντων ἐν αὑτοῖς ἀρχὴν κινήσεως καὶ μεταβολῆς· ἄλλον δὲ τρόπον ἡ μορφὴ καὶ τὸ εἶδος τὸ κατὰ τὸν λόγον. ὥσπερ γὰρ τέχνη λέγεται τὸ κατὰ τέχνην καὶ τὸ τεχνικόν, οὕτω καὶ φύσις τὸ κατὰ φύσιν λέγεται καὶ τὸ φυσικόν. οὔτε δὲ ἐκεῖ πω φαῖμεν ἂν ἔχειν κατὰ τὴν τέχνην οὐδέν, εἰ δυνάμει μόνον ἐστὶ κλίνη μήπω δ' ἔχει τὸ εἶδος τῆς κλίνης, οὐδ' εἶναι

---

[a] [*Cf. Met.* 1015 a 8 οἷον τῶν χαλκῶν ἔργων πρὸς αὐτὰ μὲν πρῶτος (proximate material) ὁ χαλκός, ὅλως δ' ἴσως ὕδωρ, εἰ πάντα τὰ τηκτὰ ὕδωρ—the doctrine of *Timaeus* 58 D (Ross *ad loc.*).—C.]

[b] [*Cf. Met.* 989 a 5, No philosopher followed popular thought in making earth the primary form of body.—C.]

[c] [*Or*, 'related as above described to more complex forms of matter.'—C.]

[d] [*Or* 'for they thought them incapable of changing *out of themselves*' (putting off their own nature).—C.]

[e] [πρώτη ὕλη is used ambiguously by Aristotle for 'proximate' or 'ultimate' matter. Here (with ἑκάστῳ) it probably

articles bear to them—if, for instance, water is the material of bronze or gold,[a] or earth of bone or timber, and so forth—then it is in the water or earth that we must look for the 'nature' and essential being of the gold and so forth. And this is why some have said that it was earth[b] that constituted the nature of things, some fire, some air, some water, and some several and some all of these elemental substances. For whichever substance or substances each thinker assumed to be primary[c] he regarded as constituting the substantive existence of all things in general, all else being mere modifications, states, and dispositions of them. Any such ultimate substance they regarded as eternal (for they did not admit the transformation of elementary substances into each other[d]), while they held that all else passed into existence and out of it endlessly.

This then is one way of regarding 'nature'—as the ultimately[e] underlying material of all things that have in themselves the principle of movement and change. But from another point of view we may think of the nature of a thing as residing rather in its form, that is to say in the 'kind' of thing it is by definition. For as we give the name of 'art' to a thing which is the product of art and is itself artistic, so we give the name of 'nature' to the products of nature which themselves are 'natural.'[f] And as, in the case of art, we should not allow that what was only potentially a bedstead and had not yet received the form of bed had in it as yet any art-formed element, or could be called 'art,' so in

means the proximate material of any given thing, as at *l.* 10.—C.]

[f] [κατὰ φύσιν, as above defined, 192 b 35.—C.].

193 b τέχνην, οὔτ' ἐν τοῖς φύσει συνισταμένοις· τὸ γὰρ δυνάμει σὰρξ ἢ ὀστοῦν οὔτ' ἔχει πω τὴν ἑαυτοῦ φύσιν, πρὶν ἂν λάβῃ τὸ εἶδος τὸ κατὰ τὸν λόγον—ὃ ὁριζόμενοι λέγομεν τί ἐστι σὰρξ ἢ ὀστοῦν—οὔτε φύσει ἐστίν. ὥστε ἄλλον τρόπον ἡ φύσις ἂν εἴη τῶν ἐχόντων ἐν αὑτοῖς κινήσεως ἀρχὴν ἡ μορφὴ καὶ τὸ εἶδος, οὐ χωριστὸν ὂν ἀλλ' ἢ κατὰ τὸν λόγον. (τὸ δ' ἐκ τούτων φύσις μὲν οὐκ ἔστι, φύσει δέ, οἷον ἄνθρωπος.) καὶ μᾶλλον αὕτη φύσις τῆς ὕλης· ἕκαστον γὰρ τότε λέγεται ὅταν ἐντελεχείᾳ ᾖ μᾶλλον ἢ ὅταν δυνάμει.

Ἔτι γίνεται ἄνθρωπος ἐξ ἀνθρώπου, ἀλλ' οὐ κλίνη ἐκ κλίνης· διὸ καί φασιν οὐ τὸ σχῆμα εἶναι τὴν φύσιν ἀλλὰ τὸ ξύλον, ὅτι γένοιτ' ἄν, εἰ βλαστάνοι, οὐ κλίνη ἀλλὰ ξύλον. εἰ δ' ἄρα τοῦτο τέχνη, καὶ ἡ μορφὴ φύσις· γίνεται γὰρ ἐξ ἀνθρώπου ἄνθρωπος.

Ἔτι δ' ἡ φύσις ἡ λεγομένη ὡς γένεσις ὁδός ἐστιν εἰς φύσιν. οὐ γὰρ ὥσπερ ἡ ἰάτρευσις λέγεται οὐκ εἰς ἰατρικὴν ὁδὸς ἀλλ' εἰς ὑγίειαν·

---

[a] The revised definition lays stress on the form while including the matter.

[b] [*Literally*. ' But if this (the artificial shape) is " art," it follows that the form (of the natural product—the timber, or the man—which *does* reproduce itself) is " nature." '—C.].

[c] So, too, in Latin *na-tura* derived from the *na* of *na-scor* and *na-tivitas*. [In fact (*g*)*natura* is derived from the same root as *gi-gno*, γί-γνομαι. *Cf. Met.* 1014 b 17 and Ross *ad loc.*—C.].

the case of natural products; what is potentially flesh or bone has not yet the 'nature' of flesh until it actually assumes the form indicated by the definition that constitutes it the thing in question, nor is this potential flesh or bone as yet a product of nature. These considerations would lead us to revise our definition of nature as follows: Nature is the distinctive form or quality of such things as have within themselves a principle of motion, such form or characteristic property not being separable from the things themselves, save conceptually.[a] (The *compositum*—a man, for example—which material and form combine to constitute, is not itself a 'nature,' but a thing that comes to be by natural process.) And this view of where to look for the nature of things is preferable to that which finds it in the material; for when we speak of the thing into the nature of which we are inquiring, we mean by its name an actuality not a potentiality merely.

Again men propagate men, but bedsteads do not propagate bedsteads; and that is why they say that the natural factor in a bedstead is not its shape but the wood—to wit, because wood and not bedstead would come up if it germinated. If, then, it is this incapacity of reproduction that makes a thing art and not nature, then the form of natural things will be their nature, as in the parallel case of art;[b] for man is generated by man, whereas a bedstead is not generated from a bedstead.

Again, *na-ture* is etymologically equivalent to *gene-sis* and (in Greek) is actually used as a synonym for it;[c] nature, then, *qua* genesis proclaims itself as the path to nature *qua* goal. Now, it is true that healing is so called, not because it is the path to the

ἀνάγκη μὲν γὰρ ἀπὸ ἰατρικῆς, οὐκ εἰς ἰατρικήν, εἶναι τὴν ἰάτρευσιν, οὐχ οὕτω δ' ἡ φύσις ἔχει πρὸς τὴν φύσιν, ἀλλὰ τὸ φυόμενον ἐκ τινὸς εἰς τὶ ἔρχεται, ᾗ φύεται. εἰς τί οὖν φύεται; οὐχὶ ἐξ οὗ, ἀλλ' εἰς ὅ. ἡ ἄρα μορφὴ φύσις. ἡ δέ γε μορφὴ καὶ ἡ φύσις διχῶς λέγεται· καὶ γὰρ ἡ στέρησις εἶδός πώς ἐστιν. εἰ δ' ἐστὶν ἡ στέρησις καὶ ἐναντίον τι περὶ τὴν ἁπλῆν γένεσιν ἢ μὴ ἔστιν, ὕστερον ἐπισκεπτέον.

CHAPTER II

ARGUMENT

[*Physics is distinguished from mathematics and from metaphysics.*

The objects of mathematics, though they do not exist apart from natural bodies, can be studied in abstraction from that power of movement by which we defined 'nature.' But Physics studies natural bodies as essentially possessing this power. (Cf. Met. E 1 *and Ross,* Aristotle, *p.* 68). *The Platonists are wrong in attempting to abstract entities* (e.g. '*man,*' '*flesh*') *whose nature involves matter* (193 b 22–194 a 12).

22 Ἐπεὶ δὲ διώρισται ποσαχῶς ἡ φύσις λέγεται,

---

[a] Which is the return to the perfect form which it started without at birth.

[b] The question is frequently discussed in other works of Aristotle, notably in *Physics* v. and in the *De gen. et cor.* A 3. The answer in substance is that there is really no such thing

healing art, but because it is the path to health, for of necessity healing proceeds from the healing art, not to the healing art itself; but this is not the relation of nature to nature, for that which is born starts as something and advances or grows towards something else. Towards what, then, does it grow? Not towards its original state at birth, but towards its final state or goal.[a] It is, then, the form that is nature; but, since 'form' and 'nature' are ambiguous terms, inasmuch as shortage is a kind of form, we shall leave to future investigation whether shortage is, or is not, a sort of contrasted term (opposed to positive form) in absolute generation.[b]

CHAPTER II

ARGUMENT (*continued*)

Physics must take account of both matter and form. In this respect it is like art: the doctor must know both the nature of health (form) and the material constituents of the body. Also, matter is related to the 'nature' (form) as means to end. Matter, moreover, is only a relative term. But Physics is concerned only with forms immersed in matter. Pure forms fall under Metaphysics or Theology (a 12–b 15).—C.]

Now that we have determined the different senses in which "nature" may be understood (as signifying

as absolute genesis and therefore no contrast to concrete entity. But we say that a concrete entity has perished when the indestructible matter which lies at the core of it has assumed forms which evade our ability to recognize its identity any longer.

193 b μετὰ τοῦτο θεωρητέον τίνι διαφέρει ὁ μαθηματικὸς τοῦ φυσικοῦ· καὶ γὰρ ἐπίπεδα καὶ στερεὰ ἔχει τὰ φυσικὰ σώματα καὶ μήκη καὶ στιγμάς, περὶ ὧν σκοπεῖ ὁ μαθηματικός. ἔτι ἡ ἀστρολογία ἑτέρα ἢ μέρος τῆς φυσικῆς· εἰ γὰρ τοῦ φυσικοῦ τὸ τί ἐστιν ἥλιος ἢ σελήνη εἰδέναι, τῶν δὲ συμβεβηκότων καθ' αὑτὰ μηδέν, ἄτοπον, ἄλλως τε καὶ ὅτι φαίνονται λέγοντες οἱ περὶ φύσεως καὶ περὶ σχήματος σελήνης καὶ ἡλίου, καὶ πότερον σφαιροειδὴς ἡ γῆ καὶ ὁ κόσμος ἢ οὔ.

Περὶ τούτων μὲν οὖν πραγματεύεται καὶ ὁ μαθηματικός, ἀλλ' οὐχ ᾗ φυσικοῦ σώματος πέρας ἕκαστον· οὐδὲ τὰ συμβεβηκότα θεωρεῖ ᾗ τοιούτοις οὖσι συμβέβηκεν. διὸ καὶ χωρίζει· χωριστὰ γὰρ τῇ νοήσει κινήσεώς ἐστι, καὶ οὐδὲν διαφέρει, οὐδὲ γίνεται ψεῦδος χωριζόντων.

Λανθάνουσι δὲ τοῦτο ποιοῦντες καὶ οἱ τὰς ἰδέας
194 a λέγοντες· τὰ γὰρ φυσικὰ χωρίζουσιν, ἧττον ὄντα χωριστὰ τῶν μαθηματικῶν. γίγνοιτο δ' ἂν τοῦτο δῆλον, εἴ τις ἑκατέρων πειρῷτο λέγειν τοὺς ὅρους, καὶ αὐτῶν καὶ τῶν συμβεβηκότων. τὸ μὲν γὰρ

---

[a] [*i.e.* ἔτι δ' εἰ ἡ ἀστρ. This reading (Tauchnitz ed. 1881) seems to have no authority.—C.]

either 'material' or 'form'), we have next to consider how the mathematician differs from the physicist or natural philosopher; for natural bodies have surfaces and occupy spaces, have lengths and present points, all which are subjects of mathematical study. And then there is the connected question [a] whether astronomy is a separate science from physics or only a special branch of it; for if the student of Nature is concerned to know what the sun and moon are, it were strange if he could avoid inquiry into their essential properties; especially as we find that writers on Nature have, as a fact, discoursed on the shape of the moon and sun and raised the question whether the earth, or the cosmos, is spherical or otherwise.

Physicists, astronomers, and mathematicians, then, all have to deal with lines, figures and the rest. But the mathematician is not concerned with these concepts *qua* boundaries of natural bodies, nor with their properties as manifested in such bodies. Therefore he abstracts them from physical conditions; for they are capable of being considered in the mind in separation from the motions of the bodies to which they pertain, and such abstraction does not affect the validity of the reasoning or lead to any false conclusions.

Now the exponents of the philosophy of 'Ideas' also make abstractions, but in doing so they fall unawares into error; for they abstract physical entities, which are not really susceptible to the process as mathematical entities are. And this would become obvious if one should undertake to define, respectively, the mathematical and the 'ideal' entities, together with their properties; for

194a περιττὸν ἔσται καὶ τὸ ἄρτιον καὶ τὸ εὐθὺ καὶ τὸ καμπύλον, ἔτι δὲ ἀριθμὸς καὶ γραμμὴ καὶ σχῆμα, ἄνευ κινήσεως· σὰρξ δὲ καὶ ὀστοῦν καὶ ἄνθρωπος οὐκέτι, ἀλλὰ ταῦτα ὥσπερ ῥὶς σιμὴ ἀλλ' οὐχ ὡς τὸ καμπύλον λέγεται. δηλοῖ δὲ καὶ τὰ φυσικώτερα τῶν μαθημάτων, οἷον ὀπτικὴ καὶ ἁρμονικὴ καὶ ἀστρολογία· ἀνάπαλιν γὰρ τρόπον τιν' ἔχουσι τῇ γεωμετρίᾳ· ἡ μὲν γὰρ γεωμετρία περὶ γραμμῆς φυσικῆς σκοπεῖ, ἀλλ' οὐχ ᾗ φυσική, ἡ δ' ὀπτικὴ μαθηματικὴν μὲν γραμμήν, ἀλλ' οὐχ ᾗ μαθηματικὴ ἀλλ' ᾗ φυσική.

Ἐπεὶ δ' ἡ φύσις διχῶς—τό τε εἶδος καὶ ἡ ὕλη—ὡς ἂν εἰ περὶ σιμότητος σκοποῖμεν τί ἐστιν, οὕτω θεωρητέον· ὥστ' οὔτ' ἄνευ ὕλης τὰ τοιαῦτα, οὔτε κατὰ τὴν ὕλην.

Καὶ γὰρ δὴ καὶ περὶ τούτου διχῶς ἀπορήσειεν ἄν τις—ἐπεὶ δύο αἱ φύσεις, περὶ ποτέρας τοῦ φυσικοῦ, ἢ περὶ τοῦ ἐξ ἀμφοῖν. ἀλλ' εἰ περὶ τοῦ ἐξ ἀμφοῖν, καὶ περὶ ἑκατέρας· πότερον οὖν τῆς αὐτῆς ἢ ἄλλης ἑκατέραν γνωρίζειν;

Εἰς μὲν γὰρ τοὺς ἀρχαίους ἀποβλέψαντι δόξειεν ἂν εἶναι τῆς ὕλης· ἐπὶ μικρὸν γάρ τι μέρος Ἐμπεδοκλῆς καὶ Δημόκριτος τοῦ εἴδους καὶ τοῦ τί ἦν εἶναι ἥψαντο. εἰ δὲ ἡ τέχνη μιμεῖται τὴν

[a] [See p. 36, note *a*.—C.]

[b] [*i.e.* Empedocles, who analysed organic substances into elements (matter) and the ratio of their mixture (form), and Democritus, who 'in a sort of way defined the hot and the cold' (*Met.* 1078 b 19, *De part. anim.* 642 a 24), had *something* to say about 'form,' but not much.—C.]

the concepts 'odd,' 'even,' 'straight,' 'curved,' will be found to be independent of movement; and so too with 'number,' 'line,' and 'figure.' But of 'flesh' and 'bone' and 'man' this is no longer true, for these are in the same case as a 'turned-up nose,'[a] not in the same case as 'curved.' The point is further illustrated by those sciences which are rather physical than mathematical, though combining both disciplines, such as optics, harmonics, and astronomy; for the relations between them and geometry are, so to speak, reciprocal; since the geometer deals with physical lines, but not *qua* physical, whereas optics deals with mathematical lines, but *qua* physical not *qua* mathematical.

Since 'nature' is used ambiguously, either for the form or for the matter, Nature, as we have seen, can be regarded from two points of view, and therefore our speculations about it may be likened to an inquiry as to what 'snubnosed-ness' is; that is to say, it can neither be isolated from the material subject in which it exists, nor is it constituted by it.

At this point, in fact, we may again raise two questions. Which of the two aspects of Nature is it that claims the attention of the physicist? Or is his subject the *compositum* that combines the two? In that case—if he is concerned with the *compositum*—he must also inquire into its two factors; and then we must ask further whether this inquiry is the same for both factors or different for each.

In reading the ancients one might well suppose that the physicist's only concern was with the material; for Empedocles and Democritus have remarkably little to say about kinds of things and what is the constituent essence of them.[b] But if art imitates

194a φύσιν, τῆς δὲ αὐτῆς ἐπιστήμης εἰδέναι τὸ εἶδος καὶ τὴν ὕλην μέχρι του (οἷον ἰατροῦ ὑγίειαν καὶ χολὴν καὶ φλέγμα, ἐν οἷς ἡ ὑγίεια, ὁμοίως δὲ καὶ οἰκοδόμου τό τε εἶδος τῆς οἰκίας καὶ τὴν ὕλην, ὅτι πλίνθοι καὶ ξύλα, ὡσαύτως δὲ καὶ ἐπὶ τῶν ἄλλων), καὶ τῆς φυσικῆς ἂν εἴη τὸ γνωρίζειν ἀμφοτέρας τὰς φύσεις.

Ἔτι τὸ οὗ ἕνεκα καὶ τὸ τέλος τῆς αὐτῆς καὶ ὅσα τούτων ἕνεκα. ἡ δὲ φύσις τέλος καὶ οὗ ἕνεκα· ὧν γάρ, συνεχοῦς τῆς κινήσεως οὔσης, ἔστι τι τέλος τῆς κινήσεως, τοῦτο ἔσχατον καὶ τὸ οὗ ἕνεκα. (διὸ καὶ ὁ ποιητὴς γελοίως προήχθη εἰπεῖν 'ἔχει τελευτήν, ἧσπερ οὕνεκ' ἐγένετο'· βούλεται γὰρ οὐ πᾶν εἶναι τὸ ἔσχατον τέλος, ἀλλὰ τὸ βέλτιστον.) ἐπεὶ καὶ ποιοῦσιν αἱ τέχναι τὴν ὕλην—αἱ μὲν ἁπλῶς, αἱ δὲ εὐεργόν—καὶ χρώμεθα ὡς ἡμῶν ἕνεκα πάντων ὑπαρχόντων· ἐσμὲν γάρ πως καὶ ἡμεῖς τέλος. διχῶς γὰρ τὸ οὗ ἕνεκα (εἴρηται δὲ
194b ἐν τοῖς Περὶ φιλοσοφίας)· δύο δὴ αἱ ἄρχουσαι τῆς ὕλης καὶ αἱ γνωρίζουσαι τέχναι—ἥ τε χρωμένη,

---

[a] [*Cf.* 199 a 15.—C.]

[b] [And consequently Physics will embrace both the form (end) and the matter (means).—C.]

[c] [Meineke, *Frag. Comic. Graec.* v. p. 123 (addenda to anonymous fragments). In a modern book this criticism of the poet would stand in a footnote.—C.]

[d] [*Cf. Met.* 1072 b 2 ἔστι γὰρ τινὶ (good for someone, *e.g.* a patient, who is like ὁ χρώμενος) τὸ οὗ ἕνεκα <καὶ> τινός (good

Nature,[a] and if in the arts and crafts it pertains to the same branch of knowledge both to study its own distinctive aspect of things and likewise (up to a point) the material in which the same is manifested (as the physician, for instance, must study health and also bile and phlegm, the state of which constitutes health ; and the builder must know what the house is to be like and also that it is built of bricks and timber ; and so in all other cases), it seems to follow that physics must take cognisance both of the formal and of the material aspect of Nature.

And further the same inquiry must embrace both the purpose or end and the means to that end.[b] And the ' nature ' is the goal for the sake of which the rest exist ; for if any systematic and continuous movement is directed to a goal, this goal is an end in the sense of the purpose to which the movement is a means. (A confusion on this point betrayed the poet [c] into the unintentionally comic phrase in reference to a man's death : ' He has reached his end, for the sake of which he was born.' For the ' goal ' does not mean any kind of termination, but only the best.) For in the arts, too, it is in view of the end that the materials are either made or suitably prepared, and we make use of all the things that we have at our command as though they existed for our sake ; for we too are, in some sort, a goal ourselves. For the expression ' that for the sake of which ' a thing exists or is done has two senses (as we have explained in our treatise On Philosophy).[d] Accordingly, the arts which control the material and possess the necessary knowledge are two : the art which uses

for the sake of some product, *e.g.* health); *De anim.* 415 b 2. Aristotle refers to his lost work Περὶ φιλοσοφίας.—C.]

194b καὶ τῆς ποιητικῆς ἡ ἀρχιτεκτονική. διὸ καὶ ἡ χρωμένη ἀρχιτεκτονική πως· διαφέρει δὲ ᾗ ἡ μὲν τοῦ εἴδους γνωριστική [ἡ ἀρχιτεκτονική],[1] ἡ δὲ ὡς ποιητική, τῆς ὕλης· ὁ μὲν γὰρ κυβερνήτης ποῖόν τι τὸ εἶδος τοῦ πηδαλίου γνωρίζει καὶ ἐπιτάττει, ὁ δ' ἐκ ποίου ξύλου καὶ ποίων κινήσεων ἔσται. ἐν μὲν οὖν τοῖς κατὰ τέχνην ἡμεῖς ποιοῦμεν τὴν ὕλην τοῦ ἔργου ἕνεκα, ἐν δὲ τοῖς φυσικοῖς ὑπάρχει οὖσα.

Ἔτι τῶν πρός τι ἡ ὕλη· ἄλλῳ γὰρ εἴδει ἄλλη ὕλη.

Μέχρι δὴ πόσου τὸν φυσικὸν δεῖ εἰδέναι τὸ εἶδος καὶ τὸ τί ἐστιν; ἢ ὥσπερ ἰατρὸν νεῦρον ἢ χαλκέα χαλκόν, μέχρι του· τινὸς γὰρ ἕνεκα

[1] [Omitting ἡ ἀρχιτεκτονική (as perhaps a marginal gloss on ἡ δ' ὡς ποιητική), the construction is: ἡ μὲν τοῦ εἴδους γνωριστική (ἐστιν), ἡ δὲ (ἀρχιτεκτονικὴ) ὡς ποιητικὴ (γνωριστική ἐστι) τῆς ὕλης, which I have translated. If ἡ ἀρχιτεκτονική is retained, it might be construed: 'the art which knows the form (is) the (true) architectonic art, whereas the other (is architectonic) as operating on the matter.' But Aristotle is not contrasting *knowing* with *operating*, but two arts which *both* have knowledge and *both* control the matter (*ll.* 2 and 5-7). I have repunctuated all this paragraph and modified the translation correspondingly.—C.]

---

[a] [This distinction is taken from Plato, *Crat.* 390 D (the steersman, who uses the helm, knows what shape it should have, and directs the carpenter who makes it) and *Polit.* 259 E (the master-builder, *architecton* by his knowledge, directs the workmen). The user's art has 'use' for its end; the manufacturer's end is the product.—C.]

[b] I think Pacius is right in following Philoponus in his interpretation of ποίων κινήσεων: ἀντὶ τοῦ ἐκ ποίων φυσικῶν ῥοπῶν καὶ δυνάμεων. [As the use of the product by ourselves is the supreme end, the art of the user (*e.g.* the helmsman) has as good a right to be called architectonic (=supreme)

the product and the art of the master-craftsman who directs the manufacture.[a] Hence the art of the user also may in a sense be called the master-art; the difference is that this art is concerned with knowing the form, the other, which is supreme as controlling the manufacture, with knowing the material. Thus, the helmsman knows what are the distinctive characteristics of the helm as such—that is to say, its form—and gives his orders accordingly; while what the other knows is out of what wood and by what manipulations the helm is produced.[b] In the crafts, then, it is we that prepare the material for the sake of the function it is to fulfil, but in natural products Nature herself has provided the material. In both cases, however, the preparation of the material is commanded by the end to which it is directed.

And again, the conception of 'material' is relative, for it is different material that is suited to receive the several forms.[c]

How far then, is the physicist concerned with the form and identifying essence of things and how far with their material? With the form primarily and essentially, as the physician is with health; with the material up to a certain point, as the physician is with sinew and the smith with bronze.[d] For his

as the art of the literal *architecton, e.g.* the master-carpenter, who directs the production of the helm.—C.]

[c] [Hence 'matter' cannot be studied apart from appropriate forms.—C.]

[d] [*Literally*, 'up to a certain point (but not to the point of abstracting form from matter), as the physician is concerned with sinew (but not to the point of losing sight of its function and considering it merely as matter).' The reading and punctuation of this sentence were debated by the ancient commentators and remain doubtful.—C.]

194 b ἕκαστον, καὶ περὶ ταῦτα ἅ ἐστι χωριστὰ μὲν εἴδει, ἐν ὕλῃ δέ. ἄνθρωπος γὰρ ἄνθρωπον γεννᾷ καὶ ἥλιος. πῶς δ' ἔχει τὸ χωριστὸν καὶ τί ἐστι, φιλοσοφίας τῆς πρώτης διορίσαι ἔργον.

---

[a] There appears to be a hiatus in the original after ἥλιος, but the meaning, as I have tried to restore it, is obvious. [*Cf. Met.* 1071 a 14 ἀνθρώπου αἴτιον τά τε στοιχεῖα . . . καὶ τὸ ἴδιον εἶδος, καὶ ἔτι τι ἄλλο ἔξω, οἷον ὁ πατήρ, καὶ παρὰ ταῦτα ὁ ἥλιος καὶ ὁ λοξὸς κύκλος . . . *De gen. et corr.* B 10 explains that the inclination of the ecliptic (' oblique circle ') to the equator, whereby the sun is further off in winter, nearer in summer, accounts for the rhythm of generation and decay. —C.]

[b] [*Met.* E 1, Physics deals with natural bodies, which have

CHAPTER III

ARGUMENT

' Because ' (=by cause *of) may answer the question of the ' how and why ' of a thing from various directions. How comes it, for instance, that we have a ' saw ' here before us ?* (1) *Because (material* aitia) *there existed suitable MATERIAL (iron, to wit) out of which to make it.* (2) *Because (formal* aitia) *the iron was given the FORM* (i.e. *distinguishing attributes) of a saw.* (3) *Because (efficient* aitia) *a voluntary AGENT (the smith, with suitable apparatus and accessories) chose to make the iron into a saw.* (4) *Because (final* aitia) *the END aimed at was to create a tool able to perform a useful function,* i.e. *dividing wood by a particular method* (194b 16-195a 26).

(These four αἰτίαι, *or essential conditions of the thing being there and being what it is, are spoken of in histories of philosophy as the four ' causes,' final, material, efficient, and formal. But it does violence to the English idiom to call the*

main concern is with the goal, which is formal; but he deals only with such forms as are conceptually, but not factually, detachable from the material in which they occur. In Nature man generates man; but the process presupposes and takes place in natural material already organized by the solar heat and so forth.[a] But how we are to take the sejunct and what it is, is a question for First Philosophy to determine.[b]

separate existence (χωριστά) and can change; Mathematics, with unchanging entities, which exist not separately but only as qualifying substances; Theology, with pure forms, unchanging and separately existing.—C.]

CHAPTER III

ARGUMENT (*continued*)

material out of which a thing is made, or the distinctive attributes which define it, its 'causes,' whereas the Greek αἰτία *and the corresponding adjective* αἴτιος *can be applied to anything that is 'guilty of' or 'responsible for' a thing, or 'to the account of which,' for praise or blame, the thing may in any sense be put down. Paraphrase or barbarism offer the only escape from using English words in a sense that they cannot really bear.)*

[*Various distinctions of 'modes' of causation cut across this classification:* (1) *A cause may be described* specifically (*'a physician'*) *or by a wider* generic *term* (*'a professional man'*). (2) Incidental *are distinguished from* proper *causes: the statue is made properly by 'the sculptor,' incidentally by 'Polycleitus.'* (3) *The* actual *and the* potential: *'Builder' may mean a man actually building or merely one who has the capacity to build* (a 26-b 30).

(*This chapter, from* 194 b 23 *to* 195 b 21 *is almost word for word identical with* Met. Δ 2, *where Mr. Ross's text and commentary may be consulted*).—C.].

194 b 16 Διωρισμένων δὲ τούτων ἐπισκεπτέον περὶ τῶν αἰτίων, ποῖά τε καὶ πόσα τὸν ἀριθμόν ἐστιν. ἐπεὶ γὰρ τοῦ εἰδέναι χάριν ἡ πραγματεία, εἰδέναι δ' οὐ πρότερον οἰόμεθα ἕκαστον πρὶν ἂν λάβωμεν τὸ διὰ τί περὶ ἕκαστον, τοῦτο δ' ἐστὶ τὸ λαβεῖν τὴν πρώτην αἰτίαν, δῆλον ὅτι καὶ ἡμῖν τοῦτο ποιητέον καὶ περὶ γενέσεως καὶ φθορᾶς καὶ πάσης τῆς φυσικῆς μεταβολῆς, ὅπως εἰδότες αὐτῶν τὰς ἀρχὰς ἀνάγειν εἰς αὐτὰς πειρώμεθα τῶν ζητουμένων ἕκαστον.

Ἕνα μὲν οὖν τρόπον αἴτιον λέγεται τὸ ἐξ οὗ γίνεταί τι ἐνυπάρχοντος, οἷον ὁ χαλκὸς τοῦ ἀνδριάντος καὶ ὁ ἄργυρος τῆς φιάλης καὶ τὰ τούτων γένη· ἄλλον δὲ τὸ εἶδος καὶ τὸ παράδειγμα· τοῦτο δ' ἐστὶν ὁ λόγος ὁ τοῦ τί ἦν εἶναι καὶ τὰ τούτου γένη, οἷον τοῦ διὰ πασῶν τὰ δύο πρὸς ἓν καὶ ὅλως ὁ ἀριθμός, καὶ τὰ μέρη τὰ ἐν τῷ λόγῳ. ἔτι ὅθεν ἡ ἀρχὴ τῆς μεταβολῆς ἡ πρώτη ἢ τῆς ἠρεμήσεως, οἷον ὁ βουλεύσας αἴτιος, καὶ ὁ πατὴρ τοῦ τέκνου, καὶ ὅλως τὸ ποιοῦν τοῦ ποιουμένου καὶ τὸ

---

[a] [ἐνυπάρχοντος. Matter (like Form) is an *immanent* 'cause,' contrasted with the external moving cause, *Met.* 1070 b 22.—C.]

[b] [καὶ τὰ τούτων γένη, the wider classes to which these terms belong; *e.g.* bronze is a species of 'matter'; statue, a species of 'image,' as explained below, 195 a 29.—C.]

[c] That is to say, it must have actually 'arrived' and realized its 'being-what-it-had-to-be.'

[d] [The ratio 2 : 1 is a wider class, including the musical

We have next to consider in how many senses 'because' may answer the question 'why.' For we aim at understanding, and since we never reckon that we understand a thing till we can give an account of its 'how and why,' it is clear that we must look into the 'how and why' of things coming into existence and passing out of it, or more generally into the essential constituents of physical change, in order to trace back any object of our study to the principles so ascertained.

Well then, (1) the existence of *material* [a] for the generating process to start from (whether specifically or generically considered) [b] is one of the essential factors we are looking for. Such is the bronze for the statue, or the silver for the phial. (Material *aitia*.) Then, naturally, (2) the thing in question cannot be there unless the material has actually received the *form* or characteristics of the type, conformity to which brings it within the definition of the thing we say it is, whether specifically or generically.[c] Thus the interval between two notes is not an octave unless the notes are in the ratio of 2 to 1; nor do they stand at a musical interval at all unless they conform to one or other of the recognized ratios.[d] (Formal *aitia*.) Then again (3), there must be something to initiate the process of the change or its cessation when the process is completed, such as the act of a voluntary agent [e] (of the smith, for instance), or the father who begets a child; or more generally the prime, conscious or unconscious,

interval of the octave, and itself included in the still wider class 'Number' (195 a 31). So τὰ μέρη τὰ ἐν τῷ λόγῳ, 'the constituent terms' (*e.g. biped, animal*) 'in the definition' (of *man*), are wider classes than the term defined.—C.]

[e] [*Or* '*e.g.* the adviser is a cause of the action' (Ross).—C.]

194 b μεταβάλλον τοῦ μεταβαλλομένου. ἔτι ὡς τὸ τέλος· τοῦτο δ' ἐστὶ τὸ οὗ ἕνεκα, οἷον τοῦ περιπατεῖν ἡ ὑγίεια· διὰ τί γὰρ περιπατεῖ; φαμέν, ἵνα ὑγιαίνῃ, καὶ εἰπόντες οὕτως οἰόμεθα ἀποδεδωκέναι τὸ αἴτιον. καὶ ὅσα δὴ κινήσαντος ἄλλου μεταξὺ γίγνεται τοῦ τέλους, οἷον τῆς 195 a ὑγιείας ἡ ἰσχνασία ἢ ἡ κάθαρσις, ἢ τὰ φάρμακα ἢ τὰ ὄργανα· πάντα γὰρ ταῦτα τοῦ τέλους ἕνεκά ἐστι, διαφέρει δ' ἀλλήλων ὡς ὄντα τὰ μὲν ἔργα τὰ δ' ὄργανα.

Τὰ μὲν οὖν αἴτια σχεδὸν τοσαυταχῶς λέγεται· συμβαίνει δὲ πολλαχῶς λεγομένων τῶν αἰτίων καὶ πολλὰ τοῦ αὐτοῦ αἴτια εἶναι οὐ κατὰ συμβεβηκός, οἷον τοῦ ἀνδριάντος καὶ ἡ ἀνδριαντοποιικὴ καὶ ὁ χαλκὸς οὐ καθ' ἕτερόν τι ἀλλ' ᾗ ἀνδριάς· ἀλλ' οὐ τὸν αὐτὸν τρόπον, ἀλλὰ τὸ μὲν ὡς ὕλη τὸ δ' ὡς ὅθεν ἡ κίνησις.

Ἔστι δέ τινα καὶ ἀλλήλων αἴτια, οἷον τὸ πονεῖν τῆς εὐεξίας καὶ αὕτη τοῦ πονεῖν· ἀλλ' οὐ τὸν αὐτὸν τρόπον, ἀλλὰ τὸ μὲν ὡς τέλος τὸ δ' ὡς ἀρχὴ κινήσεως.

---

[a] [The ' tools ' of the doctor's trade include the drugs and instruments.—C.]

agent that produces the effect and starts the material on its way to the product, changing it from what it was to what it is to be. (Efficient *aitia.*) And lastly, (4) there is the *end* or purpose, for the sake of which the process is initiated, as when a man takes exercise for the sake of his health. 'Why does he take exercise?' we ask. And the answer 'Because he thinks it good for his health' satisfies us. (Final *aitia.*) Then there are all the intermediary agents, which are set in motion by the prime agent and make for the goal, as means to the end. Such are the reduction of superfluous flesh and purgation, or drugs and surgical instruments, as means to health. For both actions and tools [a] may be means, or '*media*,' through which the efficient cause reaches the end aimed at.

This is a rough classification of the causal determinants (*aitiai*) of things; but it often happens that, when we specify them, we find a number of them coalescing as joint factors in the production of a single effect, and that not merely incidentally; for it is *qua* statue that the statue depends for its existence alike on the bronze and on the statuary. The two, however, do not stand on the same footing, for one is required as the material and the other as initiating the change.

Also, it can be said of certain things indifferently that either of them is the cause or the effect of the other. Thus we may say that a man is in fine condition 'because' he has been in training, or that he has been in training 'because' of the good condition he expected as the result. But one is the cause as aim (final *aitia*) and the other as initiating the process (efficient *aitia*).

195 a ῎Ετι δὲ τὸ αὐτὸ τῶν ἐναντίων ἐστίν· ὃ γὰρ παρὸν αἴτιον τοῦδε, τοῦτο καὶ ἀπὸν αἰτιώμεθα ἐνίοτε τοῦ ἐναντίου, οἷον τὴν ἀπουσίαν τοῦ κυβερνήτου τῆς τοῦ πλοίου ἀνατροπῆς, οὗ ἦν ἡ παρουσία αἰτία τῆς σωτηρίας.

῞Απαντα δὲ τὰ νῦν εἰρημένα αἴτια εἰς τέτταρας τρόπους πίπτει τοὺς φανερωτάτους. τὰ μὲν γὰρ στοιχεῖα τῶν συλλαβῶν καὶ ἡ ὕλη τῶν σκευαστῶν καὶ τὸ πῦρ καὶ τὰ τοιαῦτα τῶν σωμάτων καὶ τὰ μέρη τοῦ ὅλου καὶ αἱ ὑποθέσεις τοῦ συμπεράσματος ὡς τὸ ἐξ οὗ αἴτιά ἐστιν· τούτων δὲ τὰ μὲν ὡς τὸ ὑποκείμενον (οἷον τὰ μέρη), τὰ δὲ ὡς τὸ τί ἦν εἶναι—τό τε ὅλον καὶ ἡ σύνθεσις καὶ τὸ εἶδος. τὸ δὲ σπέρμα καὶ ὁ ἰατρὸς καὶ ὁ βουλεύσας καὶ ὅλως τὸ ποιοῦν, πάντα ὅθεν ἡ ἀρχὴ τῆς μεταβολῆς ἢ στάσεως ἢ κινήσεως. τὰ δ᾿ ὡς τὸ τέλος καὶ τἀγαθὸν τῶν ἄλλων· τὸ γὰρ οὗ ἕνεκα βέλτιστον καὶ τέλος τῶν ἄλλων ἐθέλει εἶναι (διαφερέτω δὲ μηδὲν εἰπεῖν αὐτὸ ἀγαθὸν ἢ φαινόμενον ἀγαθόν).

Τὰ μὲν οὖν αἴτια ταῦτα καὶ τοσαῦτά ἐστι τῷ

[a] The technical term for the propositions of a syllogism apart from their logical connexion is 'material,' so that the propositions 'mammals are animals,' 'squirrels are animals,' 'squirrels are mammals,' are the 'material' of the syllogism: 'mammals are animals, squirrels are animals, therefore squirrels are mammals.' Here all three propositions, major, minor, and conclusion, are true. Therefore the conclusion is materially sound, but formally the syllogism is not sound; for the conclusion does not follow from the premisses, though they are all true. Aristotle frequently points out that the conclusion of a syllogism is materially true, but not formally proved.

[*Cf. Anal. post.* 94 a 22, where 'the (logical) antecedent

Again, the same cause is often alleged for precisely opposite effects. For if its presence causes one thing, we lay the opposite to its account if it is absent. Thus, if the pilot's presence would have brought the ship safe to harbour, we say that he caused its wreck by his absence.

But in all cases the essential and causal determinants we have enumerated fall into four main classes. For letters are the causes of syllables, and the material is the cause of manufactured articles, and fire and the like are causes of physical bodies, and the parts are causes of the whole, and the premises are causes of the conclusion,[a] in the sense of that out of which these respectively are made[b]; but of these things some are causes in the sense of the *substratum* (*e.g.* the parts stand in this relation to the whole), others in the sense of the *essence*—the whole or the synthesis or the form. And again, the fertilizing sperm, or the physician, or briefly the voluntary or involuntary *agent* sets going or arrests the transformation or movement. And finally, there is the goal or *end* in view, which animates all the other determinant factors as the best they can attain to; for the attainment of that 'for the sake of which' anything exists or is done is its final and best possible achievement (though of course 'best' in this connexion means no more than 'taken to be the best').

These are the main classes of determinant factors

which necessitates a consequence' seems to be equated with the 'material cause.'—C.]

[b] [Mr. Ross (on *Met.* 1013 b 17) points out that Aristotle here includes both the material cause and the formal (distinguished at the beginning of the chapter) under the ἐξ οὗ (immanent or internal constituents), and then divides this again into the ὑποκείμενον and the τί ἦν εἶναι.—C.]

195 a εἴδει· τρόποι δὲ τῶν αἰτίων ἀριθμῷ μέν εἰσι πολλοί, κεφαλαιούμενοι δὲ καὶ οὗτοι ἐλάττους.

Λέγεται γὰρ αἴτια πολλαχῶς, καὶ αὐτῶν τῶν ὁμοειδῶν προτέρως καὶ ὑστέρως ἄλλο ἄλλου, οἷον ὑγιείας ἰατρὸς καὶ τεχνίτης, καὶ τοῦ διὰ πασῶν τὸ διπλάσιον καὶ ὁ ἀριθμός, καὶ ἀεὶ τὰ περιέχοντα πρὸς τὸ καθ' ἕκαστα.

Ἔτι δ' ὡς τὸ συμβεβηκὸς καὶ τὰ τούτων γένη, οἷον ἀνδριάντος ἄλλως Πολύκλειτος καὶ ἄλλως ἀνδριαντοποιός, ὅτι συμβέβηκε τῷ ἀνδριαντοποιῷ τὸ Πολυκλείτῳ εἶναι· καὶ τὰ περιέχοντα δὲ τὸ συμβεβηκός, οἷον εἰ ὁ ἄνθρωπος αἴτιος εἴη ἀν- 195 b δριάντος ἢ ὅλως ζῷον. ἔστι δὲ καὶ τῶν συμβεβηκότων ἄλλα ἄλλων πορρώτερον καὶ ἐγγύτερον, οἷον εἰ ὁ λευκὸς καὶ ὁ μουσικὸς αἴτιος λέγοιτο τοῦ ἀνδριάντος.

Παρὰ πάντα[1] δὲ καὶ τὰ οἰκείως λεγόμενα καὶ τὰ κατὰ συμβεβηκός, τὰ μὲν ὡς δυνάμενα λέγεται

[1] [παρὰ πάντα I, Simpl. 324. 5; Philop. 256. 20 (*Met.* 1014 a 7): πάντα al.—C.].

---

[a] [*Met.* 1013 b 33 has καὶ ἀεὶ τὰ περιέχοντα ὁτιοῦν τῶν καθ' ἕκαστα, 'and the classes that include any particular cause are always causes of the particular effect.' Mr. Ross notes that the antithesis between the καθ' ἕκαστον and the περιέχον

and causes; and, though within these many other distinctions may be drawn, yet they too can be reduced to a manageable number of classes.

Thus, (1) causes belonging to one and the same class may be more or less closely determined by being reduced to subordination to each other. Health may be restored by a 'physician,' or by a 'professional man' (a term which includes physicians). The octave may be described as a case of the ratio 2 : 1, or less specifically as a 'number'; and generally, an inclusive term may be used instead of a more special one.[a]

And again, (2) an agent may be described not *qua* agent, but as something that characterizes him incidentally; and in this case also the specification may be more or less inclusive. We may say that the statue was made by 'the sculptor,' or that it was made by 'Polycleitus,' in as much as the art resided in the person of Polycleitus and was incidentally associated with all his other irrelevant characteristics. Incidentally, too, the sculptor was a man, or more generally still an animal. And the 'incidental' itself may be more or less remotely related to the essential. Thus the sculptor's being a man of culture or light in complexion is even more incidental and irrelevant to his production of a statue than his being Polycleitus or a man.

(3) Besides all these ways of describing the agent, whether in his agential capacity or by attributes that fall to him incidentally, we may be speaking of his potentialities merely or of the actual exertion of his

seems to include both the antitheses, specific—generic (here) and individual—universal (below).—C.]

195 b τὰ δ' ὡς ἐνεργοῦντα, οἷον τοῦ οἰκοδομεῖσθαι οἰκίαν οἰκοδόμος ἢ οἰκοδομῶν οἰκοδόμος.

Ὁμοίως δὲ λεχθήσεται καὶ ἐφ' ὧν αἴτια τὰ αἴτια τοῖς εἰρημένοις, οἷον τουδὶ τοῦ ἀνδριάντος ἢ ἀνδριάντος ἢ καὶ ὅλως εἰκόνος, καὶ χαλκοῦ τοῦδε ἢ χαλκοῦ ἢ ὅλως ὕλης· καὶ ἐπὶ τῶν συμβεβηκότων ὡσαύτως. ἔτι δὲ συμπλεκόμενα καὶ ταῦτα κἀκεῖνα λεχθήσεται, οἷον οὐ Πολύκλειτος οὐδὲ ἀνδριαντοποιὸς ἀλλὰ Πολύκλειτος ἀνδριαντοποιός.

Ἀλλ' ὅμως ἅπαντα ταῦτά ἐστι τὸ μὲν πλῆθος ἕξ, λεγόμενα δὲ διχῶς· ἢ γὰρ ὡς τὸ καθ' ἕκαστον ἢ ὡς τὸ γένος, ἢ ὡς τὸ συμβεβηκὸς ἢ ὡς τὸ γένος τοῦ συμβεβηκότος, ἢ ὡς συμπλεκόμενα ταῦτα ἢ ὡς ἁπλῶς λεγόμενα· πάντα δὲ ἢ ἐνεργοῦντα ἢ κατὰ δύναμιν. διαφέρει δὲ τοσοῦτον, ὅτι τὰ μὲν ἐνεργοῦντα καὶ τὰ καθ' ἕκαστον ἅμα ἔστι καὶ οὐκ ἔστι καὶ ὧν αἴτια—οἷον ὅδ' ὁ ἰατρεύων τῷδε

powers. Thus, we may say that 'the builder' caused the house to be built, meaning the man who knew how to build, but it was only when he was in the act of building that he was really causing the house to be built.

And all this holds just as good for the product as for the producer. The product may be regarded as this particular statue, or as a statue in general, or still more generally as an image; or again the craftsman as working upon this particular piece of bronze, or upon bronze in general, or still more generally upon material. And these distinctions, too, may be crossed by the distinction between 'directly' and 'incidentally.' And again both the incidental and the direct may be united together, if we speak, not of 'Polycleitus,' nor of 'the sculptor,' but of 'Polycleitus the sculptor.'

All these distinctions, however, may be brought under six heads, each of which may be predicated in two different senses. For every determining factor, *qua* determinant, may be designated (1) individually, or (2) as belonging to a class; and (3) incidental coincidences may have a more specific character, or (4) a more general one; and both direct and indirect agency or passion may be indicated (5) separately or (6) in combination. And finally in every case it may be either (*a*) a potentiality that is indicated, or (*b*) an actual energizing. But they differ to this extent, that the actual energizing agent, being an individual, exists as energizing, or ceases to do so, according as that which is experiencing its energy ceases or continues so to experience it (for instance, a particular physician ceases from the actual exercise of his art at the same moment as the

τῷ ὑγιαζομένῳ, καὶ ὅδε ὁ οἰκοδομῶν τῷδε τῷ οἰκοδομουμένῳ—τὰ δὲ κατὰ δύναμιν οὐκ ἀεί· φθείρεται γὰρ οὐχ ἅμα ἡ οἰκία καὶ ὁ οἰκοδόμος.

Δεῖ δ' ἀεὶ τὸ αἴτιον ἑκάστου τὸ ἀκρότατον ζητεῖν (ὥσπερ καὶ ἐπὶ τῶν ἄλλων), οἷον ἄνθρωπος οἰκοδομεῖ ὅτι οἰκοδόμος, ὁ δ' οἰκοδόμος κατὰ τὴν οἰκοδομικήν· τοῦτο τοίνυν πρότερον τὸ αἴτιον. καὶ οὕτως ἐπὶ πάντων.

Ἔτι τὰ μὲν γένη τῶν γενῶν, τὰ δὲ καθ' ἕκαστον τῶν καθ' ἕκαστον, οἷον ἀνδριαντοποιὸς μὲν ἀνδριάντος, ὁδὶ δὲ τουδί. καὶ τὰς μὲν δυνάμεις τῶν δυνατῶν, τὰ δ' ἐνεργοῦντα πρὸς τὰ ἐνεργούμενα.

Ὅσα μὲν οὖν τὰ αἴτια καὶ ὃν τρόπον αἴτια, ἔστω ἡμῖν διωρισμένα ἱκανῶς.

CHAPTER IV

INTRODUCTORY NOTE TO CHAPTERS IV-VI

We have now learnt to regard the purpose or end (*finis*) to which a voluntary act is directed as the *final* cause of that action being taken, and the action itself as the *efficient* cause of the purpose being accomplished. Here the connexion between the efficient and final causes is direct, and the result normally follows the expectation.

But sometimes we aim at one thing and ‘ by accident ’ hit another. Thus a man may go to the market to purchase wares and there may accidentally meet a debtor and recover the debt ; or he may dig his plot for the purpose of planting it and may find a hidden treasure. In such cases the action is *directed towards* one purpose, but *accomplishes* another. And this we attribute to Chance or

particular patient ceases from being actually in the process of restoration to health; and a particular builder ceases from actual building at the same moment as the house ceases from being actually in course of erection); but in the case of potentiality it is not always so, for the potential builder and the house need not perish at the same time.

And in every case we must try to determine the point at which the causality is focused, both here and elsewhere. Thus a man builds *qua* builder, and a builder builds *qua* expert in the building art; so it is in the application of the building art to the material that the building art is focused. And so with all the rest.

Again, the general is related to the general, and the individual to the individual: a statue is produced by a sculptor, but this individual statue by such and such an individual sculptor. And the relation of the potential to the potential and of the actualized to the actualized is analogous.

Let this suffice for the definition of the different classes of determinants and the ways in which they are severally related to the result.

CHAPTER IV

INTRODUCTORY NOTE TO CHAPTERS IV–VI (*continued*)

Fortune. In these cases the final cause (*e.g.* desire for the crop) has no normal relation to the effect (discovery of a treasure), though the action incidentally caused that discovery. The efficient cause of the discovery, then, viz. the digging in that place, quite definitely and directly led

INTRODUCTORY NOTE TO CHAPTERS IV–VI (*continued*)

to the result, and is exactly the action which *would have been* dictated by the desire to secure the treasure, had the man known it was there; but the final cause that actually dictated the action was only incidentally connected with the result; for the act of digging *as determined by the desire for the crop* had no reference whatever to the treasure, and no such effect as the discovery could be expected as normally following from it.

It is in cases of this kind that Aristotle traces the causalty of what we call Fortune or Accident, and he confines his attention to accidental results which are significant and which therefore *might have been* aimed at for their own sake, but as a matter of fact were hit upon without having been aimed at.

ARGUMENT

[*Besides the four causes above distinguished, the agency of Chance or Luck has been recognized in popular and philosophic thought, and events are spoken of as occurring spontaneously ' of themselves '* (ἀπὸ ταὐτομάτου) (195 b 30-35).

Some question the existence of Chance; lucky events always have some definite cause. On the other hand, men do

195 b Λέγεται δὲ καὶ ἡ τύχη καὶ τὸ αὐτόματον τῶν αἰτίων, καὶ πολλὰ καὶ εἶναι καὶ γίνεσθαι διὰ τύχην καὶ διὰ τὸ αὐτόματον. τίνα οὖν τρόπον ἐν τούτοις ἐστὶ τοῖς αἰτίοις ἡ τύχη καὶ τὸ αὐτόματον, καὶ πότερον τὸ αὐτὸ ἡ τύχη καὶ τὸ αὐτόματον ἢ ἕτερον, καὶ ὅλως τί ἐστιν ἡ τύχη καὶ τὸ αὐτόματον, ἐπισκεπτέον.

196 a Ἔνιοι γὰρ καὶ εἰ ἔστιν ἢ μὴ ἀποροῦσιν· οὐδὲν γὰρ γίνεσθαι ἀπὸ τύχης φασίν, ἀλλὰ πάντων

INTRODUCTORY NOTE TO CHAPTERS IV–VI (*continued*)

This clue must be firmly held in following the argument through the three chapters in which it is developed. It is crossed by many repetitions and obscured by elliptical omissions. Aristotle used two words, one the more general, *αὐτόματον* (imperfectly represented by ' chance ' or ' accident '), the other of narrower range, *τύχη* (imperfectly represented by ' fortune ' or ' luck '). The English word that seems best to bring out the argument will be used in every case, without any attempt being made to secure exact correspondence.

[It will appear that, strictly, a ' chance result ' means a result which (1) is produced ' incidentally ' or ' in virtue of a concomitant ' (*κατὰ συμβεβηκός*), and also (2) is ' purpose-serving ' (*ἕνεκά του*), in that it is desirable and might have been designed either (*a*) by conscious human purpose (it is then called ' luck,' *τύχη*) or (*b*) by the unconscious purposiveness of Nature (it is then called ' chance,' *ταὐτόματον*).—C.]

ARGUMENT (*continued*)

call some events lucky, others not ; and some philosophers, who allow Chance a place in their systems, ought to have given some account of it (b 35–196 b 5).

Others see in Fortune an inscrutable divine cause. The whole question calls for examination (b 5-9).—C.]

We often allege fortune, or luck, and accident as causes, saying that something came about ' as fortune or luck would have it ' or ' accidentally.' What then is the place of fortune and accident amongst the causes we have reviewed ? Is there any distinction between them ? And, in a word, what are they ?

For some question their existence, declaring that nothing happens casually, but that everything we

196 a εἶναί τι αἴτιον ὡρισμένον, ὅσα λέγομεν ἀπ' αὐτομάτου γίγνεσθαι ἢ τύχης—οἷον τοῦ ἐλθεῖν ἀπὸ τύχης εἰς τὴν ἀγορὰν καὶ καταλαβεῖν ὃν ἐβούλετο μὲν οὐκ ᾤετο δέ, αἴτιον τὸ βούλεσθαι ἀγοράσαι ἐλθόντα· ὁμοίως δὲ καὶ ἐπὶ τῶν ἄλλων τῶν ἀπὸ τύχης λεγομένων ἀεί τι εἶναι λαβεῖν τὸ αἴτιον ἀλλ' οὐ τύχην. ἐπεὶ εἴ γέ τι ἦν ἡ τύχη, ἄτοπον ἂν φανείη ὡς ἀληθῶς, καὶ ἀπορήσειεν ἄν τις διὰ τί ποτ' οὐδεὶς τῶν ἀρχαίων σοφῶν τὰ αἴτια περὶ γενέσεως καὶ φθορᾶς λέγων περὶ τύχης οὐδὲν διώρισεν, ἀλλ', ὡς ἔοικεν, οὐδὲν ᾤοντο οὐδ' ἐκεῖνοι εἶναι ἀπὸ τύχης.

Ἀλλὰ καὶ τοῦτο θαυμαστόν· πολλὰ γὰρ καὶ γίνεται καὶ ἔστιν ἀπὸ τύχης καὶ ἀπὸ ταὐτομάτου, ἃ οὐκ ἀγνοοῦντες ὅτι ἔστιν ἐπανενεγκεῖν ἕκαστον ἐπί τι αἴτιον τῶν γινομένων (καθάπερ ὁ παλαιὸς λόγος [εἶπεν][1] ὁ ἀναιρῶν τὴν τύχην), ὅμως τούτων τὰ μὲν εἶναί φασι πάντες ἀπὸ τύχης τὰ δ' οὐκ ἀπὸ τύχης· διὸ καὶ ἁμῶς γέ πως ἦν ποιητέον αὐτοῖς μνείαν. ἀλλὰ μὴν οὐδ' ἐκείνων γέ τι ᾤοντο εἶναι τὴν τύχην, οἷον φιλίαν ἢ νεῖκος, ἢ νοῦν, ἢ πῦρ ἢ ἄλλο γέ τι τῶν τοιούτων. ἄτοπον οὖν εἴτε μὴ ὑπελάμβανον εἶναι εἴτε οἰόμενοι παρέλειπον, καὶ ταῦτ' ἐνίοτε χρώμενοι, ὥσπερ Ἐμπεδοκλῆς οὐκ ἀεὶ τὸν ἀέρα ἀνωτάτω ἀπο-

[1] [εἶπεν om. Simpl. 330. 14, Torstrik, Diels.—C.]

---

[a] [*Cf.* Leucippus, *frag.* 2 οὐδὲν χρῆμα μάτην γίνεται ἀλλὰ πάντα ἐκ λόγου τε καὶ ὑπ' ἀνάγκης. Democritus is referred to, according to Eudemus (Simplic. 330. 14; Diels, *Vors.* 55 A 68). See C. Bailey, *The Greek Atomists*, p. 121.—C.]

speak of in that way has really a definite cause.[a] For instance, if a man comes to market and there chances on someone he has been wishing to meet but was not expecting to meet there, the reason of his meeting him was that he wanted to go marketing; and so too in all other cases when we allege chance as the cause, there is always some other cause to be found, and it is never really chance. And indeed it might be urged that, if there really were such a thing as luck, it would present a genuine problem, and the question might be raised, why none of the earlier philosophers should have had anything to say about it when discussing causes in relation to genesis and perishing, so as apparently to think that nothing comes about by chance.

Maybe. But is it not equally strange that, however freely men admit that every kind of luck and everything that 'happens accidentally' can really be assigned to some definite cause, still, while accepting this venerable argument for the elimination of chance from their thoughts, they nevertheless invariably distinguish, in fact, between things that do, and the things that do not, depend upon chance or luck? So in any case the philosophers should have given some account of what are called chance happenings. If they gave none, it was not because they identified chance with any one of the causes they recognized—Love or Strife, Mind, or Fire or some other element. But surely, whether they believed or disbelieved in chance, they were bound in reason to take some note of it; especially as, on occasion, they actually had recourse to it, as when Empedocles says that air is sifted out upwards not uniformly

196 a κρίνεσθαί φησιν ἀλλ' ὅπως ἂν τύχῃ—λέγει γοῦν ἐν τῇ κοσμοποιίᾳ ὡς ' οὕτω συνέκυρσε θέων τότε, πολλάκι δ' ἄλλως '—καὶ τὰ μόρια τῶν ζῴων ἀπὸ τύχης γενέσθαι τὰ πλεῖστά φησιν.

Εἰσὶ δέ τινες οἳ καὶ τοὐρανοῦ τοῦδε καὶ τῶν κόσμων πάντων αἰτιῶνται τὸ αὐτόματον· ἀπὸ ταὐτομάτου γὰρ γίγνεσθαι τὴν δίνην καὶ τὴν κίνησιν τὴν διακρίνασαν καὶ καταστήσασαν εἰς ταύτην τὴν τάξιν τὸ πᾶν. καὶ μάλα τοῦτο θαυμάσαι ἄξιον· λέγοντες[1] γὰρ τὰ μὲν ζῷα καὶ τὰ φυτὰ ἀπὸ τύχης μήτε εἶναι μήτε γίγνεσθαι, ἀλλ' ἤτοι φύσιν ἢ νοῦν ἤ τι τοιοῦτον ἕτερον εἶναι τὸ αἴτιον (οὐ γὰρ ὅ τι ἔτυχεν ἐκ τοῦ σπέρματος ἑκάστου γίγνεται, ἀλλ' ἐκ μὲν τοῦ τοιουδὶ ἐλαία ἐκ δὲ τοῦ τοιουδὶ ἄνθρωπος), τὸν δ' οὐρανὸν καὶ τὰ θειότατα τῶν φανερῶν ἀπὸ τοῦ αὐτομάτου γενέσθαι, τοιαύτην δ' αἰτίαν μηδεμίαν εἶναι οἵαν τῶν ζῴων καὶ τῶν φυτῶν. καίτοι εἰ οὕτως ἔχει, τοῦτ' αὐτὸ ἄξιον ἐπιστάσεως, καὶ καλῶς ἔχει
196 b λεχθῆναί τι περὶ αὐτοῦ. πρὸς γὰρ τῷ καὶ ἄλλως ἄτοπον εἶναι τὸ λεγόμενον, ἔτι ἀτοπώτερον τὸ λέγειν ταῦτα ὁρῶντας ἐν μὲν τῷ οὐρανῷ οὐδὲν

[1] [The lack of a construction for λέγοντες is repaired in F by the reading καὶ μάλα τοῦτό γε αὐτὸ θαυμάσαι ἄξιον λέγοντας τὰ μὲν κτλ.—C.]

---

[a] [*Frag.* 53 ' (Air) in its course meets, now in this way, now in that (with the other elements).' At *De gen. et corr.* 334 a 1 Aristotle complains that Empedocles sometimes makes air move *upwards* ' by chance ' (quoting this verse), but seems sometimes to attribute to it a *downward* movement, which he must consequently regard as ' natural ' to air. Should we read here ἄνω ἢ κάτω (for the strange superlative ἀνωτάτω, though read by Simplicius): ' E. speaks of the air as

but as it happens—for he says of it, in his *Cosmogony*,[a]

Thus did it chance to hit one while, but other whiles not thus—

and he says that the members of animals, for the most part, came out haphazard.[b]

Some[c] indeed attribute our Heaven and all the worlds[d] to chance happenings, saying that the vortex and shifting that disentangled the chaos and established the cosmic order came by chance. This is surely most amazing—for these people actually to say that, whereas neither animals nor plants are, or come to be, by chance, but are all caused by Nature or Mind or what else (for it is not a matter of chance what springs from a given sperm, since an olive comes from such an one, and a man from such another),[e] yet the heaven and the divinest things that our sight reveals come anyhow and have no such causes as animals and plants have. But if this really were so, that very fact ought to give us pause and convince us that the matter needs investigation. For, in addition to the inherently paradoxical nature of such an assertion, we may note that it is exactly in the movements of the heavenly bodies that we never observe what we call

separated out, not always up or always down, but as it happens'?—C.]

[b] [*Fragg.* 57-61. In the period when Love is gaining on Strife, animals are formed by the casual coming together of limbs separately produced. *Cf.* 198 b 29.—C.]

[c] [Democritus (Simplic. 331. 16; Diels, *Vors.* 55 A 67, 69). See C. Bailey, *The Greek Atomists*, p. 139. *Cf.* also *De part. anim.* 641 b 15, Plato, *Soph.* 265 c, *Laws* 889 B.—C.]

[d] [The Atomists believed in an indefinite number of worlds in infinite space. Diels, *Vors.* 54 A 21.—C.]

[e] [*Cf. De gen. et cor.* 333 b 3 ff.—C.]

196 b ἀπὸ ταὐτομάτου γιγνόμενον, ἐν δὲ τοῖς οὐκ ἀπὸ τύχης πολλὰ συμβαίνοντα ἀπὸ τύχης· καίτοι εἰκός γε ἦν τοὐναντίον γίγνεσθαι.

Εἰσὶ δέ τινες οἷς δοκεῖ εἶναι αἰτία μὲν ἡ τύχη, ἄδηλος δὲ ἀνθρωπίνῃ διανοίᾳ ὡς θεῖόν τι οὖσα καὶ δαιμονιώτερον.

Ὥστε σκεπτέον καὶ τί ἑκάτερον, καὶ εἰ ταὐτὸν ἢ ἕτερον τό τε αὐτόματον καὶ ἡ τύχη, καὶ πῶς εἰς τὰ διωρισμένα αἴτια ἐμπίπτουσιν.

CHAPTER V

ARGUMENT

The phrases we are examining are never used in connexion with normal or expected sequences of things.

Now man has purposes, and Nature has trends, which normally determine the course of deliberate action or of natural processes ; but in either case such actions or processes may incidentally involve unexpected results outside their normal progress towards the goal they are making for ;

Πρῶτον μὲν οὖν, ἐπειδὴ ὁρῶμεν τὰ μὲν ἀεὶ ὡσαύτως γινόμενα τὰ δὲ ὡς ἐπὶ πολύ, φανερὸν ὅτι οὐδετέρου τούτων αἰτία ἡ τύχη λέγεται οὐδὲ τὸ ἀπὸ τύχης, οὔτε τοῦ ἐξ ἀνάγκης καὶ ἀεὶ οὔτε τοῦ ὡς ἐπὶ πολύ. ἀλλ' ἐπειδὴ ἔστιν ἃ γίγνεται καὶ παρὰ ταῦτα, καὶ ταῦτα πάντες φασὶν εἶναι ἀπὸ

[a] [Aet. i. 29. 7 Ἀναξαγόρας καὶ Δημόκριτος καὶ οἱ Στωικοὶ ἄδηλον αἰτίαν ἀνθρωπίνῳ λογισμῷ (τὴν τύχην). Diels, *Dox.* 326 ; *Vors.* 46 A 66. *Cf.* also *Eud. Eth.* 1247 b 6.—C.]

[b] [*Cf. Anal. post.* 96 a 8, Some occurrences are universal (for they are as they are, or occur as they do, always

casual or accidental variations, whereas in all that these people tell us is exempt from chance such things are common. Of course it ought to be just the other way.

Some, moreover, hold that fortune is a genuine cause of things, but one that has a something divine and mysterious about it, that makes it inscrutable to the human intelligence.[a]

So we must obviously investigate the whole matter, and must see whether the phrases in question can be classified or distinguished from each other, and how they fall in with the causes and determinants we have defined.

CHAPTER V

ARGUMENT (*continued*)

it is such results that we attribute to chance. Every concrete thing has incidentally a number of attributes which are not essential to its being the thing it is. And the same is true of agents and actions regarded as causes. (196 b 10–197 a 8.)

Hence the sense of vagueness that attaches to incidental or accidental causation. (a 8–35.)

To begin with, then, we note that some things follow upon others uniformly or generally,[b] and it is evidently not such things that we attribute to chance or luck. Necessary or customary successions, therefore, are excluded from our present inquiry. On the other hand irregular and exceptional consequences do occasionally occur, and since it is precisely to this class of actual happenings that we ourselves

and in every case); others occur only as a general rule; *e.g.*, a man generally grows a beard, but not always.—C.]

τύχης, φανερὸν ὅτι ἔστι τι ἡ τύχη καὶ τὸ αὐτόματον· τά τε γὰρ τοιαῦτα ἀπὸ τύχης καὶ τὰ ἀπὸ τύχης τοιαῦτα ὄντα ἴσμεν.

Τῶν δὲ γινομένων τὰ μὲν ἕνεκά του γίγνεται τὰ δ' οὔ· τούτων δὲ τὰ μὲν κατὰ προαίρεσιν τὰ δ' οὐ κατὰ προαίρεσιν, ἄμφω δ' ἐν τοῖς ἕνεκά του, ὥστε δῆλον ὅτι καὶ ἐν τοῖς παρὰ τὸ ἀναγκαῖον καὶ τὸ ὡς ἐπὶ πολὺ ἔστιν ἔνια περὶ ἃ ἐνδέχεται ὑπάρχειν τὸ ἕνεκά του. ἔστι δ' ἕνεκά του ὅσα τε ἀπὸ διανοίας ἂν πραχθείη καὶ ὅσα ἀπὸ φύσεως. τὰ δὴ τοιαῦτα ὅταν κατὰ συμβεβηκὸς γένηται, ἀπὸ τύχης φαμὲν εἶναι. ὥσπερ γὰρ καὶ ὄν ἐστι τὸ μὲν καθ' αὑτὸ τὸ δὲ κατὰ συμβεβηκός, οὕτω καὶ αἴτιον ἐνδέχεται εἶναι, οἷον οἰκίας καθ' αὑτὸ μὲν αἴτιον τὸ οἰκοδομικόν, κατὰ συμβεβηκὸς δὲ τὸ λευκὸν ἢ τὸ μουσικόν. τὸ μὲν οὖν καθ' αὑτὸ αἴτιον ὡρισμένον, τὸ δὲ κατὰ συμβεβηκὸς ἀόριστον· ἄπειρα γὰρ ἂν τῷ ἑνὶ συμβαίη.

Καθάπερ οὖν ἐλέχθη, ὅταν ἐν τοῖς ἕνεκά του γιγνομένοις τοῦτο γένηται, τότε λέγεται ἀπὸ ταὐτο-

---

[a] And if it happened that the cultivated tastes of the builder, which had not influenced our choice, brought us a pleasant acquaintance, our selection of him would have been directed to one purpose primarily, but would have served another purpose incidentally, and that would be a 'lucky chance.'

apply such terms as 'luck' or 'lucky' and of this class that we think when anyone else uses these terms, it follows that what we call luck or chance corresponds to some reality and is not a mere fiction.

Or taking it from another side: some things make for a purpose and some not; and of the 'ends' actually achieved some are those we aimed at, but some are not. In this latter case our actions were as a fact making for the uncontemplated result just as much as if it had been intended; so that an action may actually 'serve a purpose' which does not necessarily or normally follow upon it, and which it was not intended to serve. In this sense any action may be regarded as purpose-serving if it leads to a result that *might have been* voluntarily sought, or to a result which stands in the corresponding relation to the movements of Nature. Now when such results accrue incidentally, we say that they come 'by chance.' 'Incidentally,' I say; for as a thing takes its name, and is what it is, in virtue of certain essential attributes, but incidentally has other attributes, not essential to it as that thing, so too it may be with causes. Thus the essential efficient cause of a house being built is the application of the builder-craft to the task by a builder; but if the man in whom that builder-craft is embodied is a pale-complexioned or cultivated person, then these characteristics are incidentally part and parcel with the direct cause of the building.[a] Thus direct causation is determinate and calculable, but incidental causation indeterminate; for one and the same person or thing may have an indefinite number of incidental qualifications.

As we said, then, what we mean by luck or chance (for at present we are attempting no distinction

196 b ματου καὶ ἀπὸ τύχης (αὐτῶν δὲ πρὸς ἄλληλα τὴν διαφορὰν τούτων ὕστερον διοριστέον, νῦν δὲ τοῦτο ἔστω φανερόν, ὅτι ἄμφω ἐν τοῖς ἕνεκά τού ἐστιν) οἷον ἕνεκα τοῦ ἀπολαβεῖν τὸ ἀργύριον ἦλθεν ἄν, κομιζομένου[1] τὸν ἔρανον, εἰ ᾔδει· ἦλθε δ' οὐ τούτου ἕνεκα, ἀλλὰ συνέβη αὐτῷ ἐλθεῖν καὶ ποιῆσαι τοῦτο τοῦ κομίσασθαι ἕνεκα· τοῦτο δὲ οὔθ' ὡς ἐπὶ τὸ 197 a πολὺ φοιτῶν εἰς τὸ χωρίον οὔτ' ἐξ ἀνάγκης· ἔστι δὲ τὸ τέλος—ἡ κομιδή—οὐ τῶν ἐν αὐτῷ αἰτίων, ἀλλὰ τῶν προαιρετῶν καὶ ἀπὸ διανοίας. καὶ λέγεταί γε τότε ἀπὸ τύχης ἐλθεῖν· εἰ δὲ προελόμενος καὶ τούτου ἕνεκα, ἢ ἀεὶ φοιτῶν, ἢ ὡς ἐπὶ τὸ πολὺ κομιζόμενος, οὐκ ἀπὸ τύχης. δῆλον ἄρα ὅτι ἡ τύχη αἰτία κατὰ συμβεβηκὸς ἐν τοῖς κατὰ προαίρεσιν τῶν ἕνεκά του. διὸ περὶ τὸ αὐτὸ

[1] [The reading κομιζομένου is guaranteed by the ancient commentaries. E_2 (according to Prantl) has κομισαμένου. The variants κομιζόμενος, κομισόμενος may be due to the notion that the same transaction is meant as by κομίσασθαι (l. 36). —C.]

---

[a] [The definite article ('*the* sum subscribed') can be explained by supposing that Aristotle, as often, is alluding to a scene in some well-known comedy, already used as an illustration by those who denied the existence of chance, 196 a 1. —C.]

between different terms within this general purport) is the incidental production of some significant result by a cause that took its place in the causal chain incidentally, and without the result in question being contemplated. I call actions purpose-serving, and speak of them as accomplishing ' ends,' when the result is such as *would have been* recognized as a purpose and *would have* determined the action, had it been anticipated. Thus the man who came to the market-place for some other reason would have come there on purpose to recover his debt if he had known that he would there meet his debtor in the act of receiving the [a] sum subscribed by his friends; and though he did not come for that reason, yet his coming there incidentally made for that end, though directed to another. But we must suppose that he did not habitually go to the place (frequented occasionally by his debtor), and still less that he was compelled to go there habitually on some other business; so that in his case the achieved end of recovering his debt was a result not normally involved in his action, but was yet of the class of things that may be deliberately determined upon and purposed. In this case he would in fact be said to have come there 'by luck'; whereas, if that had been the purpose he contemplated, or if he always went to market, or if he generally recovered a debt when he did, we should not say that the result came by luck. Clearly then luck itself, regarded as a cause, is the name we give to causation which incidentally inheres in deliberately purposeful action taken with respect to some other end but leading to the event we call fortunate. And the significant results of such causes we say 'come by luck.' Thus, since

197 a διάνοια καὶ τύχη· ἡ γὰρ προαίρεσις οὐκ ἄνευ διανοίας.

Ἀόριστα μὲν οὖν τὰ αἴτια ἀνάγκη εἶναι, ἀφ' ὧν ἂν γένοιτο τὸ ἀπὸ τύχης. ὅθεν καὶ ἡ τύχη τοῦ ἀορίστου εἶναι δοκεῖ καὶ ἄδηλος ἀνθρώπῳ, καὶ ἔστιν ὡς οὐδὲν ἀπὸ τύχης δόξειεν ἂν γίγνεσθαι. πάντα γὰρ ταῦτα ὀρθῶς λέγεται, ὅτι εὐλόγως. ἔστι μὲν γὰρ ὡς γίνεται ἀπὸ τύχης· κατὰ συμβεβηκὸς γὰρ γίνεται, καὶ ἔστιν αἴτιον ὡς συμβεβηκὸς ἡ τύχη, ὡς δ' ἁπλῶς οὐδενός· οἷον οἰκίας οἰκοδόμος μὲν αἴτιος κατὰ συμβεβηκὸς δὲ αὐλητής, καὶ τοῦ ἐλθόντα κομίσασθαι τὸ ἀργύριον, μὴ τούτου ἕνεκα ἐλθόντα, ἄπειρα τὸ πλῆθος (καὶ γὰρ ἰδεῖν τινα βουλόμενος καὶ διώκων καὶ φεύγων καὶ θεασάμενος).[1] καὶ τὸ φάναι εἶναί τι παράλογον τὴν τύχην ὀρθῶς· ὁ γὰρ λόγος ἢ τῶν ἀεὶ ὄντων ἢ τῶν ὡς ἐπὶ τὸ πολύ, ἡ δὲ τύχη ἐν τοῖς γιγνομένοις

[1] [The translator omitted καὶ θεασάμενος, which is not in E, but was read by Simplic. (lemma 341. 19) and Themistius (53. 9). FI have καὶ θεασόμενος (before καὶ φεύγων,) of which another trace appears in Simplicius's paraphrase (340. 26) καὶ θέαν τινὰ ὀψόμενος.—C.]

---

[a] [This serves to distinguish τύχη from ταὐτόματον. τύχη is to purposive thought as ταὐτόματον is to the unconscious purposiveness of Nature (Ross on *Met.* 1065 a 31).—C.]

choice implies intention, it follows that luck and intention are concerned with the same field of objects.[a]

The incidentally causative forces are, in the nature of the case, indefinite as to number. This is why luck appears to have something evasive about it and to be inscrutable by man; and why, on the other hand (since everything really *has* a definite cause) there is a sense in which it might seem that nothing at all really goes by luck or chance. For all these opinions have some justification in the facts; inasmuch as there is a sense in which things do go by luck, when they come to pass incidentally to some other chain of causation, and 'luck' is the name we give to causes that act incidentally; but in the absolute sense, without this qualification, luck is not the cause of anything. For instance, the builder, as such, is the efficient cause of the house being built, but his skill in flute-playing is incidental, and therefore might equally well attach itself to any other man, not a builder. In like manner the man's meeting his debtor was a by-product incidentally determined by the purpose, whatever it was, that brought him to the market-place; and it would have attached itself in the same incidental way to any other of the countless reasons that might have brought him there at that moment, such as the desire to meet a friend, or legal business as prosecutor or defendant. It is only if he had come on purpose to meet his debtor that his meeting him would have followed not incidentally but primarily from his purposeful action. And this is why we are justified in saying that luck cannot be calculated; for we can calculate only from necessary or normal sequences, and luck acts outside such. So

197 a παρὰ ταῦτα. ὥστ' ἐπειδὴ ἀόριστα τὰ οὕτως αἴτια, καὶ ἡ τύχη ἀόριστον.

Ὅμως δ' ἐπ' ἐνίων ἀπορήσειεν ἄν τις, ἆρ' οὖν τὰ τυχόντα αἴτι' ἂν γένοιτο τῆς τύχης. οἷον ὑγιείας ἢ πνεῦμα ἢ εἵλησις ἀλλ' οὐ τὸ ἀποκεκάρθαι· ἔστι γὰρ ἄλλα ἄλλων ἐγγύτερα τῶν κατὰ συμβεβηκὸς αἰτίων.

Τύχη δὲ ἀγαθὴ μὲν λέγεται ὅταν ἀγαθόν τι ἀποβῇ, φαύλη δὲ ὅταν φαῦλόν τι, εὐτυχία δὲ καὶ δυστυχία ὅταν μέγεθος ἔχοντα ταῦτα· διὸ καὶ τὸ παρὰ μικρὸν κακὸν ἢ ἀγαθὸν μέγα λαβεῖν ἢ εὐτυχεῖν ἢ ἀτυχεῖν ἐστιν, ὅτι ὡς ὑπάρχον λέγει ἡ διάνοια· τὸ γὰρ παρὰ μικρὸν ὥσπερ οὐδὲν ἀπέχειν δοκεῖ.

Ἔτι ἀβέβαιον ἡ εὐτυχία εὐλόγως· ἡ γὰρ τύχη ἀβέβαιος· οὔτε γὰρ ἀεὶ οὔθ' ὡς ἐπὶ τὸ πολὺ οἷόν τ' εἶναι τῶν ἀπὸ τύχης οὐθέν.

Ἔστι μὲν οὖν ἄμφω αἴτια (καθάπερ εἴρηται) κατὰ συμβεβηκός—καὶ ἡ τύχη καὶ τὸ αὐτόματον—ἐν τοῖς ἐνδεχομένοις γίγνεσθαι μὴ ἁπλῶς μηδ' ὡς ἐπὶ τὸ πολύ, καὶ τούτων ὅσ' ἂν γένοιτο ἕνεκά του.

---

[a] [*Literally*, 'whether any and every accidental cause can be considered responsible for the lucky event.'—C.]

[b] [Philoponus and Simplicius interpret as follows: A sick man *happens* to have his hair cut (not as part of his treatment, but for the usual reason), and the cold wind or hot sun to which he is consequently exposed *happen* to cure him. Shall we say that this lucky result is due to the nearer accidents (heat or cold) and not to the more remote (hair-cutting)? The question is not answered (Simplic.).—C.]

[c] ['Purposeful action' includes here the unconscious purposiveness of Nature, and so covers *ταὐτόματον*. This last sentence would be better placed at the beginning of the

the indeterminate nature of these incidental lines of causation makes luck indeterminate.

But when we have said all this, the question still remains in some cases how far we are to carry back our search for this incidental causation.[a] If a man has his head shaved for some special treatment,[b] and afterwards goes out for some reason indifferent to his cure, and the air and sun cure him, is the exposure to air and sun (which is the efficient cause of his cure) incident to his going out? Or must we go further back and say that it is incident to his being shaved, without reference to going out, his going out itself being without reference to the cure? The only answer to such questions is that, here as elsewhere, 'incidence' may be more or less proximate or remote.

We speak of 'good luck' when luck brings us something good, and 'bad luck' in the opposite event, or, in serious cases, of 'good fortune' or 'misfortune'; and accordingly if we just miss some important good thing or just escape a bad one, we call that also bad or good fortune, because by anticipation we regard the good or ill as having been actually present, so close did it seem.

Yet again, we may well say that good fortune is unstable; for so is all luck, inasmuch as nothing that is constant or normal can be attributed to luck.

Both luck and chance, then, as we have said, are causes that come into play incidentally and produce effects that possibly, but not necessarily or generally, follow from the purposeful action to which in this case they are incident, though the action might have been taken directly and primarily for their sake.[c]

next chapter, which introduces the distinction between *τύχη* and *ταὐτόματον*.—C.]

CHAPTER VI

ARGUMENT

As the greater part of this chapter is devoted to the definition of two Greek words to which there are no exact English equivalents, I have preserved the Greek words themselves and only attempted to translate the commentary. If the English reader cares to substitute ' luck ' or ' fortune ' for tyche, *and ' accident ' or ' chance result ' for* automaton,[a] *it may make it easier to understand the general drift of Aristotle's analysis, but he must not expect it to fit the English words with any approach to precision.*

Note that Aristotle, swayed by a false etymology, inclines throughout to treat the incidentally emergent result as superseding the intended one, rather than as additionally accruing to it. And this is the more remarkable inasmuch as his examples and illustrations signally fail to conform to this prepossession ; since they are drawn either from human actions that may very easily achieve their contemplated result as well as the uncontemplated one that incidentally emerges ;

197 a 36 Διαφέρει δ' ὅτι τὸ αὐτόματον ἐπὶ πλεῖόν ἐστι· τὸ μὲν γὰρ ἀπὸ τύχης πᾶν ἀπὸ ταὐτομάτου, τοῦτο
197 b δ' οὐ πᾶν ἀπὸ τύχης.

Ἡ μὲν γὰρ τύχη καὶ τὸ ἀπὸ τύχης ἐστὶν ὅσοις καὶ τὸ εὐτυχῆσαι ἂν ὑπάρξειεν καὶ ὅλως πρᾶξις. διὸ καὶ ἀνάγκη περὶ τὰ πρακτὰ εἶναι τὴν τύχην —σημεῖον δ' ὅτι δοκεῖ ἤτοι ταὐτὸν εἶναι τῇ εὐδαιμονίᾳ ἡ εὐτυχία ἢ ἐγγύς· ἡ δ' εὐδαιμονία πρᾶξίς

[a] Automaton may be translated by ' of itself' ; *e.g.*, ' it happened of itself.'

[b] [Aristotle refers both to the etymological sense of *eudaemonia*, ' having a good *daemon* (destiny, fortune) ' and to his own doctrine that happiness consists in the highest *activity* (πρᾶξις, ἐνέργεια) of the soul, conditioned by virtue.

CHAPTER VI

ARGUMENT *(continued)*

or else from natural processes which, as such, are in no way deflected by their incidental results to human beings, however casual or unusual. It is only quite at the end of the analysis, as a sort of supplement, that he notices, as a different kind of thing from what he has been investigating, the case of a warp in nature, where the fulfilment of the normal 'intention' of the process is really precluded by the incidental deflection.

In the concluding section of the chapter Aristotle unmasks his battery, and we see why he has been so insistent on chance being always incident to purposeful activities, whether of man or of Nature, and on its always resulting in something that might have been directly aimed at. It is because it enables him to bring out the necessary antecedence of purposeful to casual causation, by insisting that, as substantive existence must always underlie attributive or relational existence, so direct must always underlie incidental causation, direct causation being always purposeful.

In Greek *tyche* and *automaton* differ in this, that *automaton* is the more general term and includes *tyche* as a special class.

For (1) *tyche* itself, as a cause, and the results that accrue by the action of *tyche*, are only spoken of in connexion with beings capable of enjoying good fortune, or more generally of 'doing well' or 'doing ill,' in the sense either of 'faring' or of 'acting' so. Therefore *tyche* must always be connected with our doings and farings—a truth indicated by the common belief that good fortune (*tyche*) is the same, or much the same, thing as 'happiness' [b]; and to be 'happy'

in beings capable of deliberate choice and moral *conduct*.—C.]

197 b τις (εὐπραξία γὰρ)—ὥσθ' ὁπόσοις μὴ ἐνδέχεται πρᾶξαι, οὐδὲ τὸ ἀπὸ τύχης τι ποιῆσαι. καὶ διὰ τοῦτο οὔτε ἄψυχον οὐδὲν οὔτε θηρίον οὔτε παιδίον οὐδὲν ποιεῖ ἀπὸ τύχης, ὅτι οὐκ ἔχει προαίρεσιν· οὐδ' εὐτυχία οὐδ' ἀτυχία ὑπάρχει τούτοις, εἰ μὴ καθ' ὁμοιότητα, ὥσπερ ἔφη Πρώταρχος εὐτυχεῖς εἶναι τοὺς λίθους ἐξ ὧν οἱ βωμοί, ὅτι τιμῶνται, οἱ δὲ ὁμόζυγες αὐτῶν καταπατοῦνται. τὸ δὲ πάσχειν ἀπὸ τύχης ὑπάρξει πως καὶ τούτοις, ὅταν ὁ πράττων τι περὶ αὐτὰ πράξῃ ἀπὸ τύχης· ἄλλως δὲ οὐκ ἔστιν.

Τὸ δ' αὐτόματον καὶ τοῖς ἄλλοις ζῴοις καὶ πολλοῖς τῶν ἀψύχων· οἷον ὁ ἵππος αὐτόματος (φαμέν) ἦλθεν, ὅτι ἐσώθη μὲν ἐλθών, οὐ τοῦ σωθῆναι δὲ ἕνεκα ἦλθεν. καὶ ὁ τρίπους αὐτόματος κατέπεσεν· ἔστη μὲν γὰρ τοῦ καθῆσθαι ἕνεκα, ἀλλ' οὐ τοῦ καθῆσθαι ἕνεκα κατέπεσεν.

Ὥστε φανερὸν ὅτι ἐν τοῖς ἁπλῶς ἕνεκά του γινομένοις, ὅταν μὴ τοῦ συμβάντος ἕνεκα γένηται οὗ ἔξω τὸ αἴτιον, τότε ἀπὸ ταὐτομάτου λέγομεν· ἀπὸ τύχης δέ, τούτων ὅσα ἀπὸ ταὐτομάτου γίνεται τῶν προαιρετῶν τοῖς ἔχουσι προαίρεσιν.

---

[a] [Perhaps the Protarchus, son of Callias, who appears in Plato's *Philebus*. (Wilamowitz, *Platon*, i. p. 629.)—C.]

[b] [*Or* 'we say the horse came "of itself" (spontaneously)' —though, in Aristotle's view, the cause in this case was *external* to the horse.—C.]

[c] [*Literally*, 'It is clear that, in the field of occurrences which are in a general way for some purpose, we use the expression 'chance result' in any case where a thing of which the cause is external occurs not for the sake of the

is to have 'done well' in life; so that 'doing well' or 'ill' by *tyche* is impossible to creatures that have no self-direction. That is why neither inanimate things nor brute beasts nor infants can ever accomplish anything by *tyche*, since they exercise no deliberate choice; nor can such be said to have good or bad *tyche*, except by a figure of speech, as when Protarchus [a] speaks of the 'fortunate' stones that have been built into altars and are treated with reverence, while their fellows are trampled under foot. But even such things may be brought, as passive agents, under the action of *tyche*, if a rational agent does something with them which turns out by *tyche* to affect his well-being; but not otherwise.

(2) *Automaton*, on the other hand, may be used to describe the behaviour of brute beasts and even of many inanimate things. For instance, we attribute it to *automaton* [b] if a horse escapes a danger by coming accidentally to a place of safety. Or again, if a tripod chances to fall on its feet for a man to sit down upon, this is due to *automaton*, for though a man would put it on its feet *with a view to* its being a seat, the forces of Nature that controlled its fall had no such aim.

It is clear, then, that when *any* causal agency incidentally produces a significant result outside its aim, we attribute it to *automaton* [c]; and in the special cases where such a result springs from deliberate action (though not aimed at it) on the part of a being capable of choice, we may say that it comes by *tyche*.

actual result' (as in the instance of the horse, which was not brought to a place of safety by an *internal* purpose of its own).—C.]

197b (Σημεῖον δὲ τὸ μάτην, ὅτι λέγεται ὅταν μὴ γένηται τὸ οὗ ἕνεκα ἀλλ' ὃ ἐκείνου ἕνεκα,[1] οἷον τὸ βαδίσαι λαπάξεως ἕνεκά ἐστιν· εἰ δὲ μὴ ἐγένετο βαδίσαντι, μάτην φαμὲν βαδίσαι καὶ ἡ βάδισις ματαία, ὡς τοῦτο ὂν τὸ μάτην, τὸ πεφυκὸς ἄλλου ἕνεκα, ὅταν μὴ περαίνῃ ἐκεῖνο οὗ ἕνεκα ἐπεφύκει· ἐπεὶ εἴ τις λούσασθαι φαίη μάτην ὅτι οὐκ ἐξέλιπεν ὁ ἥλιος, γελοῖος ἂν εἴη· οὐ γὰρ ἦν τοῦτο ἐκείνου ἕνεκα. οὕτω δὴ τὸ αὐτόματον καὶ κατὰ τὸ ὄνομα, ὅταν αὐτὸ μάτην γένηται· κατέπεσε γὰρ οὐ τοῦ πατάξαι ἕνεκα ὁ λίθος· ἀπὸ τοῦ αὐτομάτου ἄρα κατέπεσεν ὁ λίθος, ὅτι πέσοι ἂν ὑπό τινος καὶ τοῦ πατάξαι ἕνεκα.)

Μάλιστα δ' ἐστὶ χωριζόμενον τοῦ ἀπὸ τύχης ἐν τοῖς φύσει γινομένοις· ὅταν γὰρ γένηταί τι παρὰ φύσιν, τότε οὐκ ἀπὸ τύχης ἀλλὰ μᾶλλον ἀπὸ ταὐτομάτου γεγονέναι φαμέν. ἔστι δὲ καὶ τοῦτο ἕτερον· τοῦ μὲν γὰρ ἔξω τὸ αἴτιον, τοῦ δ' ἐντός.

[1] [τὸ οὗ ἕνεκα ἀλλ' ὃ ἐκείνου ἕνεκα, a reading recorded by Simplicius (349. 6): τὸ ἕνεκα ἄλλου ἐκείνου ἕνεκα codd.—C.]

---

[a] [Prellwitz (*Etym. Wörterbuch*, 1905) supports the derivation from μάτην. Boisacq (*Dict. Etym.* 1910) prefers a root meaning 'think': 'one who thinks (and acts) for himself.'—C.]

[b] ['To no purpose' has the same ambiguity as μάτην: (1) *failure* of purpose (walking which fails to produce evacuation); (2) *absence* of purpose (a stone falls with effect which might have been, but was not, purposed).—C.]

[c] [*Literally*, 'the distinction (of the chance result, *automaton*) from the lucky result is best seen. . . .' The connexion of thought (broken by the parenthesis on the etymology of *automaton*) is with the last paragraph but one: the chance result with an *external* cause. If we might be tempted to call such a result 'lucky,' we can see the dis-

(The etymology of *automaton* indicates this[a]; for the expression *maten*—'for nothing,' 'to no purpose'[b] —is used in cases where the end or purpose is not realized, but only the means to it. Walking, for instance, is a means to evacuation; if a man takes a walk and this natural effect does not follow, we say that he took his walk 'for nothing' and the walk was 'to no purpose,' meaning by this phrase what fails to accomplish that purpose to which it is naturally a means. 'Naturally a means,' for it would be ridiculous to say one had taken a bath 'to no purpose' because an eclipse of the sun did not follow: an eclipse is not a natural consequence of taking a bath. So then *automaton*, as the form of the word implies, means an occurrence that is *in itself* (*auto*) *to no purpose* (*maten*). A stone falls and hits someone, but it does not fall for the purpose of hitting him; the fall accordingly was 'in-itself-to-no-purpose'—a chance result—because the fall might have been caused by someone who had the purpose of hitting the man.)

It would be most inappropriate of all to speak of *tyche*[c] in cases where Nature herself produces unnatural monstrosities; and accordingly in these cases we may attribute it to an 'accident' (*automaton*) in Nature, but can hardly say that a piece of bad 'luck' (*tyche*) has come to her. But this case is different from that of the horse; for the horse's escape was due to an external cause, but the causes of Nature's miscarriage are internal to her own processes.

tinction better by taking the case of monstrosities, where the cause is *internal*, but the effect is anything but 'lucky' or such as Nature might have purposed; for the purpose of Nature is actually thwarted (παρὰ φύσιν) by a monstrous birth.—C.]

198 a Τί μὲν οὖν ἐστι τὸ αὐτόματον καὶ τί ἡ τύχη, εἴρηται, καὶ τί διαφέρουσιν ἀλλήλων. τῆς δ' αἰτίας τῶν τρόπων ἐν τοῖς ὅθεν ἡ ἀρχὴ τῆς κινήσεως ἑκάτερον αὐτῶν· ἢ γὰρ τῶν φύσει τι ἢ τῶν ἀπὸ διανοίας αἴτιων ἀεί ἐστιν· ἀλλὰ τούτων τὸ πλῆθος ἀόριστον.

Ἐπεὶ δ' ἐστὶ τὸ αὐτόματον καὶ ἡ τύχη αἴτια ὧν ἂν ἢ νοῦς γένοιτο αἴτιος ἢ φύσις, ὅταν κατὰ συμβεβηκὸς αἴτιόν τι γένηται τούτων αὐτῶν, οὐδὲν δὲ κατὰ συμβεβηκός ἐστι πρότερον τῶν καθ' αὑτό, δῆλον ὅτι οὐδὲ τὸ κατὰ συμβεβηκὸς αἴτιον πρότερον τοῦ καθ' αὑτό. ὕστερον ἄρα τὸ αὐτόματον καὶ ἡ τύχη καὶ νοῦ καὶ φύσεως· ὥστ' εἰ ὅτι μάλιστα τοῦ οὐρανοῦ αἴτιον τὸ αὐτόματον, ἀνάγκη πρότερον νοῦν καὶ φύσιν αἰτίαν εἶναι καὶ ἄλλων πολλῶν καὶ τοῦδε τοῦ παντός.

CHAPTER VII

ARGUMENT

This chapter, together with much repetition, points out that, whereas the efficient cause must to some extent coincide with the 'formal' determinant of the result, the coincidence between the formal determinant and the 'final' cause appears to be complete. It would seem, therefore, that the final 'because' might be left out of consideration, as already included in the formal 'because.' And, as a matter of fact, natural philosophers do generally concern themselves with the starting-point, the actual result, and the moving forces that push from the one to the other. Nevertheless 'form,' considered under its aspect of 'aim,' is a distinct factor in causation that can least of all be neglected, for 'purpose' is the ultimate 'because' of all physical change.

What *automaton* and *tyche* are, and how they differ, has now been stated; and it follows that both are causes of the 'efficient' order, that set processes in motion; for they are always attached to efficient causes either of the natural or volitional order, such attachments being indefinite in number.

And since the results of *automaton* and *tyche* are always such as might have been aimed at by mind or Nature, though in fact they emerged incidentally, and since there can be nothing incidental unless there is something primary for it to be incidental to, it follows that there can be no incidental causation except as incident to direct causation. Chance and fortune, therefore, imply the antecedent activity of mind and Nature as causes; so that, even if the cause of the heavens were ever so casual, yet mind and Nature must have been causes antecedently, not only of many other things we could mention, but of this universe itself.

CHAPTER VII

ARGUMENT (*continued*)

Since this question of aim is the point at which Physics leads directly up to Metaphysics or Theology, some references are here introduced to this higher branch of philosophy (parenthetically, to the disturbance of the main progress of the discourse, and with repetitions that suggest the confluence of two distinct recensions); and a new classification is incidentally introduced, constituted by Metaphysics (dealing with the immaterial and unchanging) on the one hand, and two departments of Physics on the other hand, dealing, respectively, the one with Astronomy (the study of things that move, indeed, but are not subject to other change or to decay), and the other covering all sublunary phenomena (in which both decay and generation reign).

198 a 14 Ὅτι δ᾿ ἔστιν αἴτια, καὶ ὅτι τοσαῦτα τὸν ἀριθμὸν ὅσα φαμέν, δῆλον. τοσαῦτα γὰρ τὸν ἀριθμὸν τὸ διὰ τί περιείληφεν· ἢ γὰρ εἰς τὸ τί ἐστιν ἀνάγεται τὸ διὰ τί ἔσχατον ἐν τοῖς ἀκινήτοις (οἷον ἐν τοῖς μαθήμασιν· εἰς ὁρισμὸν γὰρ τοῦ εὐθέος ἢ συμμέτρου ἢ ἄλλου τινὸς ἀνάγεται ἔσχατον), ἢ εἰς τὸ κινῆσαν πρῶτον (οἷον διὰ τί ἐπολέμησαν; ὅτι ἐσύλησαν), ἢ τίνος ἕνεκα; (ἵν᾿ ἄρξωσιν), ἢ ἐν τοῖς γινομένοις ἡ ὕλη.

Ὅτι μὲν οὖν τὰ αἴτια ταῦτα καὶ τοσαῦτα, φανερόν. ἐπεὶ δ᾿ αἱ αἰτίαι τέτταρες, περὶ πασῶν τοῦ φυσικοῦ εἰδέναι, καὶ εἰς πάσας ἀνάγων τὸ διὰ τί ἀποδώσει φυσικῶς—τὴν ὕλην, τὸ εἶδος, τὸ κινῆσαν, τὸ οὗ ἕνεκα. ἔρχεται δὲ τὰ τρία εἰς ἓν πολλάκις· τὸ μὲν γὰρ τί ἐστι καὶ τὸ οὗ ἕνεκα ἕν ἐστι, τὸ δ᾿ ὅθεν ἡ κίνησις πρῶτον τῷ εἴδει ταὐτὸ τούτοις· ἄνθρωπος γὰρ ἄνθρωπον γεννᾷ. καὶ ὅλως ὅσα κινούμενα κινεῖ· (ὅσα δὲ μή, οὐκέτι φυσικῆς· οὐ γὰρ ἐν αὑτοῖς ἔχοντα κίνησιν οὐδ᾿ ἀρχὴν κινήσεως κινεῖ,

---

[a] [The illustration is more fully stated in *Anal. post.* 94 a 36. The Athenians were involved in war with Persia because they raided Sardis with the Eretrians.—C.]

[b] [*Literally*, ' and in referring to them all, he will be answering the question " why ? " as a natural philosopher ' (keeping within his province).—C.]

It is clear, then, that there are such things as causes, and that they can be classified under the four heads that have been enumerated. For these are the four ways of apprehending the 'how and why' of things: we may refer it either (1) to the essential nature of the thing in question, in the sphere of unchanging objects (as in mathematics, where the conclusions ultimately depend upon the definitions of straight line, or commensurability, or whatever it may be); or (2) to that which first initiated the movement (as in: 'Why did they go to war? Because the others had raided them')[a]; or (3) to the result aimed at (as: 'To gain an empire'); or (4) in the case of anything that comes into existence out of something already there, to the material.

Clearly, then, the 'becauses' being such and so classified, it behoves the natural philosopher to understand all four, and to be able to indicate, in answer to the question 'how and why,' the material, the form, the moving force, and the goal or purpose, so far as they come within the range of Nature.[b] But in many cases three of these 'becauses' coincide; for the essential nature of a thing and the purpose for which it is produced are often identical (so that the final cause coincides with the formal), and moreover the efficient cause must bear some resemblance in 'form' to the effect (so that the efficient cause too must, so far, coincide with the formal); for instance, man is begotten by man. And this applies universally to all things that cause motion and are themselves moved. (Where that is not the case, we are no longer in the domain of Physics at all, since we are dealing, not with things that move other things in virtue of their own motion or because

ἀλλ᾽ ἀκίνητα ὄντα. διὸ τρεῖς αἱ πραγματεῖαι, ἡ μὲν περὶ ἀκίνητον, ἡ δὲ περὶ κινούμενον μὲν ἄφθαρτον δέ, ἡ δὲ περὶ τὰ φθαρτά). ὥστε τὸ διὰ τί καὶ εἰς τὴν ὕλην ἀνάγοντι ἀποδίδοται, καὶ εἰς τὸ τί ἐστι, καὶ εἰς τὸ πρῶτον κινῆσαν. περὶ γενέσεως γὰρ μάλιστα τοῦτον τὸν τρόπον τὰς αἰτίας σκοποῦσι, τί μετὰ τί γίνεται, καὶ τί πρῶτον ἐποίησεν ἢ τί ἔπαθε, καὶ οὕτως ἀεὶ τὸ ἐφεξῆς.

Διτταὶ δὲ αἱ ἀρχαὶ αἱ κινοῦσαι φυσικῶς, ὧν ἡ ἑτέρα οὐ φυσική· οὐ γὰρ ἔχει κινήσεως ἀρχὴν ἐν αὑτῇ. τοιοῦτον δ᾽ ἐστὶν εἴ τι κινεῖ μὴ κινούμενον, ὥσπερ τό τε παντελῶς ἀκίνητον καὶ τὸ πάντων πρῶτον καὶ τὸ τί ἐστι καὶ ἡ μορφή· τέλος γὰρ καὶ οὗ ἕνεκα· ὥστε ἐπεὶ ἡ φύσις ἕνεκά του, καὶ ταύτην εἰδέναι δεῖ. καὶ πάντως ἀποδοτέον τὸ διὰ τί, οἷον ὅτι ἐκ τοῦδε ἀνάγκη τόδε (τὸ δὲ ἐκ τοῦδε ἢ ἁπλῶς ἢ ὡς ἐπὶ τὸ πολύ)· καὶ εἰ μέλλει τοδὶ ἔσεσθαι, ὥσπερ ἐκ τῶν προτάσεων τὸ συμπέρασμα·

[a] [See note on 195 a 18.—C.]

they have the principle of motion in themselves, but with something that moves other things, though itself motionless. So that we have three fields of inquiry, concerned respectively with (1) things motionless, (2) things that, though in motion, are imperishable, and (3) things perishable.) Thus to give an account of the how and why of anything is to trace it to its material, to its essential characteristics, and to its provoking cause ; for in investigating the genesis of a thing men are chiefly concerned with the nature of what emerges from the process, with the impulse that initiates the process, and with what was already there to undergo the process, at the start ; together with all the successive steps that lie between the starting-point and the end.

But the principles which direct physical movement or change are of two orders, one of which is not itself physical, for it is not in motion, nor has it in itself the principle of motion. Such would be anything that should move other things while itself motionless, as being absolutely unchanging and primary, and such the essential characteristic or form in its capacity of constituting the end and aim to be reached, and therefore, since Nature is purposeful, demanding to be recognized by the natural philosopher. In short, under all four aspects, we must give an account of the how and why, so as to show (1) that from this *efficient* cause this result must follow, or if the cause in question does not absolutely involve a certain definite result, we must show that it will lead to it normally. We must also show (2) that, if such and such a thing is to exist, there must be a *material* substrate, related to it as the premisses to the conclusion,[a] and (3) that the result manifests the

198b καὶ ὅτι τοῦτ' ἦν τὸ τί ἦν εἶναι· καὶ διότι βέλτιον οὕτως—οὐχ ἁπλῶς ἀλλὰ τὸ πρὸς τὴν ἑκάστου οὐσίαν.

CHAPTER VIII

ARGUMENT

[In support of final causality in Nature, Aristotle reviews and criticizes the tacit assumptions and explicit arguments of those who deny its existence ; in particular, Empedocles' theory of the origin of animal species. (Empedocles held (1) that in the period when a world is formed by the action of Love, gaining upon Strife, the parts of animals arise separately and are united by Love in all sorts of casual ways.

Λεκτέον δὴ πρῶτον μὲν διότι ἡ φύσις τῶν ἕνεκά του αἰτίων· ἔπειτα περὶ τοῦ ἀναγκαίου, πῶς ἔχει ἐν τοῖς φυσικοῖς. εἰς γὰρ ταύτην τὴν αἰτίαν ἀνάγουσι πάντες, ὅτι ἐπειδὴ τὸ θερμὸν τοιονδὶ πέφυκε καὶ τὸ ψυχρὸν καὶ ἕκαστον δὴ τῶν τοιούτων, ταδὶ ἐξ ἀνάγκης ἔστι καὶ γίνεται· καὶ γὰρ ἐὰν ἄλλην αἰτίαν εἴπωσιν, ὅσον ἁψάμενοι χαίρειν ἐῶσιν, ὁ μὲν τὴν φιλίαν καὶ τὸ νεῖκος, ὁ δὲ τὸν νοῦν.

Ἔχει δ' ἀπορίαν τί κωλύει τὴν φύσιν μὴ ἕνεκά του ποιεῖν μηδ' ὅτι βέλτιον ἀλλ' ὥσπερ ὕει ὁ Ζεύς, οὐχ ὅπως τὸν σῖτον αὐξήσῃ ἀλλ' ἐξ ἀνάγκης—τὸ γὰρ ἀναχθὲν ψυχθῆναι δεῖ καὶ τὸ ψυχθὲν ὕδωρ

essential nature aimed at by the process, and (4) why it was *better* thus—not absolutely, but relatively to the being of the thing in question.

CHAPTER VIII

ARGUMENT (*continued*)

Monsters like man-headed oxen, ox-headed men, etc., perish; those forms survive which happen to be fitted to survive. (2) *In the opposite period, when Strife is gaining on Love* (*as in our world*), *animal species are developed from undifferentiated* (' *whole-natured* ') *forms produced by fire rising from the earth* ' *to meet its like* ' *and mixing with water*).—C.]

We must now consider why Nature is to be ranked among causes that are final, that is to say purposeful; and further we must consider what is meant by ' necessity ' when we are speaking of Nature. For thinkers are for ever referring things to necessity as a cause, and explaining that, since hot and cold and so forth are what they are, this or that exists or comes into being ' of necessity ' ; for even if one or another of them alleges some other cause, such as ' Sympathy and Antipathy ' or ' Mind,' he straight away drops it again, after a mere acknowledgement.

So here the question rises whether we have any reason to regard Nature as making for any goal at all, or as seeking any one thing as preferable to any other. Why not say, it is asked, that Nature acts as Zeus drops the rain, not to make the corn grow, but of necessity (for the rising vapour must needs be condensed into water by the cold, and must then

198 b γενόμενον κατελθεῖν, τὸ δ' αὐξάνεσθαι τούτου γενομένου τὸν σῖτον συμβαίνει—· ὁμοίως δὲ καὶ εἴ τῳ ἀπόλλυται ὁ σῖτος ἐν τῇ ἅλῳ, οὐ τούτου ἕνεκα ὕει ὅπως ἀπόληται, ἀλλὰ τοῦτο συμβέβηκεν. ὥστε τί κωλύει οὕτω καὶ τὰ μέρη ἔχειν ἐν τῇ φύσει, οἷον τοὺς ὀδόντας ἐξ ἀνάγκης ἀνατεῖλαι τοὺς μὲν ἐμπροσθίους ὀξεῖς, ἐπιτηδείους πρὸς τὸ διαιρεῖν, τοὺς δὲ γομφίους πλατεῖς καὶ χρησίμους πρὸς τὸ λεαίνειν τὴν τροφήν, ἐπεὶ οὐ τούτου ἕνεκα γενέσθαι ἀλλὰ συμπεσεῖν; ὁμοίως δὲ καὶ περὶ τῶν ἄλλων μερῶν, ἐν ὅσοις δοκεῖ ὑπάρχειν τὸ ἕνεκά του. ὅπου μὲν οὖν ἅπαντα συνέβη ὥσπερ κἂν εἰ ἕνεκά του ἐγίνετο, ταῦτα μὲν ἐσώθη ἀπὸ τοῦ αὐτομάτου συστάντα ἐπιτηδείως· ὅσα δὲ μὴ οὕτως, ἀπώλετο καὶ ἀπόλλυται, καθάπερ Ἐμπεδοκλῆς λέγει τὰ βουγενῆ ἀνδρόπρωρα.

Ὁ μὲν οὖν λόγος, ᾧ ἄν τις ἀπορήσειεν, οὗτος, καὶ εἴ τις ἄλλος τοιοῦτός ἐστιν· ἀδύνατον δὲ τοῦτον ἔχειν τὸν τρόπον. ταῦτα μὲν γὰρ καὶ πάντα τὰ φύσει ἢ ἀεὶ οὕτω γίνεται ἢ ὡς ἐπὶ τὸ πολύ, τῶν δ' 199 a ἀπὸ τύχης καὶ τοῦ αὐτομάτου οὐδέν. οὐ γὰρ ἀπὸ τύχης οὐδ' ἀπὸ συμπτώματος δοκεῖ ὕειν πολλάκις τοῦ χειμῶνος, ἀλλ' ἐὰν ὑπὸ κύνα· οὐδὲ καύματα ὑπὸ κύνα, ἀλλ' ἂν χειμῶνος. εἰ οὖν ἢ ὡς ἀπὸ συμπτώματος δοκεῖ ἢ ἕνεκά του εἶναι, εἰ μὴ οἷόν τε ταῦτ' εἶναι μήτε ἀπὸ συμπτώματος μήτ' ἀπὸ

---

[a] [*i.e.*, if we attribute the bad result (destruction of the corn) to accident or necessity, why attribute the good result (growth of the crop) to benevolence ?—C.]

[b] [*Cf. De part. anim.* 661 b 7, Xen. *Mem.* i. 4, where the shape of the teeth is alleged among other evidences of design.

descend, and incidentally, when this happens, the corn grows), just as, when a man loses his corn on the threshing-floor, it did not rain on purpose to destroy the crop, but the result was merely incidental to the raining?[a] So why should it not be the same with natural organs like the teeth? Why should it not be a coincidence that the front teeth come up with an edge, suited to dividing the food, and the back ones flat and good for grinding it, without there being any design in the matter?[b] And so with all other organs that seem to embody a purpose. In cases where a coincidence brought about such a combination as might have been arranged on purpose, the creatures, it is urged, having been suitably formed by the operation of chance, survived; otherwise they perished, and still perish, as Empedocles says of his 'man-faced oxen.'[c]

Such and suchlike are the arguments which may be urged in raising this problem; but it is impossible that this should really be the way of it. For all these phenomena and all natural things are either constant or normal, and this is contrary to the very meaning of luck or chance. No one assigns it to chance or to a remarkable coincidence if there is abundant rain in the winter, though he would if there were in the dog-days; and the other way about, if there were parching heat. Accordingly, if the only choice is to assign these occurrences either to coincidence or to purpose, and if in these cases chance coincidence is out of the question, then it must be pur-

W. Theiler, *Zur Geschichte der teleologischen Naturbetrachtung* (Zürich u. Leipzig, 1925), p. 25, argues that both Xenophon and Aristotle borrowed from Diogenes of Apollonia. —C.]

[c] [*Frag.* 61.—C.]

199 a ταὐτομάτου, ἕνεκά του ἂν εἴη. ἀλλὰ μὴν φύσει γ' ἐστὶ τὰ τοιαῦτα πάντα, ὡς κἂν αὐτοὶ φαῖεν οἱ ταῦτα λέγοντες. ἔστιν ἄρα τὸ ἕνεκά του ἐν τοῖς φύσει γινομένοις καὶ οὖσιν.

Ἔτι ἐν ὅσοις τέλος ἐστί τι, τούτου ἕνεκα πράττεται τὸ πρότερον καὶ τὸ ἐφεξῆς. οὐκοῦν ὡς πράττεται, οὕτω πέφυκε, καὶ ὡς πέφυκεν, οὕτω πράττεται ἕκαστον, ἂν μή τι ἐμποδίζῃ. πράττεται δ' ἕνεκά του· καὶ πέφυκεν ἄρα τούτου ἕνεκα. οἷον εἰ οἰκία τῶν φύσει γινομένων ἦν, οὕτως ἂν ἐγίνετο ὡς νῦν ὑπὸ τῆς τέχνης· εἰ δὲ τὰ φύσει μὴ μόνον φύσει ἀλλὰ καὶ τέχνῃ γίγνοιτο, ὡσαύτως ἂν γίνοιτο ᾗ πέφυκεν. ἕνεκα ἄρα θατέρου θάτερον. ὅλως τε ἡ τέχνη τὰ μὲν ἐπιτελεῖ ἃ ἡ φύσις ἀδυνατεῖ ἀπεργάσασθαι, τὰ δὲ μιμεῖται. εἰ οὖν τὰ κατὰ τὴν τέχνην ἕνεκά του, δῆλον ὅτι καὶ τὰ κατὰ τὴν φύσιν· ὁμοίως γὰρ ἔχει πρὸς ἄλληλα ἐν τοῖς κατὰ τέχνην καὶ ἐν τοῖς κατὰ φύσιν τὰ ὕστερα πρὸς τὰ πρότερα.

Μάλιστα δὲ φανερὸν ἐπὶ τῶν ζῴων τῶν ἄλλων, ἃ οὔτε τέχνῃ οὔτε ζητήσαντα οὔτε βουλευσάμενα ποιεῖ· ὅθεν διαποροῦσί τινες πότερον νῷ ἢ τινι ἄλλῳ ἐργάζονται οἵ τ' ἀράχναι καὶ οἱ μύρμηκες καὶ τὰ τοιαῦτα. κατὰ μικρὸν δ' οὕτω προϊόντι

---

[a] [At *Met.* 980 b 22, bees and other such species are called 'intelligent' (φρόνιμα) but unteachable, as having no hearing. *De part. an.* 648 a 5, Bees, etc., are more intelligent than some sanguineous animals, because of the thin and cold nature of the fluid corresponding to blood in them. 'Some similar faculty,' perhaps the unreasoning intuition of instinct, which resembles the human faculty of intuition (νόησις).—C.]

pose. But, as our opponents themselves would admit, these occurrences are all natural. There is purpose, then, in what is, and in what happens, in Nature.

Further, in any operation of human art, where there is an end to be achieved, the earlier and successive stages of the operation are performed for the purpose of realizing that end. Now, when a thing is produced by Nature, the earlier stages in every case lead up to the final development in the same way as in the operation of art, and *vice versa*, provided that no impediment balks the process. The operation is directed by a purpose; we may, therefore, infer that the natural process was guided by a purpose to the end that is realized. Thus, if a house were a natural product, the process would pass through the same stages that it in fact passes through when it is produced by art; and if natural products could also be produced by art, they would move along the same line that the natural process actually takes. We may therefore say that the earlier stages are for the purpose of leading to the later. Indeed, as a general proposition, the arts either, on the basis of Nature, carry things further than Nature can, or they imitate Nature. If, then, artificial processes are purposeful, so are natural processes too; for the relation of antecedent to consequent is identical in art and in Nature.

This principle comes out most clearly when we consider the other animals. For their doings are not the outcome of art (design) or of previous research or deliberation; so that some raise the question whether the works of spiders and ants and so on should be attributed to intelligence or to some similar faculty.[a] And then, descending step by

καὶ ἐν τοῖς φυτοῖς φαίνεται τὰ συμφέροντα γινόμενα πρὸς τὸ τέλος, οἷον τὰ φύλλα τῆς τοῦ καρποῦ ἕνεκα σκέπης. ὥστ᾽ εἰ φύσει τε ποιεῖ καὶ ἕνεκά του ἡ χελιδὼν τὴν νεοττιὰν καὶ ὁ ἀράχνης τὸ ἀράχνιον, καὶ τὰ φυτὰ τὰ φύλλα ἕνεκα τῶν καρπῶν καὶ τὰς ῥίζας οὐκ ἄνω ἀλλὰ κάτω ἕνεκα τῆς τροφῆς, φανερὸν ὅτι ἔστιν ἡ αἰτία ἡ τοιαύτη ἐν τοῖς φύσει γινομένοις καὶ οὖσιν.

Καὶ ἐπεὶ ἡ φύσις διττή, ἡ μὲν ὡς ὕλη ἡ δ᾽ ὡς μορφή, τέλος δ᾽ αὕτη, τοῦ τέλους δ᾽ ἕνεκα τἆλλα, αὕτη ἂν εἴη ἡ αἰτία ἡ οὗ ἕνεκα.

Ἁμαρτία δὲ γίγνεται καὶ ἐν τοῖς κατὰ τέχνην—ἔγραψε γὰρ οὐκ ὀρθῶς ὁ γραμματικός, καὶ ἐπότισεν οὐκ ὀρθῶς ὁ ἰατρὸς τὸ φάρμακον—ὥστε δῆλον ὅτι ἐνδέχεται καὶ ἐν τοῖς κατὰ φύσιν. εἰ δὴ ἔστιν ἔνια κατὰ τέχνην ἐν οἷς τὸ ὀρθῶς ἕνεκά του, ἐν δὲ τοῖς ἁμαρτανομένοις ἕνεκα μέν τινος ἐπιχειρεῖται ἀλλ᾽ ἀποτυγχάνεται, ὁμοίως ἂν ἔχοι καὶ ἐν τοῖς φυσικοῖς, καὶ τὰ τέρατα ἁμαρτήματα ἐκείνου τοῦ ἕνεκά του. καὶ ἐν ταῖς ἐξ ἀρχῆς ἄρα συστάσεσι τὰ βουγενῆ, εἰ μὴ πρός τινα ὅρον καὶ τέλος δυνατὰ ἦν ἐλθεῖν, διαφθειρομένης ἂν ἀρχῆς τινος ἐγίνετο, ὥσπερ νῦν τοῦ σπέρματος. ἔτι ἀνάγκη σπέρμα γενέσθαι πρῶτον, ἀλλὰ μὴ εὐθὺς τὰ ζῷα· καὶ τὸ ‘ οὐλοφυὲς μὲν πρῶτα ’ σπέρμα ἦν.

[a] [*Frag.* 62 οὐλοφυεῖς μὲν πρῶτα τύποι χθονὸς ἐξανέτελλον, ‘ whole-natured forms first arose from the earth ’ (Burnet),

step, we find that plants too produce organs subservient to their perfect development—leaves, for instance, to shelter the fruit. Hence, if it is by nature and also for a purpose that the swallow makes her nest and the spider his web, and that plants make leaves for the sake of the fruit and strike down (and not up) with their roots in order to get their nourishment, it is clear that causality of the kind we have described is at work in things that come about or exist in the course of Nature.

Also, since the term ' nature ' is applied both to material and to form, and since it is the latter that constitutes the goal, and all else is for the sake of that goal, it follows that the form is the final cause.

Now there are failures even in the arts (for writers make mistakes in writing and physicians administer the wrong dose); so that analogous failures in Nature may evidently be anticipated as possible. Thus, if in art there are cases in which the correct procedure serves a purpose, and attempts that fail are aimed at a purpose but miss it, we may take it to be the same in Nature, and monstrosities will be like failures of purpose in Nature. So if, in the primal combinations, such ' ox-creatures ' as could not reach an equilibrium and goal, should appear, it would be by the miscarriage of some principle, as monstrous births are actually produced now by abortive developments of sperm. Besides, the sperm must precede the formation of the animal, and Empedocles' ' primal all-generative '[a] is no other than such sperm.

rohgeballte Erdklumpen (Diels); primitive forms of life, in which sex and species were not yet differentiated, composed of water and fire.—C.]

199 b Ἔτι καὶ ἐν τοῖς φυτοῖς ἔνεστι τὸ ἕνεκά του, ἧττον δὲ διήρθρωται. πότερον οὖν καὶ ἐν τοῖς φυτοῖς ἐγίνετο, ὥσπερ τὰ βουγενῆ ἀνδρόπρωρα, οὕτω καὶ ἀμπελογενῆ ἐλαιόπρωρα, ἢ οὔ; ἄτοπον γάρ· ἀλλὰ μὴν ἔδει γε, εἴπερ καὶ ἐν τοῖς ζῴοις. ἔτι ἔδει ἐν τοῖς σπέρμασι γίνεσθαι ὅπως ἔτυχεν.

Ὅλως δ' ἀναιρεῖ ὁ οὕτω λέγων τὰ φύσει τε καὶ φύσιν· φύσει γάρ, ὅσα ἀπό τινος ἐν αὑτοῖς ἀρχῆς συνεχῶς κινούμενα ἀφικνεῖται εἴς τι τέλος· ἀφ' ἑκάστης δὲ οὐ τὸ αὐτὸ ἑκάστοις οὐδὲ τὸ τυχόν, ἀεὶ μέντοι ἐπὶ τὸ αὐτό, ἂν μή τι ἐμποδίσῃ. τὸ δὲ οὗ ἕνεκα, καὶ ὃ τούτου ἕνεκα, γένοιτο ἂν καὶ ἀπὸ τύχης· οἷον λέγομεν ὅτι ἀπὸ τύχης ἦλθεν ὁ ξένος καὶ λυσάμενος ἀπῆλθεν, ὅταν ὥσπερ ἕνεκα τούτου ἐλθὼν πράξῃ, μὴ ἕνεκα δὲ τούτου ἔλθῃ. καὶ τοῦτο κατὰ συμβεβηκός—ἡ γὰρ τύχη τῶν κατὰ συμβεβηκὸς αἰτίων, καθάπερ καὶ πρότερον εἴπομεν—ἀλλ' ὅταν τοῦτο ἀεὶ ἢ ὡς ἐπὶ τὸ πολὺ

---

[a] [In this final argument Aristotle brings together all three kinds of 'causation': Luck, Nature, Art. By Luck a desirable result which serves a purpose is achieved incidentally and once in a way. In Nature, since desirable results are produced normally or invariably, *a fortiori* they must be purpose-serving, though we cannot see the purposing agency. Even in Art we can imagine the purpose-serving process carrying itself out without the previous conscious design which is certainly present.—C.]

[b] [ἀφ' ἑκάστης δὲ κτλ. might possibly mean: 'The final development reached from any one principle (*e.g.*, human seed) is neither (exactly) the same for every individual (for no two men are exactly alike), nor yet is it any random result (*e.g.* dog or horse); there is, however, in each species, always a tendency towards an identical result (*e.g.* the perfect human type that Nature strives, but never successfully, to produce), if nothing interferes.'—C.]

In plants, too, though they are less elaborately articulated, there are manifest indications of purpose. Are we to suppose, then, that as there were 'ox-creatures man-faced' so also there were 'vine-growths olive-bearing'? Incongruous as such a thing seems, it ought to follow if we accept the principle in the case of animals. Moreover, it ought still to be a matter of chance what comes up when you sow this seed or that.

In general,[a] the theory does away with the whole order of Nature, and indeed with Nature's self. For natural things are exactly those which do move continuously, in virtue of a principle inherent in themselves, towards a determined goal; and the final development which results from any one such principle is not identical for any two species, nor yet is it any random result; but in each there is always a tendency towards an identical result, if nothing interferes with the process.[b] A desirable result and the means to it may also be produced by chance, as for instance we say it was 'by luck' that the stranger came and ransomed the prisoner before he left,[c] where the ransoming is done as if the man had come for that purpose, though in fact he did not. In this case the desirable result is incidental; for, as we have explained, chance is an incidental cause. But when the desirable result is effected invariably or normally,

[c] [W. translated λουσάμενος, but λυσάμενος is the better reading. The allusion cannot be to the Μισούμενος of Menander (*Oxyr. Pap.* No. 1013), to which Simplic. 384. 14 refers. Philoponus 324. 21 suggests the ransoming of Plato at Aegina (Diels compares Diog. L. iii. 20 λυτροῦται δὴ αὐτὸν κατὰ τύχην παρὼν Ἀννίκερις ὁ Κυρηναῖος).—C.]

γίγνηται, οὐ συμβεβηκὸς οὐδ' ἀπὸ τύχης· ἐν δὲ τοῖς φυσικοῖς ἀεὶ οὕτως, ἂν μή τι ἐμποδίσῃ. ἄτοπον δὲ τὸ μὴ οἴεσθαι ἕνεκά του γίνεσθαι, ἐὰν μὴ ἴδωσι τὸ κινοῦν βουλευσάμενον. καίτοι καὶ ἡ τέχνη οὐ βουλεύεται· καὶ γὰρ εἰ ἐνῆν ἐν τῷ ξύλῳ ἡ ναυπηγική, ὁμοίως ἂν φύσει ἐποίει· ὥστ' εἰ ἐν τῇ τέχνῃ ἔνεστι τὸ ἕνεκά του, καὶ ἐν φύσει. μάλιστα δὲ δῆλον, ὅταν τις ἰατρεύῃ αὐτὸς ἑαυτόν· τούτῳ γὰρ ἔοικεν ἡ φύσις.

Ὅτι μὲν οὖν αἰτία ἡ φύσις, καὶ οὕτως ὡς ἕνεκά του, φανερόν.

CHAPTER IX

ARGUMENT

[*Most philosophers recognize in the process of natural generation only an 'unconditional,' mechanical necessity, whereby the forces inherent in matter blindly produce their effects. Aristotle elsewhere* (*e.g.* De part. anim. 642 a 1) *admits that many things are produced by the 'simple' necessity of material and efficient causes alone ; but he argues that we must also recognize a 'hypothetical' necessity, seen in the processes of Nature, as in those of art, where purpose fixes a goal to which the mechanical chain of causation shall lead, rather than in other directions.* If *a given desirable*

Τὸ δ' ἐξ ἀνάγκης πότερον ἐξ ὑποθέσεως ὑπάρχει

---

[a] [*Cf. Eth. Nic.* iii. 3. In some arts (*e.g.* writing) there is no deliberation ; the right spelling of a word is not debatable. In others (*e.g.* medicine) we deliberate about means (how to heal), not about the end (healing).—C.]

[b] [The 'hypothetically' necessary is described as 'the *sine qua non* of a desirable result' (*οὗ οὐκ ἄνευ τὸ εὖ*, *Met.* 1072 b 12)—necessary, *if* that result is to be achieved.—C.]

it is not an incidental or chance occurrence; and in the course of Nature the result always is achieved either invariably or normally, if nothing hinders. It is absurd to suppose that there is no purpose because in Nature we can never detect the moving power in the act of deliberation. Art, in fact, does not deliberate either,[a] and if the shipbuilding art were incorporate in the timber, it would proceed by nature in the same way in which it now proceeds by art. If purpose, then, is inherent in art, so is it in Nature also. The best illustration is the case of a man being his own physician, for Nature is like that—agent and patient at once.

That Nature is a cause, then, and a goal-directed cause, is above dispute.

CHAPTER IX

ARGUMENT (*continued*)

result is to be achieved, such and such matter is necessary. The existence of the matter and its forces is a condition sine qua non *of the realization of such ends, but will not account for the process of realization, which is explained and directed by the end to be brought about.* (*Cf. the fuller discussion in* De part. anim. i. 1).

The physicist is primarily concerned with the form, definable essence, or end; but where the form 'necessitates' a certain kind of matter, his definitions will include mention of the appropriate matter.—C.]

The phrase 'must of necessity' may be used of what is unconditionally necessary or of what is 'necessary to this or that.'[b] Of which kind is the

199 b 35 ἢ καὶ ἁπλῶς; νῦν μὲν γὰρ οἴονται τὸ ἐξ ἀνάγκης
200 a εἶναι ἐν τῇ γενέσει, ὥσπερ ἂν εἴ τις τὸν τοῖχον ἐξ ἀνάγκης γεγενῆσθαι νομίζοι, ὅτι τὰ μὲν βαρέα κάτω πέφυκε φέρεσθαι τὰ δὲ κοῦφα ἐπιπολῆς, διὸ οἱ λίθοι μὲν κάτω καὶ τὰ θεμέλια, ἡ δὲ γῆ ἄνω διὰ κουφότητα, ἐπιπολῆς δὲ μάλιστα τὰ ξύλα· κουφότατα γάρ.

Ἀλλ' ὅμως οὐκ ἄνευ μὲν τούτων γέγονεν, οὐ μέντοι διὰ ταῦτα—πλὴν ὡς δι' ὕλην—ἀλλ' ἕνεκα τοῦ κρύπτειν ἄττα καὶ σώζειν. ὁμοίως δὲ καὶ ἐν τοῖς ἄλλοις πᾶσιν ἐν ὅσοις τὸ ἕνεκά τού ἐστιν, οὐκ ἄνευ μὲν τῶν ἀναγκαίαν ἐχόντων τὴν φύσιν, οὐ μέντοι γε διὰ ταῦτα—ἀλλ' ἢ ὡς ὕλην—ἀλλ' ἕνεκά του. οἷον διὰ τί ὁ πρίων τοιοσδί; ὅπως τοδὶ καὶ ἕνεκα τουδί· τοῦτο μέντοι τὸ οὗ ἕνεκα ἀδύνατον γενέσθαι, ἂν μὴ σιδηροῦς ᾖ· ἀνάγκη ἄρα σιδηροῦν εἶναι, εἰ πρίων ἔσται καὶ τὸ ἔργον αὐτοῦ. ἐξ ὑποθέσεως δὴ τὸ ἀναγκαῖον, ἀλλ' οὐχ ὡς τέλος· ἐν γὰρ τῇ ὕλῃ τὸ ἀναγκαῖον, τὸ δ' οὗ ἕνεκα ἐν τῷ λόγῳ.

Ἔστι δὲ τὸ ἀναγκαῖον ἔν τε τοῖς μαθήμασι καὶ

---

[a] [γῆ, unbaked brick, of which house walls were commonly made.—C.]

[b] [*Literally*, 'So the "necessity" here is necessity in the hypothetical sense, but not in the sense in which an end (is a "necessary" consequence of its antecedents); for it resides in the matter, whereas the end is contained in the notion (or definition of the thing's essential nature, *e.g.* "what it is to be a saw").' *Cf. De part. anim.* 639 b.—C.]

necessity that exists in Nature? For people ignore this distinction and talk about things being 'necessarily generated,' much as if they thought that a wall would come up in the necessary course of things because what is heavy naturally descends and what is light is naturally on the top, so that stones go down and make the foundations, while the lighter brick[a] rises above them, and the timber, lightest of all, roofs them above.

No doubt it is a fact that the building cannot dispense with these materials, and in that sense they 'must be there'; but they do not themselves 'make' the building in the sense of constructing it, but only in that of constituting its material. What causes the building to be made is the purpose of protecting and preserving certain goods. And so in all other cases where a purpose can be traced. It cannot be accomplished without materials that have the required nature; but it is not they that 'make' the purpose-fulfilling instrument, except materially. For what makes it, in the formative sense, is the purposeful intention of the maker. For instance: 'Why is a saw like this?' 'In order that it may have the essential character of a saw and serve for sawing.' This purpose, however, could not be served if it were not made of iron. So if it is to be a saw, and to do its work, it 'must necessarily' be made of iron. The necessity, then, is conditional, or hypothetical. The purpose, mentally conceived, demands the material as necessary to its accomplishment; but the nature of the material, as already existing, does not 'necessarily' lead to the accomplishment of the purpose.[b]

Note, too, that in a certain respect, there is a kind

200 a ἐν τοῖς κατὰ φύσιν γινομένοις τρόπον τινὰ παραπλησίως· ἐπεὶ γὰρ τὸ εὐθὺ τοδί ἐστιν, ἀνάγκη τὸ τρίγωνον δύο ὀρθαῖς ἴσας ἔχειν· ἀλλ᾿ οὐκ ἐπεὶ τοῦτο, ἐκεῖνο· ἀλλ᾿ εἴ γε τοῦτο μὴ ἔστιν, οὐδὲ τὸ εὐθὺ ἔστιν. ἐν δὲ τοῖς γινομένοις ἕνεκά του ἀνάπαλιν, εἰ τὸ τέλος ἔσται ἢ ἔστι, καὶ τὸ ἔμπροσθεν ἔσται ἢ ἔστιν· εἰ δὲ μή, ὥσπερ ἐκεῖ μὴ ὄντος τοῦ συμπεράσματος ἡ ἀρχὴ οὐκ ἔσται, καὶ ἐνταῦθα τὸ τέλος καὶ τὸ οὗ ἕνεκα· ἀρχὴ γὰρ καὶ αὕτη, οὐ τῆς πράξεως ἀλλὰ τοῦ λογισμοῦ (ἐκεῖ δὲ τοῦ λογισμοῦ· πράξεις γὰρ οὐκ εἰσίν). ὥστ᾿

---

[a] [*Or*, ‘There is, in a way, some resemblance between the working of (logical) necessity in mathematics and the necessity involved in things coming into existence by process of nature. Thus, if the straight line is defined as it is, it is a necessary consequence that the angles of the triangle should be equal to two right angles. On the other hand we cannot prove the truth of the definition from this property of the triangle; though we can say that, if the triangle has not this property, then nothing answering to the definition of "straight line" exists.’ (The property of the (rectilinear) triangle, demonstrated in Euclid i. 32, follows from (1) the definition of straight line (and of other terms), (2) the "hypothesis" that such things as those defined *exist*, (3) other assumptions. No one of these assumptions can be proved from the consequence of them all. The ultimate principles are, in fact, indemonstrable. On the other hand, if the conclusion turned out to be false, we could infer that some hypothesis, such as the existence of the straight line, was false. *Cf. Anal. post.* i. 10; Heath, *Thirteen Books of Euclid's Elements* i. 117 ff.) ‘But in the production of things for a purpose (of art or of Nature), the necessity works the other way: the assumption is that the *end* is to exist or does exist, and this necessitates that the antecedents shall exist or do exist; otherwise, as in mathematics the failure of the conclusion necessitates the non-existence of the premiss we start from (straight line), here the non-existence of the end

of inverted parallelism between mathematics and the things that take place under the control of nature. For a straight line being what it is, if (i) a triangle is straight-sided, then (ii) its angles are equal to two right angles; but it does not follow that if (ii) its angles are equal to two right angles, then (i) the triangle must be straight-sided, though it does follow that if (ii) is not, *i.e.* its angles are not equal to two right angles, then (i) is not, *i.e.* the triangle cannot be straight-sided.[a] But in purposeful constructions it is the other way about; for if (ii) the end is to be secured or actually has been secured, then (i) the antecedent conditions must be, or must have been, present; but just as in the case of the triangle the failure of the *consequent* (its angles are equal to two right angles) necessarily involves the failure of the *antecedent* (that the triangle be straight-sided), so here the failure of the material *antecedent* (bricks or iron) necessarily carries with it the failure of the proposed *end* (house or saw). But in both cases it is the 'principle' or 'primary factor' that imposes the necessity upon the accessary one; for in the case of the triangle it is the premise that imposes a logical necessity on the conclusion (for the whole construc-

or purpose will be necessitated. For the end or purpose also is a principle or starting-point, not indeed of the process of realization, but of the process of reasoning; whereas in mathematics there is only the starting-point, or premiss, of the reasoning, no action being involved.'

Cf. Eth. Nic. 1112 b 11 ff. In action we reason backwards from the proposed end ('If I am to bring this about, what must I do first?') along the chain of means, till we arrive at some action we can at once perform. We perform it, and so start the train of means from its ἀρχή (the beginning of the action and the conclusion of the reasoning) to the τέλος (the end of the action and the starting-point of the reasoning).—C.]

200 a εἰ ἔσται οἰκία, ἀνάγκη ταῦτα γενέσθαι ἢ ὑπάρχειν ἢ εἶναι, ἢ ὅλως τὴν ὕλην τὴν ἕνεκά του (οἷον πλίνθους καὶ λίθους, εἰ οἰκία)· οὐ μέντοι διὰ ταῦτά ἐστι τὸ τέλος—ἀλλ' ἢ ὡς ὕλην—οὐδ' ἔσται διὰ ταῦτα· ὅλως μέντοι μὴ ὄντων οὐκ ἔσται οὔτε ἡ οἰκία οὔθ' ὁ πρίων, ἡ μὲν εἰ μὴ οἱ λίθοι, ὁ δ' εἰ μὴ ὁ σίδηρος (οὐδὲ γὰρ ἐκεῖ αἱ ἀρχαί, εἰ μὴ τὸ τρίγωνον δύο ὀρθαῖς).

Φανερὸν δὴ ὅτι τὸ ἀναγκαῖον ἐν τοῖς φυσικοῖς τὸ ὡς ὕλη λεγόμενον καὶ αἱ κινήσεις αἱ ταύτης. καὶ ἄμφω μὲν τῷ φυσικῷ λεκτέαι αἱ αἰτίαι, μᾶλλον δὲ ἡ τινὸς ἕνεκα· αἴτιον γὰρ τοῦτο τῆς ὕλης, ἀλλ' οὐχ αὕτη τοῦ τέλους. καὶ τὸ τέλος τὸ οὗ ἕνεκα, καὶ ἡ ἀρχὴ ἀπὸ τοῦ ὁρισμοῦ καὶ τοῦ λόγου,

---

[a] [*Or*, '(as in the mathematical illustration, entities whose existence is assumed at the outset will not exist, if it is not true that the angles of a triangle are equal to two right angles).' Aristotle makes confusion by at once comparing and contrasting logical and teleological necessity. His real point is that logical necessity works forward from premiss to conclusion, but in Nature there is not only mechanical necessity, working blindly forward from cause to effect, but 'hypothetical' necessity, seen in the processes of Nature or of art, where a 'final cause' fixes the goal to which the mechanical chain shall lead, rather than in other directions. The assumption, by the artist or by Nature, that this end shall be brought about, starts the whole process and accounts for its course.—C.]

[b] [*Literally* 'what is necessary in the things of Nature is what is so called in the sense of (the necessary) matter and its motions' (its inherent behaviour). These are a condition

tion is logical, and there is no action at all in question), whereas in the other cases, though the purpose (which is the principle that imposes the necessity on the material) has to do with action, yet it is only a logical necessity that it can impose, for it does not actually create or constitute the material. Thus, if there is to be a house, then stones, bricks and so forth, if not already on hand, must be produced (for at any rate they must be there). Or, to put it generally, the purpose-serving materials must be there, if the purpose is to be accomplished ; but since they do not themselves fulfil the purpose, except materially, it does not follow that, if they are there, the purpose will be fulfilled. To sum up : the materials will not account for the existence of the house or of the saw, though, if they are simply not there—no stones for the house, no iron for the saw—there will be no house and no saw (just as in the mathematical illustration straight-sidedness does not inhere in biorthogonality, but cannot exist without it).[a]

It is clear, then, that when physicists speak of necessity absolutely, they should limit the term to what is inherent in the material, and should recognize purposeful movement imposed upon the material as a distinct addition to its inherent qualities.[b] And though the physicist has to deal with both material and purpose, he is more deeply concerned with the latter ; for purpose directs the moving causes that act upon the material, not the reverse. It is the goal that determines the purpose, and the principle of causation is derived from the definition and rationale of the end, in Nature just as much as in artificial

sine qua non, but will not of themselves produce the desirable 'end.'—C.]

200 b ὥσπερ ἐν τοῖς κατὰ τέχνην, ἐπεὶ ἡ οἰκία τοιόνδε, τάδε δεῖ γίγνεσθαι καὶ ὑπάρχειν ἐξ ἀνάγκης, καὶ ἐπεὶ ἡ ὑγίεια τοδί, τάδε δεῖ γίγνεσθαι ἐξ ἀνάγκης καὶ ὑπάρχειν. οὕτως καὶ εἰ ἄνθρωπος τοδί, ταδί· εἰ δὲ ταδί, ταδί. ἴσως δὲ καὶ ἐν τῷ λόγῳ ἐστὶ τὸ ἀναγκαῖον. ὁρισαμένῳ[1] γὰρ τὸ ἔργον τοῦ πρίειν ὅτι διαίρεσις τοιαδί· αὕτη δ' οὐκ ἔσται, εἰ μὴ ἕξει ὀδόντας τοιουσδί· οὗτοι δ' οὔ, εἰ μὴ σιδηροῦς. ἔστι γὰρ καὶ ἐν τῷ λόγῳ ἔνια μόρια ὡς ὕλη τοῦ λόγου.

[1] [ὁρισαμένῳ F. Bekker: ὁρισάμενοι E: ὁρισαμένου I.—C.]

---

[a] [Or, 'Though, perhaps, we may also say that what is

constructions. *E.g.* since this is what a house is, these things must necessarily be ready or be produced; and since this is what health is, such and such things must be provided to secure it. And in like manner, if this is what 'man' is, then there must antecedently be such and such things, and that they again may be there, such and such other things. And the same kind of 'necessity' may be traced on the conceptual side.[a] For if we define 'sawing' as such and such a method of dividing, it follows that it can only be accomplished by teeth of such and such a character, and these teeth can exist only if the saw be made of iron. For a definition, too, contains constituent terms which are, as it were, its materials.

" necessary " in this way is actually contained in the definition.']

BOOK III

INTRODUCTION

[In the last Book 'Nature' was defined as a principle of 'movement' (including change of all kinds). The next step is to define movement, and to study certain conceptions which are held to be allied to movement: continuity, infinity, place, time, and void. Books III. and IV. consist of a series of Essays on these subjects.

Chapters i.-iii. define movement or change as the passage from potential to actual existence: given something that is actually *a* and potentially *b*, change is the process which ends in the realization of its capacity of being *b*. Hence a change is something that is essentially incomplete so long as it lasts; the potential is all the while progressively losing its character; but when it is transformed into complete actuality (ἐνέργεια), the change is over. This definition is substantiated by comparison with the inadequate conceptions of Plato and the Pythagoreans; and finally it is argued that, although a change or movement is the actualization of two potentialities—that of the mover and that of the moved—it occurs in the moved. This preliminary account of movement will be amplified in Book V.

In Chapter iv. the essay on Infinity opens with a survey of the part played by this conception in earlier systems, and a statement of the problems it involves. Among these, the physicist is most concerned with the question whether there exists an infinite material body. Chap. v.: That no such body can actually exist is proved, chiefly by means of the doctrine that sensible bodies have 'natural places' (*cf. De caelo* i. 7-8). Chap. vi.: On the other hand, infinity must exist in some sense; otherwise time would not be unlimited in duration, there would be indivisible magnitudes (for it would not be possible to carry the process of division beyond a certain point), and the series of numbers would be limited. In spatial magnitude there is the illimitable potentiality of division that can never be completed. Conversely, there is the illimitable series converging to a finite sum. Chap. vii.: Mathematical demonstrations do not require a line of infinite length, but only a finite line as long as the mathematician chooses to suppose. There is no actual straight line greater than the diameter of the physical world-sphere. Chap. viii.: Time and number, on the other hand, are illimitable divergent series. It follows from the illimitable divisibility of magnitude that number can be increased without limit (see Gen. Introd. pp. lxxxix f.); but number is limited in the other direction, *i.e.* there is a smallest, but not a greatest number. The illimitability of Time is a consequence of the illimitability of motion, which Time measures. What is conceivable is not limited by what can be.

The result of this analysis is that 'illimitable' means 'the open possibility of more,' which can never be completed.—C.]

For the sake of clarity I have added chapter headings and also an extract from chapter viii and have substituted both here and in the text and notes of Bk. III. and elsewhere the terms 'illimitable' and its derivatives for 'unlimited,' 'Infinite,' and 'infinitely,' and also 'completed' for 'realized' here and for 'actualized' in the note on p. 192.

The term 'illimitable' implies that there is no assignable limit beyond which the argument is inapplicable, whereas 'infinite' is beyond the realm of 'nature,' which is the subject matter of the *Physics*.—R. W.

Γ

CHAPTER I

ARGUMENT

[*Nature having been defined as a principle of movement or change, we must first investigate the nature of movement. This will lead on to further studies of the associated conceptions of continuity, infinity, place, void, and time* (200 b 12-25).

There are four classes of things that can change or move, and four corresponding kinds of change : (1) *of substance,* (2) *of quality,* (3) *of quantity,* (4) *of place.* (Cf. *the longer discussion in* De gen. et corr. i. 4.) *Change is the process of actually realizing what exists potentially, in so far as it exists potentially* (b 25-201 a 29).

The qualification ' in so far as it exists potentially ' is

200 b 12 Ἐπεὶ δ᾽ ἡ φύσις μέν ἐστιν ἀρχὴ κινήσεως καὶ μεταβολῆς, ἡ δὲ μέθοδος ἡμῖν περὶ φύσεώς ἐστι, δεῖ μὴ λανθάνειν τί ἐστι κίνησις· ἀναγκαῖον γὰρ ἀγνοουμένης αὐτῆς ἀγνοεῖσθαι καὶ τὴν φύσιν. διορισαμένοις δὲ περὶ κινήσεως πειρατέον τὸν αὐτὸν ἐπελθεῖν τρόπον περὶ τῶν ἐφεξῆς. δοκεῖ δ᾽ ἡ κίνησις εἶναι τῶν συνεχῶν, τὸ δ᾽ ἄπειρον ἐμφαίνεται πρῶτον ἐν τῷ συνεχεῖ· διὸ καὶ τοῖς

BOOK III

CHAPTER I

ARGUMENT (*continued*)

added because movement is not the actualization of the whole character (say) of the bricks used for the process of building a house, but only of that part of their character which consists in their capacity to serve as materials for the house. The movement or process of building is not the realization of the building materials as bricks (they are bricks before the process begins), nor of their potentiality of being a house (the house is the realization of that), but of their potentiality of serving as building materials (ᾗ οἰκοδομητόν) (a 29-b 15.)

(Extracts from this chapter appear in *Metaphysics* K 9, where Mr. Ross's commentary may be consulted.)—*C.*]

Since Nature is the principle of movement and change, and it is Nature that we are studying, we must understand what ' movement ' is ; for, if we do not know this, neither do we understand what Nature is. When we have defined the meaning of movement or progress from this to that, we must attempt in the same way a discussion of the associated conceptions to which it leads. Now, movement is clearly one of the things we think of as ' continuous,' and it is in connexion with continuity that we first encounter the concept of the ' illimitable.' And this

200 b ὁριζομένοις τὸ συνεχὲς συμβαίνει προσχρήσασθαι πολλάκις τῷ λόγῳ τῷ τοῦ ἀπείρου, ὡς τὸ εἰς ἄπειρον διαιρετὸν συνεχὲς ὄν. πρὸς δὲ τούτοις ἄνευ τόπου καὶ κενοῦ καὶ χρόνου κίνησιν ἀδύνατον εἶναι. δῆλον οὖν ὡς διά τε ταῦτα καὶ διὰ τὸ πάντων εἶναι κοινὰ καὶ καθόλου ταῦτα πᾶσι, σκεπτέον προχειρισαμένοις περὶ ἑκάστου τούτων· ὑστέρα γὰρ ἡ περὶ τῶν ἰδίων θεωρία τῆς περὶ τῶν κοινῶν ἐστιν.

Καὶ πρῶτον, καθάπερ εἴπαμεν, περὶ κινήσεως. ἔστι δή τι τὸ μὲν ἐντελεχείᾳ μόνον, τὸ δὲ δυνάμει καὶ ἐντελεχείᾳ, τὸ μὲν τόδε τι, τὸ δὲ τοσόνδε, τὸ δὲ τοιόνδε, καὶ ἐπὶ τῶν ἄλλων τῶν τοῦ ὄντος κατηγοριῶν ὁμοίως. τοῦ δὲ πρός τι τὸ μὲν καθ' ὑπεροχὴν λέγεται καὶ κατ' ἔλλειψιν, τὸ δὲ κατὰ τὸ ποιητικὸν καὶ παθητικὸν καὶ ὅλως κινητικόν τε καὶ κινητόν· τὸ γὰρ κινητικὸν κινητικὸν τοῦ κινητοῦ, καὶ τὸ κινητὸν κινητὸν ὑπὸ τοῦ κινητικοῦ.

Οὐκ ἔστι δέ τις κίνησις παρὰ τὰ πράγματα.

---

[a] [Viz., pure Intelligences. After *μόνον* (*l.* 26) probably <*τὸ δὲ δυνάμει*> should be inserted (as in *Met.* 1065 b 5; so Spengel and others), so as to include in the classification potentialities never completed, such as the illimitable. The third class (*δυνάμει καὶ ἐντελεχείᾳ*) contains natural objects (matter + form).—C.] But *cf.* Introd. pp. 186 ff. and pp. 203-205; also Ross, commentary on *Met. loc. cit.*

[b] [*κινητικόν*, as distinguished from *κινοῦν* at 202 a 16 *κινητικὸν μὲν γάρ ἐστι τῷ δύνασθαι, κινοῦν δὲ τῷ ἐνεργεῖν*.—C.]

is why in definitions of continuity this concept of the 'illimitable' frequently occurs, as when we say that the continuous is that which is susceptible of division without limit. Further, movement (it is said) cannot occur except in relation to place, void, and time. Evidently, then, for these reasons and because these four things—movement, place, void, and time—are universal conditions common to all natural phenomena, we must consider each of them on the threshold of our inquiry; for the treatment of peculiar properties must come after that of properties common to all natural things.

We must begin, then, as already said, with movement in general or progress from this to that. Now, some potentialities never exist apart, but always reveal themselves as actualized;[a] others, while they are something actually, are capable of becoming something else than they are, that is to say, have potentialities not realized at the moment; and these potentialities may concern their substantive being (what they are) or their quantity or their qualities; and so on with the other categories of existence. And under the category of 'relation' may be relations between the 'more' and the 'less,' or between that which is active and that which is acted on, and generally between that which 'moves' (or changes) something as the agent and that which is moved (or changed) by it as the patient. For that which has the power of producing a change can only act in reference to a thing capable of being changed; and that which is capable of being changed can only suffer change under the action of that which has the power to change it.[b]

Now, motion and change cannot exist in them-

200 b μεταβάλλει γὰρ τὸ μεταβάλλον ἀεὶ ἢ κατ' οὐσίαν ἢ κατὰ ποσὸν ἢ κατὰ ποιὸν ἢ κατὰ τόπον· κοινὸν δ' ἐπὶ τούτων οὐδὲν ἔστι λαβεῖν (ὥς φαμεν) ὃ
201 a οὔτε τόδε οὔτε ποσὸν οὔτε ποιὸν οὔτε τῶν ἄλλων κατηγορημάτων οὐθέν· ὥστ' οὐδὲ κίνησις οὐδὲ μεταβολὴ οὐθενὸς ἔσται παρὰ τὰ εἰρημένα, μηδενός γε ὄντος παρὰ τὰ εἰρημένα.

Ἕκαστον δὲ διχῶς ὑπάρχει πᾶσιν, οἷον τὸ τόδε (τὸ μὲν γὰρ μορφὴ αὐτοῦ, τὸ δὲ στέρησις), καὶ κατὰ τὸ ποιόν (τὸ μὲν γὰρ λευκὸν τὸ δὲ μέλαν), καὶ κατὰ τὸ ποσὸν τὸ μὲν τέλειον τὸ δ' ἀτελές· ὁμοίως δὲ καὶ κατὰ τὴν φορὰν τὸ μὲν ἄνω τὸ δὲ κάτω ἢ τὸ μὲν κοῦφον τὸ δὲ βαρύ. ὥστε κινήσεως καὶ μεταβολῆς ἐστιν εἴδη τοσαῦτα ὅσα τοῦ ὄντος.

Διῃρημένου δὲ καθ' ἕκαστον γένος τοῦ μὲν ἐντελεχείᾳ τοῦ δὲ δυνάμει, ἡ τοῦ δυνάμει ὄντος ἐντελέχεια, ᾗ τοιοῦτον, κίνησίς ἐστιν· οἷον τοῦ μὲν ἀλλοιωτοῦ, ᾗ ἀλλοιωτόν, ἀλλοίωσις, τοῦ δὲ αὐξητοῦ καὶ τοῦ ἀντικειμένου φθιτοῦ (οὐδὲν γὰρ ὄνομα κοινὸν ἐπ' ἀμφοῖν) αὔξησις καὶ φθίσις, τοῦ δὲ γενητοῦ καὶ φθαρτοῦ γένεσις καὶ φθορά, τοῦ δὲ φορητοῦ φορά. ὅτι δὲ τοῦτό ἐστιν ἡ κίνησις,

---

[a] [Aristotle seems to be attacking Plato for discussing Change or Motion as one of the 'widest kinds' (*Soph.* 254 B ff.). There is no such thing as Change in itself, apart from something that changes.—C.]

[b] [τούτων, literally, 'these classes of entity'—substance, quality, etc.—which are ultimate irreducible heads under one (and only one) of which anything that exists can be classified. *Cf. Met.* 1065 b 7.—C.]

[c] [*Literally*, 'each (of these categories) belongs to all (its subjects) in two ways.'—C.]

[d] See Book V. ch. i.

selves apart from what moves and changes.[a] For, wherever anything changes, it always changes either from one thing to another, or from one magnitude to another, or from one quality to another, or from one place to another ; but there is nothing that embraces all these kinds of change[b] in common, and is itself neither substantive nor quantitive nor qualitive nor pertaining to any of the other categories, but existing in detachment ; so neither can movement or change exist independently of these, for there is nothing independent of them.

Again, in each of these four cases, there are two poles between which the change moves ;[c] in substantive existence, for example, form and shortage from form ; in quality, white and black ; in quantity, the perfectly normal and an achievement short of perfection[d] ; and so, too, in the case of vection, up and down, or the action of levity and gravity. So there are as many kinds of change as there are categories of existence.

Reverting, therefore, to the universal distinction already established between ‘ being-at-the-goal ’ in actuality and being in potentiality ‘ such-as-is-capable-of-attaining-the-goal,’ we can now define motion or change as the progress of the realizing of a potentiaality, *qua* potentiality, *e.g.* the actual progress of qualitive modification in any modifiable thing *qua* modifiable ; the actual growing or shrinking (for we have no single word to include them both) of anything capable of expanding or contracting ; the process of coming into existence or passing out of it of that which is capable of so coming and passing ; the actual moving of the physical body capable of changing its place. That this is what we really mean

201 a ἐντεῦθεν δῆλον· ὅταν γὰρ τὸ οἰκοδομητόν, ᾗ τοιοῦτον αὐτὸ λέγομεν εἶναι, ἐντελεχείᾳ ᾖ, οἰκοδομεῖται, καὶ ἔστι τοῦτο οἰκοδόμησις· ὁμοίως δὲ καὶ μάθησις καὶ ἰάτρευσις καὶ κύλισις καὶ ἅλσις καὶ ἅδρυνσις καὶ γήρανσις.

Ἐπεὶ δ' ἔνια ταὐτὰ καὶ δυνάμει καὶ ἐντελεχείᾳ ἐστίν—οὐχ ἅμα δέ, ἢ οὐ κατὰ τὸ αὐτό, ἀλλ' οἷον θερμὸν μὲν δυνάμει ψυχρὸν δὲ ἐντελεχείᾳ—πολλὰ ἤδη ποιήσει καὶ πείσεται ὑπ' ἀλλήλων· ἅπαν γὰρ ἔσται ἅμα ποιητικὸν καὶ παθητικόν. ὥστε καὶ τὸ κινοῦν φυσικῶς κινητόν· πᾶν γὰρ τὸ τοιοῦτον κινεῖ κινούμενον καὶ αὐτό. δοκεῖ μὲν οὖν τισιν ἅπαν κινεῖσθαι τὸ κινοῦν· οὐ μὴν ἀλλὰ περὶ τούτου μὲν ἐξ ἄλλων ἔσται δῆλον ὅπως ἔχει—ἔστι γάρ τι κινοῦν καὶ ἀκίνητον—ἡ δὲ τοῦ δυνάμει

---

[a] [*i.e.*, things in the sublunary world. (The Heavens, whose lowest sphere is conterminous with the uppermost stratum (fire) of the sublunary world, act upon this world, but are in no way affected in return. *De gen. et corr.* A 6.) The clause ἅπαν γὰρ . . . παθητικόν seems to mean that every such thing has both potentialities at once.—C.]

[b] The important and easily overlooked point here is that Aristotle had no conception of inertia, the fatal defect which runs through his physics. Having no conception of inertia, he could not conceive why a projectile goes forward after it leaves the hand that is hurling or thrusting it forward. He gives obviously inadequate alternatives as explanations of this, which he is never quite willing to father, but which he allows to pass with some uneasiness (see Book V. ch. vi). But in other respects he accepts the principles which the differences of idiom easily allow us to think we understand when we do not really do so. If one brick by being thrown upon another sets a whole row of bricks on the fall one after another, Aristotle takes this to show not only that every fresh motion must have been started by a moving body already moving, but also that every moving

by motion or change may be shown thus: building material is actualizing the potentialities in virtue of which we call it 'building material' when it is in the act of being built into a structure, and this act is the process or 'movement' of 'building'; and so too with other processes—learning, healing, rolling, jumping, maturing, aging.

And since in certain cases the same thing may have both an actuality and a potentiality (not indeed at the same time or not in the same respect, but potentially hot, for instance, and actually cold), it follows that many things act on, and are acted on by, each other; for anything will be at once capable of acting and of being acted upon.[a] And so it happens that every physical body which causes motion must be capable of being moved; for whenever it causes motion it is itself under the action of some other body which is keeping it in motion.[b] But the inference that has sometimes been drawn, that there is *no* cause of a thing being in motion, which cause is not itself in motion, is false. The truth in this matter will be explained later on;[c] suffice it now to say that there *is* a cause of things being in motion, which cause is itself immovable; but motion is the

body during the whole time that it is in motion must be being kept in motion by some other body, which other body is itself not only in motion but kept in motion by some other body for the whole time during which it is in motion. Few of us are conscious of the wide place that the law of inertia —that a body in motion will go on of itself at the same speed and in the same direction until something else affects it— takes in our actual thinking. We all assume its constant action subconsciously without ever thinking of it at all. But Aristotle, who had no conception of it, was always conscious of having to do without it.

[c] [In Book VIII. 1-6.—C.]

201 a ὄντος, ὅταν ἐντελεχείᾳ ὂν ἐνεργῇ οὐχ ᾗ αὐτὸ ἀλλ' ᾗ κινητόν, κίνησίς ἐστιν.

Λέγω δὲ τὸ ᾗ ὡδί. ἔστι γὰρ ὁ χαλκὸς δυνάμει ἀνδριάς, ἀλλ' ὅμως οὐχ ἡ τοῦ χαλκοῦ ἐντελέχεια, ᾗ χαλκός, κίνησίς ἐστιν· οὐ γὰρ τὸ αὐτὸ τὸ χαλκῷ εἶναι καὶ δυνάμει τινὶ κινητῷ, ἐπεὶ εἰ ταὐτὸν ἦν ἁπλῶς καὶ κατὰ τὸν λόγον, ἦν ἂν ἡ τοῦ χαλκοῦ, ᾗ χαλκός, ἐντελέχεια κίνησις· οὐκ ἔστι δὲ ταὐτόν, ὡς εἴρηται. (δῆλον δ' ἐπὶ τῶν ἐναντίων· τὸ μὲν 201 b γὰρ δύνασθαι ὑγιαίνειν καὶ δύνασθαι κάμνειν ἕτερον—καὶ γὰρ ἂν τὸ κάμνειν καὶ τὸ ὑγιαίνειν ταὐτὸν ἦν—τὸ δὲ ὑποκείμενον καὶ τὸ ὑγιαῖνον καὶ τὸ νοσοῦν, εἴθ' ὑγρότης εἴθ' αἷμα, ταὐτὸν καὶ ἕν.) ἐπεὶ δ' οὐ ταὐτόν (ὥσπερ οὐδὲ χρῶμα ταὐτὸν καὶ ὁρατόν), ἡ τοῦ δυνατοῦ, ᾗ δυνατόν, ἐντελέχεια φανερὸν ὅτι κίνησίς ἐστιν.

Ὅτι μὲν οὖν ἐστιν αὕτη, καὶ ὅτι συμβαίνει τότε κινεῖσθαι ὅταν ἡ ἐντελέχεια ᾖ αὔτη,[1] καὶ οὔτε πρότερον οὔτε ὕστερον, δῆλον. ἐνδέχεται γὰρ ἕκαστον ὁτὲ μὲν ἐνεργεῖν ὁτὲ δὲ μή, οἷον τὸ

[1] [αὐτή conj. Christ at *Met.* 1065 b 35 : αὔτη codd. (in both places).—C.]

---

[a] [*Sc.* ἐντελέχεια, which Bonitz inserted after ὄντος. But see *Met.* 1065 b 21, where this clause follows a sentence from which the word can be supplied.—C.]

[b] [W. here translated the Berlin text ; but for ᾗ αὐτὸ ᾗ ἄλλο most editors read (with one MS. here and with E and others at *Met.* 1065 b 22) οὐχ ᾗ αὐτὸ ἀλλ' ᾗ κινητόν. This, as Ross remarks on *Met.*, *loc. cit.*, is shown to be the true reading by the explanation of ᾗ which follows (in both places). Ross renders : ' The complete reality of that which exists potentially, when it is completely real and actual, not *qua* itself, but *qua* movable, is movement.'—C.]

[c] [The parallel passage (with slight verbal variations) in

functioning[a] of a movable thing, all the time that it is bringing its potentiality into act, not *qua* itself, but *qua* movable.[b]

To illustrate what I mean by '*qua*' this or that. The bronze is potentially the statue, but neither to be the statue nor to move or change in any respect is the self-realizing of the bronze *qua* bronze; for it is not the same thing to be bronze and to be potentially movable or changeable. Were it the same thing, absolutely and by definition, then indeed moving would be its self-realization; but it is not the same. (It is clear in the case of opposites: the potentiality of health and the potentiality of disease are different things—otherwise being diseased and being healthy would be identical—but whatever it is, humour or blood, that is subject to the healthy or unhealthy condition is one and the same thing in both cases.) And since it is not the same (any more than colour and visibility, for instance, are the same), it is clear that motion must be the realization of the specific potentiality in question and of the subject only *qua* seat of this specific potentiality.

Clearly, then, this is the nature of movement, and a thing is moving just as long as it is actually functioning in this particular way, and neither before nor after. For anything capable of this special kind of functioning may be exercising it at one time, but not at another; for instance,[c] the building materials are

Met. 1066 a 1 is translated more literally by Mr. Ross: '*e.g.*, the "buildable," *qua* "buildable"; and the actuality of the "buildable" *qua* "buildable" is building. For the actuality is either this—building—or the house. But when the *house* exists, it is no longer "buildable"; the "buildable" is what *is being* built. The actuality then must be the *building*, and the building is a movement.'—C.]

201 b οἰκοδομητόν, καὶ ἡ τοῦ οἰκοδομητοῦ ἐνέργεια, ᾗ οἰκοδομητόν, οἰκοδόμησίς ἐστιν (ἢ γὰρ οἰκοδόμησις ἡ ἐνέργεια τοῦ οἰκοδομητοῦ ἢ ἡ οἰκία· ἀλλ᾽ ὅταν οἰκία ᾖ, οὐκέτ᾽ οἰκοδομητόν ἐστιν, οἰκοδομεῖται δὲ τὸ οἰκοδομητόν· ἀνάγκη ἄρα τὴν οἰκοδόμησιν τὴν ἐνέργειαν εἶναι), ἡ δ᾽ οἰκοδόμησις κίνησίς τίς ἐστιν. ἀλλὰ μὴν ὁ αὐτὸς ἐφαρμόσει λόγος καὶ ἐπὶ τῶν ἄλλων κινήσεων.

CHAPTER II

ARGUMENT

[*The definition of motion is compared with the conceptions of the Pythagoreans and Plato, which were inadequate but prompted by a dim perception of the truth that movement is something incomplete* (201 b 15-202 a 3).

In the sublunary world the action of the moving body is accompanied by action upon it, so that the mover is also

Ὅτι δὲ καλῶς εἴρηται, δῆλον καὶ ἐξ ὧν οἱ ἄλλοι περὶ αὐτῆς λέγουσι, καὶ ἐκ τοῦ μὴ ῥᾴδιον εἶναι διορίσαι ἄλλως αὐτήν. οὔτε γὰρ τὴν κίνησιν καὶ τὴν μεταβολὴν ἐν ἄλλῳ γένει θεῖναι δύναιτ᾽ ἄν τις. δῆλον δὲ σκοποῦσιν ὡς τιθέασιν αὐτὴν ἔνιοι,

[a] [οὔτε has no grammatical correlate.—C.]

functioning *as materials for building* only so long as they are in process of being built with ; for as soon as the edifice itself is actually raised, the functioning of what were materials for a house is merged in the functioning of the house itself ; but as long as they are being built with, they are functioning as materials for a house. The act of building, then, is the energizing or bringing into actuality of the potentiality of the materials *qua* materials ; and the passage of the materials of a house into the texture of the house itself, so long as it is in progress, is their ' movement ' *qua* materials of building. And this is the theory of all the other ' movements ' equally.

CHAPTER II

ARGUMENT (*continued*)

moved. This justifies our defining motion as the actualization of what is capable of being moved (*not, as one might expect, of* causing motion) (a 3-9).

The moving cause carries with it some element of form, which will serve as a principle of movement (a 9-12). (Extracts from this chapter, *Met.* 1066 a 7-27.)—*C.*]

That we have found the true definition becomes clear when we note what others have said about it and the difficulties presented by all their definitions. For, in the first place,[a] it is not possible to place movement and change under any other heading, nor have any previous definitions justified themselves. This comes out clearly enough when we consider

201 b ἑτερότητα καὶ ἀνισότητα καὶ τὸ μὴ ὂν φάσκοντες εἶναι τὴν κίνησιν· ὧν οὐδὲν ἀναγκαῖον κινεῖσθαι—οὔτ' ἂν ἕτερα ᾖ, οὔτ' ἂν ἄνισα, οὔτ' ἂν οὐκ ὄντα—ἀλλ' οὐδ' ἡ μεταβολὴ οὔτ' εἰς ταῦτα οὔτ' ἐκ τούτων μᾶλλόν ἐστιν ἢ τῶν ἀντικειμένων. αἴτιον δὲ τοῦ εἰς ταῦτα τιθέναι ὅτι ἀόριστόν τι δοκεῖ εἶναι ἡ κίνησις, τῆς δὲ ἑτέρας συστοιχίας αἱ ἀρχαὶ διὰ τὸ στερητικαὶ εἶναι ἀόριστοι· οὔτε γὰρ τόδε οὔτε τοιόνδε οὐδεμία αὐτῶν ἐστιν, [ὅτι][1] οὐδὲ τῶν ἄλλων κατηγοριῶν.

Τοῦ δὲ δοκεῖν ἀόριστον εἶναι τὴν κίνησιν αἴτιον ὅτι οὔτε εἰς δύναμιν τῶν ὄντων οὔτε εἰς ἐνέργειαν ἔστι θεῖναι αὐτὴν ἁπλῶς· οὔτε γὰρ τὸ δυνατὸν ποσὸν εἶναι κινεῖται ἐξ ἀνάγκης, οὔτε τὸ ἐνεργείᾳ ποσόν, ἥ τε κίνησις ἐνέργεια μέν τις εἶναι δοκεῖ, ἀτελὴς δέ· αἴτιον δ' ὅτι ἀτελὲς τὸ δυνατὸν οὗ

[1] [ὅτι del. Bonitz.—C.]

---

[a] [The Pythagoreans, who in their list of ten contrary principles in parallel columns (συστοιχίαι, *Met.* 986 a 22 ff.), headed by 'Limit' and 'Unlimited,' included κινούμενον under the Unlimited; and Plato, whose pupil Hermodorus (*ap.* Simplic. 247. 33) says that all things related to one another as 'great' to 'small' admit degrees of 'more' or 'less,' so that they can go on to infinity in either direction—towards the still more, or the still less. But things described as the *equal* or as that which is *at rest* or *in tune* do not admit degrees of more or less. There is always something more unequal than what is unequal, moving more than what is moving, more out of tune than what is out of tune. So what is of this nature is unstable, formless, unlimited, and may be called '*not-being*.' *Cf.* note on 187 a 17.—C.]

[b] [*Literally*, 'the principles in the second column are regarded as indefinite because of their negative character.' 'Not only is συστοιχία used of the Pythagorean list of opposites, but Aristotle himself recognizes a positive συστοιχία or column including such terms as being, unity, substance,

how movement has been defined by some[a] as 'otherwiseness' or 'unequalness' or non-existence. For none of these—being 'other' or 'unequal' or 'non-existent,' to wit—necessarily carries movement with it; nor is change either into or out of these any more than into or out of their opposites. The reason why motion was brought under these negative definitions was because it seemed to share the indefinite and elusive character of the negative limb of every antithesis.[b] For neither motion nor any of these negations seems to be a 'thing' or a 'quality' or to come under any other category.

Now the indefinite character attributed to motion is due to our inability to place it frankly either amongst the potentialities or amongst the realized functions of any kind of 'thing' as such; for there is no necessity in anything's changing its magnitude (for instance) either because it is potentially or because it is actually of this or that specified magnitude. [c] And the fact turns out to be that movement is a realization, but an uncompleted one; because a

and a negative συστοιχία including not-being, plurality, not-substance, . . In each case the negative is known, not *per se* but as the negation of the positive term' (Ross on *Met.* 1072 a 31).—C.]

[c] [*Literally*, 'And movement appears to be an actuality, but incomplete; the reason is that the potential, of which it is the actuality, is incomplete.' *Cf. Met.* 1048 b 29, 'Every movement is incomplete, for it is not true that at the same time we are taking a walk and have taken a walk, are being moved and have been moved; whereas it is true that the same thing at the same time has seen and is seeing, or has thought and is thinking. The latter sort of process is an actuality (ἐνέργεια), the former a movement (κίνησις).' Thus a movement is a process which is never complete till it is over; while an activity, *e.g.* seeing, is complete at every moment of its duration.—C.]

201 b ἐστιν ἐνέργεια. καὶ διὰ τοῦτο δὴ χαλεπὸν αὐτὴν λαβεῖν τί ἐστιν· ἢ γὰρ εἰς στέρησιν ἀναγκαῖον θεῖναι ἢ εἰς δύναμιν ἢ εἰς ἐνέργειαν ἁπλῆν, τούτων
202 a δ' οὐδὲν φαίνεται ἐνδεχόμενον. λείπεται τοίνυν ὁ εἰρημένος τρόπος, ἐνέργειαν μέν τινα εἶναι, τοιαύτην δ' ἐνέργειαν οἵαν εἴπαμεν—χαλεπὴν μὲν ἰδεῖν, ἐνδεχομένην δ' εἶναι.

Κινεῖται δὲ καὶ τὸ κινοῦν ὥσπερ εἴρηται, πᾶν τὸ δυνάμει ὂν κινητὸν καὶ οὗ ἡ ἀκινησία ἠρεμία ἐστίν (ᾧ γὰρ ἡ κίνησις ὑπάρχει, τούτῳ ἡ ἀκινησία ἠρεμία)· τὸ γὰρ πρὸς τοῦτο ἐνεργεῖν, ᾗ τοιοῦτον, αὐτὸ τὸ κινεῖν ἐστι· τοῦτο δὲ ποιεῖ θίξει, ὥστε ἅμα καὶ πάσχει. διὸ ἡ κίνησις ἐντελέχεια τοῦ

---

[a] [*Or*: ' Now that which causes motion in the way above mentioned (ὥσπερ εἴρηται alludes to φυσικῶς in the passage referred to, 201 a 23 ὥστε καὶ τὸ κινοῦν φυσικῶς κινητόν), —namely anything that is a potential movable or whose motionlessness can be described as " rest " (for " rest " means the absence of motion in a thing to which motion properly belongs)—is also itself in motion : for the act of causing motion is precisely acting upon that which is thus movable, as such; and this effect is produced by contact, so that the mover is acted upon at the same time that it acts. Accordingly, motion is the actualization of the movable, *qua* movable, but this occurs by contact with the thing capable of moving it, so that this latter thing is simultaneously affected (by some further thing moving it). But the thing which is actually causing motion (unlike the moved, which is potential and incomplete while the motion lasts) will always bring with it some element of (actual) form. . . .'—C.]

[b] [Contact and action-passion between bodies are exhaustively discussed in *De gen. et corr.* i. 6. The Heavens ' touch ' the sublunary world and communicate motion to it,

potentiality, as long as it is such, is by its nature uncompleted, and therefore its actual functioning —which motion is—must stop short of the completion: on the attainment of the end, the motion towards it no longer exists, but is merged in the reality. Hence the difficulty of seizing its nature. For what can it be, one may ask, save a shortage, or a potentiality, or a full attainment? But none of these seems possible. It remains then, as has been said, to define it as a kind of realization or attainment, but the kind we have described, difficult to pin down, it is true, but not impossible to be.

Now[a] everything that is capable of motion and which is at rest when not moving, is itself in motion whenever it produces motion in anything else. For what we mean by 'station' or 'being at rest' is the absence of motion in that to which motion is possible, and its actualizing of this specific potentiality is just that we mean by the very act of moving. But this is effected by the body which has the potentiality of being in motion receiving the impact of a body which is in actual motion, so that the body that becomes active, by entering into motion, at the same time becomes passive to the impact that keeps it in motion; but this is incidental, for primarily the movement is the realization of the thing's capacity for being in motion.[b]

but are not affected by it; whereas bodies in the sublunary world have *reciprocal* contact, which involves that they alter and are altered by one another. The Heavens are *κινοῦντα*, but not *δυνάμει κινητὰ καὶ ὧν ἡ ἀκινησία ἠρεμία ἐστίν*. Hence this qualification is added here. This paragraph leads up to the problem discussed in the next chapter: Where does the actualization reside—in the mover or in the moved?—C.]

202 a κινητοῦ, ᾗ κινητόν· συμβαίνει δὲ τοῦτο θίξει τοῦ κινητικοῦ, ὥσθ' ἅμα καὶ πάσχει.

Εἶδος δὲ ἀεὶ οἴσεταί τι τὸ κινοῦν—ἤτοι τόδε ἢ τοιόνδε ἢ τοσόνδε—ὃ ἔσται ἀρχὴ καὶ αἴτιον τῆς κινήσεως, ὅταν κινῇ· οἷον ὁ ἐντελεχείᾳ ἄνθρωπος ποιεῖ ἐκ τοῦ δυνάμει ὄντος ἀνθρώπου ἄνθρωπον.

The apparent contradiction between the passage from *Met.* 1048 b 29 quoted in note *c*, p. 203 and that in *Phys.* VI. vi. (Vol. II. p. 153) is resolved when we realize that in *Met.* 1048 b 29 Aristotle is referring to the whole movement towards a particular goal; which movement no longer exists when the goal is reached.

Whereas in the *Physics* he is thinking not only of the whole movement towards this final goal, but also of movement towards a point on its course, which movement automatically ceases to exist as the mobile passes, *without*

CHAPTER III

ARGUMENT

[*It follows from the definition of motion that the movement, which is the actualization of both potentialities—that of the mover and that of the moved—occurs in the moved. It is one actualization, with two conceptually distinct aspects* (202 a 13-21 = *Met.* 1066 a 26-34).

Against this there is a specious objection: that there must be two distinct actualizations, active and passive, and if the active is in the agent, the mover will itself be moved; if both are in the patient, we shall have two different changes in the same thing producing a single result (a 21-36). *Whereas if there is only one actualization, how can two con-*

Καὶ τὸ ἀπορούμενον δὲ φανερόν, ὅτι ἐστὶν ἡ κίνησις ἐν τῷ κινητῷ· ἐντελέχεια γάρ ἐστι τούτου, καὶ ὑπὸ τοῦ κινητικοῦ. καὶ ἡ τοῦ κινητικοῦ δὲ

But since, as we have said, the motion of one body is initiated and sustained by the impact and pressure of another body, and is therefore incidentally passive to that body, it is natural that the moving cause of the motion should carry over to the thing moved certain characteristics of its own that fall under such categories as substance or quality or quantity, which become principles and determinants of the character of the change (within the potentialities of the subject) which it produces by its casual action. Thus the man in actuality produces the actual man out of something which has the potentiality of becoming man.

stopping at that point. Thus movement towards an intermediate point has ceased while movement towards the final goal is still going on.—R. W.

CHAPTER III

ARGUMENT (*continued*)

ceptually different things have the same actualization? And will not to learn and to teach be the same thing (a 36-b 5)?

These objections are shown not to be substantial (b 5-22).

The formula by which we have defined movement can easily be applied to each special sort of movement. The treatment of movement is concluded with a final definition (b 22-29).—*C.*]

The question where the actual movement is realized is answered easily, for it obviously consists in that which was in potentiality capable of motion being now in the realized 'act of motion'; and this realized 'act of motion' is caused by what was in potentiality 'capable of causing' it now realizing that potentiality by actually moving it. So that the one 'act of moving' in the movable is the realization,

202 a ἐνέργεια οὐκ ἄλλη ἐστίν. δεῖ μὲν γὰρ εἶναι ἐντελέχειαν ἀμφοῖν· κινητικὸν μὲν γάρ ἐστι τῷ δύνασθαι, κινοῦν δὲ τῷ ἐνεργεῖν, ἀλλ' ἔστιν ἐνεργητικὸν τοῦ κινητοῦ, ὥστε ὁμοίως μία ἡ ἀμφοῖν ἐνέργεια ὥσπερ τὸ αὐτὸ διάστημα ἓν πρὸς δύο καὶ δύο πρὸς ἕν, καὶ τὸ ἄναντες καὶ τὸ κάταντες· ταῦτα γὰρ ἓν μὲν ἔστιν, ὁ μέντοι λόγος οὐχ εἷς. ὁμοίως δὲ καὶ ἐπὶ τοῦ κινοῦντος καὶ κινουμένου.

Ἔχει δ' ἀπορίαν λογικήν· ἀναγκαῖον γὰρ ἴσως εἶναί τινα ἐνέργειαν ἄλλην τοῦ ποιητικοῦ καὶ τοῦ παθητικοῦ· τὸ μὲν δὴ ποίησις, τὸ δὲ πάθησις, ἔργον δὲ καὶ τέλος τοῦ μὲν ποίημα, τοῦ δὲ πάθος. ἐπεὶ οὖν ἄμφω κινήσεις, εἰ μὲν ἕτεραι, ἐν τίνι; ἢ γὰρ ἄμφω ἐν τῷ πάσχοντι καὶ κινουμένῳ, ἢ ἡ μὲν ποίησις ἐν τῷ ποιοῦντι ἡ δὲ πάθησις ἐν τῷ πάσχοντι (εἰ δὲ δεῖ καὶ ταύτην ποίησιν καλεῖν, ὁμώνυμος ἂν εἴη). ἀλλὰ μὴν εἰ τοῦτο, ἡ κίνησις

---

[a] [*Literally*, ' for it must be the realization of both.'—C.]

[b] The substrate (ὑποκείμενον) of these two ratios is the same interval. But the ratios are not the same. And again the same slope is the substrate of both an upward and a downward slope, but these two are different.

not only of its own capacity of being moved, but of the capacity of the potential mover to be actually moving it. For potential mover and potential moved must each have its proper realization or 'being in act,'[a] for to be a potential mover is to be capable of moving something, but to be an actual mover is to be in the 'act of moving' the something. Now this act is accomplished in the movement of the thing moved, so that the actualizing of the two potentialities coincides (objectively and materially) in the one movement; just as the interval between 1 and 2 is the same as that between 2 and 1,[b] and the slope of a line is the same whether you regard it as sloping up or down; for in these cases we are dealing with one and the same thing which may be regarded or defined from two different approaches. So too with the mover and the moved.[c]

A specious difficulty may, however, be urged. It seems hard to deny that the energizing of a potential activity must be different from that of a potential passivity; so the one will be a *doing* and the other a *suffering* ('passion'), and the actuality and goal of the one is a deed and of the other an experience. Since, then, both are movements, (*a*) if they are two movements and not one, where are they? Obviously one of them is in the moved. Either then (1) they are both in the moved, or 'patient,' or (2) the one that is a *doing* is in the agent and the one that is a *suffering* in the patient (and it will not do to say that this too is a doing, for that would be using words equivocally). In this case (2) the 'movement' (which we must regard as a 'suffering,' wherever it is seated, or it would not be a movement

[c] See Gen. Introd. p. xxviii, on magnet and needle.

202a ἐν τῷ κινοῦντι ἔσται· ὁ γὰρ αὐτὸς λόγος ἐπὶ κινοῦντος καὶ κινουμένου. ὥστ' ἢ πᾶν τὸ κινοῦν κινήσεται, ἢ ἔχον κίνησιν οὐ κινήσεται. εἰ δ' ἄμφω ἐν τῷ κινουμένῳ καὶ πάσχοντι—καὶ ἡ ποίησις καὶ ἡ πάθησις—καὶ ἡ δίδαξις καὶ ἡ μάθησις δύο οὖσαι ἐν τῷ μανθάνοντι, πρῶτον μὲν ἡ ἐνέργεια ἡ ἑκάστου οὐκ ἐν ἑκάστῳ ὑπάρξει, εἶτα ἄτοπον τὸ δύο κινήσεις ἅμα κινεῖσθαι· τίνες γὰρ ἔσονται ἀλλοιώσεις δύο τοῦ ἑνὸς καὶ εἰς ἓν εἶδος; ἀλλ' ἀδύνατον, ἀλλὰ μία ἔσται ἡ ἐνέργεια. 202b ἀλλ' ἄλογον δύο ἑτέρων τῷ εἴδει τὴν αὐτὴν καὶ μίαν εἶναι ἐνέργειαν· καὶ ἔσται, εἴπερ ἡ δίδαξις καὶ ἡ μάθησις ταὐτὸ καὶ ἡ ποίησις καὶ ἡ πάθησις, καὶ τὸ διδάσκειν τῷ μανθάνειν ταὐτὸ καὶ τὸ ποιεῖν τῷ πάσχειν, ὥστε τὸν διδάσκοντα ἀνάγκη ἔσται πάντα μανθάνειν καὶ τὸν ποιοῦντα πάσχειν.

Ἢ οὔτε τὸ τὴν ἄλλου ἐνέργειαν ἐν ἑτέρῳ εἶναι

---

[a] See Book VIII. ch. v.

[b] [*Literally*, 'how can there be two particular qualitative changes of one thing terminating in a single form?' referring to the illustration μάθησις. *Cf. De anim.* 417 a 31 ὁ διὰ μαθήσεως ἀλλοιωθείς, though (as Philoponus there remarks) μάθησις might better be regarded as a γένεσις than as an ἀλλοίωσις.—C.]

at all) in which the agent realizes its potentiality must reside in the mover. Thus every mover must, as such, be in motion[a]—unless we are to say that a thing may have motion and yet be motionless; which is not the case. If, on the other hand, (1) the two movements—that which actualizes the potentiality of the agent and that which actualizes the potentiality of the patient—both take place in the moved or patient; if, for instance, the process which realizes the power of teaching and that which realizes the power of learning, being two distinct processes, both take place in the learner, then, in the first place, this would be to say that the two 'potentials' are not realizing their several potentialities each in itself; and, in the next place, it would involve the paradox of the same thing experiencing two movements or progressions at the same time—which is only possible if the progressions differ generically, *e.g.* one being from place to place and the other from quality to quality.[b] That is impossible: the actualization must be one. But (*b*) if it is one, then it may be urged that it is monstrous to assert that two capacities that are conceptually different can find one and the same realization; and if the actualizing of teaching and the actualizing of learning, or that of doing and that of suffering, coincide, then to be teaching and to be learning will be the same thing, and to be acting will be the same as being subject to another's action. At that rate every one who was teaching would have to be learning all the things he taught, and every one who was doing anything would be having the same thing done to him all the while.

But this is only a quibble. For there is nothing

202 b ἄτοπον (ἔστι γὰρ ἡ δίδαξις ἐνέργεια τοῦ διδασκαλικοῦ, ἐν τινὶ μέντοι, καὶ οὐκ ἀποτετμημένη, ἀλλὰ τοῦδε ἐν τῷδε)· οὔτε μίαν δυοῖν τὴν αὐτὴν εἶναι κωλύει, μὴ ὡς τὸ εἶναι τὸ αὐτὸ ἀλλ' ὡς ὑπάρχει τὸ δυνάμει ὂν πρὸς τὸ ἐνεργοῦν. οὔτ' ἀνάγκη τὸν διδάσκοντα μανθάνειν, οὐδ' εἰ τὸ ποιεῖν καὶ πάσχειν τὸ αὐτό ἐστι, μὴ μέντοι ὡς τὸν λόγον εἶναι ἕνα τὸν τὸ τί ἦν εἶναι λέγοντα (ὡς λώπιον καὶ ἱμάτιον) ἀλλ' ὡς ἡ ὁδὸς ἡ Θήβηθεν Ἀθήναζε καὶ ἡ Ἀθήνηθεν εἰς Θήβας, ὥσπερ εἴρηται καὶ πρότερον· οὐ γὰρ ταὐτὰ πάντα ὑπάρχει τοῖς ὁπωσοῦν τοῖς αὐτοῖς, ἀλλὰ μόνον οἷς τὸ εἶναι τὸ αὐτό. οὐ μὴν ἀλλ' οὐδ' εἰ ἡ δίδαξις τῇ μαθήσει τὸ αὐτό, καὶ τὸ μανθάνειν τῷ διδάσκειν· ὥσπερ οὐδ' εἰ ἡ διάστασις μία τῶν διεστηκότων, καὶ τὸ διίστασθαι ἐνθένδε ἐκεῖσε κἀκεῖθεν δεῦρο ἓν καὶ τὸ αὐτό. ὅλως δ' εἰπεῖν οὐδ' ἡ δίδαξις τῇ μαθήσει οὐδ' ἡ ποίησις τῇ παθήσει τὸ αὐτὸ κυρίως, ἀλλ' ᾧ ὑπάρχει ταῦτα—ἡ κίνησις· τὸ γὰρ

---

[a] [At 202 a 18.—C.]

paradoxical in the actualizing of the potentiality of one thing having its effective seat in another (for the potentiality of an agent—the ' teacher ' for instance —is actualized not *in vacuo* or in isolation, but in the ' taught ' : it is the actualizing *of* this *in* that) ; nor is there anything to prevent the same one thing being the actualizing of two potentialities—not in the sense of unqualified identity, but in the sense of being in the one relation of actuality to potentiality with regard to the two potentialities. So it does not follow that the teacher must be learning, even though we admit that action and passion coincide, not in the sense of being the same in the definitions that determine them (like ' garments ' and ' clothes '), but in the sense in which (to vary the illustrations already given [a]) the road from Thebes to Athens is the same as the road from Athens to Thebes ; for things need not be identical in all respects because they are the same in some, but only if they are identical in what they actually are—in a word if they are not two ' things ' at all, but only two names or definitions of the same thing. Nor yet, even if the actualizing of teaching is identical with the actualizing of learning, does it follow that to be learning is the same thing as to be teaching ; any more than, if the interval between two distant points *A* and *B* is one interval, it follows that being distant from *B* when you are at *A* is the same as being distant from *A* when you are at *B*. To put it generally ; there is, in the strict sense, no absolute identity even between the actualization of teaching and that of learning, or between action and passion ; but only the thing which has these two aspects—the motion—is one and the same thing ; for there is a conceptual

202 b τοῦδε ἐν τῷδε καὶ τὸ τοῦδε ὑπὸ τοῦδε ἐνέργειαν εἶναι ἕτερον τῷ λόγῳ.

Τί μὲν οὖν ἐστι κίνησις εἴρηται καὶ καθόλου καὶ κατὰ μέρος—οὐ γὰρ ἄδηλον ὡς ὁρισθήσεται τῶν εἰδῶν ἕκαστον αὐτῆς· ἀλλοίωσις μὲν γὰρ ἡ τοῦ ἀλλοιωτοῦ, ᾗ ἀλλοιωτόν, ἐντελέχεια—ἔτι δὲ γνωριμώτερον, ἡ τοῦ δυνάμει ποιητικοῦ καὶ παθητικοῦ, ᾗ τοιοῦτον, ἁπλῶς τε καὶ πάλιν καθ' ἕκαστον, ἢ[1] οἰκοδόμησις ἢ ἰάτρευσις. τὸν αὐτὸν δὲ λεχθήσεται τρόπον καὶ περὶ τῶν ἄλλων κινήσεων ἑκάστης.

CHAPTER IV

ARGUMENT

[*The study of Infinity, as involved in space, motion, and time, belongs to Physics* (202 b 30-36).

Earlier philosophers all discuss the Unlimited. (1) *The Pythagoreans and Plato treat 'the Infinite itself' as having an independent substantive existence* (b 36-203 a 16). (2) *The physicists believe in an infinite element (the Ionian monists), or an infinite mass of particles. These, according to Anaxagoras, originally formed an infinite mixture before differentiation began. Democritus' atoms are eternally distinct, but the 'body' of which they are all composed is a sort of material*

Ἐπεὶ δ' ἐστὶν ἡ περὶ φύσεως ἐπιστήμη περὶ μεγέθη καὶ κίνησιν καὶ χρόνον, ὧν ἕκαστον ἀναγκαῖον ἢ ἄπειρον ἢ πεπερασμένον εἶναι—εἰ

[1] [For ἢ E has ὅτι (according to Bekker).—C.]

distinction between the aspects—between 'being the actualizing of this in that' and 'being the actualizing of that by this.'

We have now explained the nature both of movement in general and of the special kinds of movement (for it is easy to see how each special kind is to be defined ; qualitative alteration, for example, as the actualization of what has the potentiality of this kind of alteration, in so far as it has that potentiality); or to put it still more clearly : motion is the actualization of the potentially active and of the potentially passive, as such—as a general formula and again as applied to any particular case, the process of building or of healing. Each of the other kinds of motion will be described in similar terms.

CHAPTER IV

ARGUMENT (*continued*)

principle, indeterminate when considered in abstraction from atomic forms (a 16-b 2).

All agree that the Infinite is a 'beginning' or principle, which the earliest physicists regarded as a divine, self-moving, element (b 3-15).

Five considerations which lead to a belief in something that is infinite (b 15-30).

The problems that arise about its nature (b 30-204 a 2).

The senses in which 'infinite' is used (a 2-7 = *Met.* 1066 a 35-b 1).—*C.*]

The study of Nature is concerned with extension, motion, and time ; and since each one of these three must be either limited or unlimited (which is not to

202 b καὶ μὴ πᾶν ἔστιν ἄπειρον ἢ πεπερασμένον, οἷον πάθος ἢ στιγμή· τῶν γὰρ τοιούτων ἴσως οὐδὲν ἀναγκαῖον ἐν θατέρῳ τούτων εἶναι—προσῆκον ἂν εἴη τὸν περὶ φύσεως πραγματευόμενον θεωρῆσαι περὶ ἀπείρου, εἰ ἔστιν ἢ μή, καὶ εἰ ἔστι, τί ἐστιν.

203 a Σημεῖον δ' ὅτι τῆς ἐπιστήμης οἰκεία ἡ περὶ αὐτὸ θεωρία· πάντες γὰρ οἱ δοκοῦντες ἀξιολόγως ἧφθαι τῆς τοιαύτης φιλοσοφίας πεποίηνται λόγον περὶ τοῦ ἀπείρου, καὶ πάντες ὡς ἀρχήν τινα τιθέασι τῶν ὄντων,—

Οἱ μέν, ὥσπερ οἱ Πυθαγόρειοι καὶ Πλάτων, καθ' αὑτό, οὐχ ὡς συμβεβηκός τινι ἑτέρῳ ἀλλ' οὐσίαν αὐτὸ ὂν τὸ ἄπειρον. πλὴν οἱ μὲν Πυθαγόρειοι ἐν τοῖς αἰσθητοῖς (οὐ γὰρ χωριστὸν ποιοῦσι τὸν ἀριθμόν), καὶ εἶναι τὸ ἔξω τοῦ οὐρανοῦ τὸ ἄπειρον· Πλάτων δὲ ἔξω μὲν οὐδὲν εἶναι σῶμα, οὐδὲ τὰς ἰδέας, διὰ τὸ μηδέ που εἶναι αὐτάς, τὸ μέντοι ἄπειρον καὶ ἐν τοῖς αἰσθητοῖς καὶ ἐν ἐκείναις εἶναι. καὶ οἱ μὲν τὸ ἄπειρον εἶναι τὸ ἄρτιον· τοῦτο γὰρ ἐναπολαμβανόμενον καὶ ὑπὸ τοῦ περιττοῦ περαινόμενον παρέχειν τοῖς οὖσι τὴν ἀπειρίαν· σημεῖον δ' εἶναι τούτου τὸ συμβαῖνον ἐπὶ τῶν ἀριθμῶν· περιτιθεμένων γὰρ τῶν γνωμόνων περὶ τὸ ἓν καὶ

---

[a] [*Or*, 'contained in sensible things' as a constituent of them; for 'the elements of Number (Even = Unlimited, Odd = Limit or Limited) are the elements of all things,' *Met.* 986 a 1 and 17.—C.]

say that *everything* must be limited or unlimited, for it does not preclude us from denying the necessity of such things as a ‘ point ’ or an ‘ experience ’ being either one or the other), it follows that the student of Nature must consider the question of the unlimited, with a view to determining whether it exists at all, and, if so, what is its nature.

That this inquiry is really germane to our subject is indicated by the fact that all philosophers of repute who have dealt with Physics have discussed the ‘ unlimited,’ and all have regarded it as in some sense a ‘ principle ’ of actually existing things.

Some, such as the Pythagoreans and Plato, have regarded the unlimited or undetermined as existing in itself, and not as being a condition incident to something else, but having its own independent substantive existence. But the Pythagoreans regard it as something cognizable by the senses [a] (since they do not regard ‘ number ’ as having a detached existence apart from the objects of the senses), and hold that what lies beyond the heaven is ‘ the unlimited ’; whereas Plato holds that no material body lies outside the heaven at all—nor any ‘ Ideas ’ either, for they cannot be said so much as to be ‘ located ’ anywhere ; but, for all that, he finds the ‘ undetermined ’ both in the objects of sense and in the ‘ Ideas.’ Again, the Pythagoreans identify the ‘ undetermined ’ with the ‘ even.’ For evenness, when it has been enclosed and determined by the odd unit, is still the principle of the undetermined aspect of things. And this they find reflected in the properties of numbers ; for successive additions to the unit of the odd numbers (even numbers gnomonized by the extra unit) preserve the definite quad-

203 a χωρὶς ὁτὲ μὲν ἄλλο ἀεὶ γίγνεσθαι τὸ εἶδος, ὁτὲ

---

[a] [The Pythagoreans represented numbers by patterns or figures (εἴδη) of pebbles or counters, called 'terms' (ὅροι), as being like terminal stones marking out a field or space, which without them would be 'unlimited' and 'void.' Each pebble stands for an *indivisible* unit (monad). The unit is not a number. Numbers are aggregates of two or more monads; so 2 is the first even number, 3 the first odd.

(1) τὸ ἄπειρον εἶναι τὸ ἄρτιον. *Cf. Met.* 986 a 1 and 17, the elements of number (and hence of all things) are the Even, which is Unlimited, and the Odd, which is Limited. According to the authorities quoted by Simplic. (455. 15), the Even is unlimited because the Even is divisible into two equal parts *ad infinitum*, but the addition of the odd sets a limit to this division. This and similar statements are explained graphically by Heidel (*Arch. f. Gesch. Phil.* xiv. 390 ff.). Take any even number, say 10 (figured thus ⁚⁚⁚⁚⁚→). The process of halving, represented by the arrow, goes on indefinitely; but it would be arrested at any point by the addition of an odd unit, *e.g.* in 11 (⁚⁚⁚⁚⁚→·). (*Cf.* Ross on *Met.* 986 a 18 and references there.)

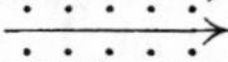

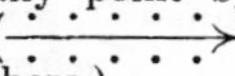

(2) 11 τοῦτο γὰρ . . . 12 ἀπειρίαν. In this clause τοῦτο may mean 'this principle, the Even = Unlimited,' and τὸ περιττόν its opposite, 'the Odd = Limit.' As applied to numbers, the Unlimited is the empty space *enclosed* between two Limits (units) set in a row. But, by some Pythagoreans at least, the process of 'generating' numbers was not distinguished from the generation of 'things'—cosmogony,—for 'things' (bodies existing in actual space) *are* 'numbers' (aggregates of monads or atomic bodies). So this clause may be compared with the Pythagorean cosmogonical process as described at *Met.* 1091 a 14, 'when the One had been constructed . . . the nearest part of the Unlimited began to be constrained and limited by the Limit.' Here the first monad, working outwards from the centre of an unformed mass, probably, of 'air' (or 'void'), introduces shape or limit, and forms a world. This void would (in the terms of our

rate form intact, whereas, if the determining unit is withheld, the succession of even numbers yields a perpetually varying and undetermined figure.[a]

passage) ' when enclosed by the Odd (Limit) contribute to things the element of unlimitedness ' (the vacancy contained by the planes bounding a body).

(3) περιτιθεμένων . . . ὁτὲ δὲ ἕν. The gnomon is the carpenter's tool, shaped , for measuring a right angle. When the *uneven* numbers (3, 5, 7, etc.) are ' put round ' the unit in figures of this shape, the result is always the same figure, viz., a square, thus: 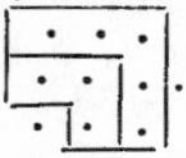. Arithmetically

$$1+3+5+\ .\ .\ .\ +(2n-1)=n^2.$$

(4) καὶ χωρὶς . . . ἄλλο ἀεὶ γίγνεσθαι εἶδος is explained as meaning that, if *even* numbers are put in gnomons round the first even, 2, the resulting figures are oblongs, all dissimilar in form—an infinity of forms 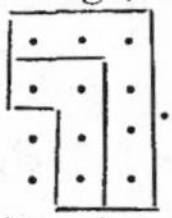. The problem is, how to get this meaning out of καὶ χωρίς. Heath (*Gk. Maths.* i. 83) and others take χωρίς to mean ' in the separate (*or* other) case.' Taylor's explanation (*C.R.* xl. 150) is unsatisfactory. Burnet (*E. G. Ph.*[3] 103) understands ' χωρὶς τοῦ ἑνός, *i.e.* starting from 2, not from 1.' But why say ' without the unit ' instead of περὶ τὰ δύο? Possibly ' putting gnomons round without the unit ' means putting *two* minimum gnomons *round one another*, thus . Each succeeding gnomon will then contain an even number (6, 8, 10, etc.), and we shall obtain the traditional series of varying oblongs.—C.]

203 a δὲ ἕν. Πλάτων δὲ δύο τὰ ἄπειρα, τὸ μέγα καὶ τὸ μικρόν.

Οἱ δὲ περὶ φύσεως ἅπαντες[1] ὑποτιθέασιν ἑτέραν τινὰ φύσιν τῷ ἀπείρῳ τῶν λεγομένων στοιχείων, οἷον ὕδωρ ἢ ἀέρα ἢ τὸ μεταξὺ τούτων. τῶν δὲ πεπερασμένα ποιούντων στοιχεῖα οὐθεὶς ἄπειρα ποιεῖ· ὅσοι δ' ἄπειρα ποιοῦσι τὰ στοιχεῖα, καθάπερ Ἀναξαγόρας καὶ Δημόκριτος, ὁ μὲν ἐκ τῶν ὁμοιομερῶν, ὁ δ' ἐκ τῆς πανσπερμίας τῶν σχημάτων, τῇ ἁφῇ συνεχὲς τὸ ἄπειρον εἶναί φασιν.

Καὶ ὁ μὲν ὁτιοῦν τῶν μορίων εἶναι μῖγμα ὁμοίως τῷ παντί, διὰ τὸ ὁρᾶν ὁτιοῦν ἐξ ὁτουοῦν γιγνόμενον. ἐντεῦθεν γὰρ ἔοικε καὶ ὁμοῦ ποτε πάντα χρήματα φάναι εἶναι· οἷον ἥδε ἡ σὰρξ καὶ τόδε τὸ ὀστοῦν, καὶ οὕτως ὁτιοῦν· καὶ πάντα ἄρα· καὶ ἅμα τοίνυν· ἀρχὴ γὰρ οὐ μόνον ἐν ἑκάστῳ ἐστὶ τῆς διακρίσεως, ἀλλὰ καὶ πάντων. ἐπεὶ γὰρ τὸ γιγνόμενον ἐκ τοῦ τοιούτου γίνεται σώματος, πάντων δ' ἐστὶ γένεσις (πλὴν οὐχ ἅμα), καί τινα ἀρχὴν δεῖ εἶναι τῆς γενέσεως. αὕτη δ' ἐστὶ μία,

[1] [ἅπαντες] ἀεὶ πάντες E, ἅπαντες ἀεὶ *al.* Simplic. 458. 17 (lemma) and 459. 8 omits ἀεί.—C.]

---

[a] [See 187 a 17, 201 b 20, notes.—C.]

[b] [*i.e.*, whereas Plato and the Pythagoreans talked simply of ' the Unlimited ' in the abstract, the physicists have an unlimited *something*. (1) The Ionian monists make this *something* water, air, etc. Of the pluralists, (2) Empedocles has no unlimited, but a finite number of elements, each finite in quantity; (3) Anaxagoras and Democritus have an infinite number of elementary particles, and ' the Infinite ' means the infinite aggregate of these particles in contact with one another.—C.]

Plato, for his part, recognizes two aspects of the 'unlimited,' the great and the small.[a]

The physicists, on the other hand, all make some other nature—one of their so-called elements, water or air or the intermediate between these—a subject of which 'unlimited' is predicate;[b] and only those of them, such as Anaxagoras and Democritus, who hold the elements themselves to be unlimited, admit the principle of the unlimited at all. The others, who regard the 'elements' as limited, will not allow that anything at all is unlimited; whereas Anaxagoras finds the unlimited in his 'parts-after-their-kind,' and Democritus in his hotch-potch of multi-shaped atoms; while both of them regarded it as a mass having continuity by contact of unlike particles.

Anaxagoras, moreover, held that any part whatsoever was as much a mixture as the whole, because he saw that any one thing comes into being out of any other thing. For this seems to be the original ground of his assertion that 'all things are together' at some moment. Take, for instance, this piece of flesh and this piece of bone; either can come out of the other; and so with any one thing; and accordingly, *all* things must come out of *all* things, and therefore all things must be together at the same moment; for there must be a beginning of the separating out, not only in each thing, but of all things in general. For since the thing that comes into separate existence comes out of that in which it exists unseparated, and everything thus comes out of the unseparated (though successively, not all at once), there must be some initiating principle of 'coming out of the undistinguished' in general.

203a ὃν ἐκεῖνος καλεῖ νοῦν· ὁ δὲ νοῦς ἀπ᾽ ἀρχῆς τινος ἐργάζεται νοήσας, ὥστε ἀνάγκη ὁμοῦ ποτε πάντα εἶναι καὶ ἄρξασθαί ποτε κινούμενα. Δημόκριτος δ᾽ οὐδὲν ἕτερον ἐξ ἑτέρου γίγνεσθαι τῶν πρώτων
203b φησίν· ἀλλ᾽ ὅμως γε αὐτῶν τὸ κοινὸν σῶμα πάντων ἐστὶν ἀρχή, μεγέθει κατὰ μόρια καὶ σχήματι διαφέρον.

Ὅτι μὲν οὖν προσήκουσα τοῖς φυσικοῖς ἡ θεωρία, δῆλον ἐκ τούτων. εὐλόγως δὲ καὶ ἀρχὴν αὐτὸ τιθέασι πάντες· οὔτε γὰρ μάτην αὐτὸ οἷόν τε εἶναι, οὔτε ἄλλην ὑπάρχειν αὐτῷ δύναμιν πλὴν ὡς ἀρχήν· ἅπαντα γὰρ ἢ ἀρχὴ ἢ ἐξ ἀρχῆς, τοῦ δὲ ἀπείρου οὐκ ἔστιν ἀρχή· εἴη γὰρ ἂν αὐτοῦ πέρας. ἔτι δὲ καὶ ἀγένητον καὶ ἄφθαρτον ὡς ἀρχή τις οὖσα· τό τε γὰρ γενόμενον ἀνάγκη τέλος λαβεῖν, καὶ τελευτὴ πάσης ἐστὶ φθορᾶς. διὸ (καθάπερ λέγομεν) οὐ ταύτης ἀρχή, ἀλλ᾽ αὕτη τῶν ἄλλων εἶναι δοκεῖ καὶ περιέχειν ἅπαντα καὶ πάντα κυβερνᾶν, ὥς φασιν ὅσοι μὴ ποιοῦσι παρὰ τὸ ἄπειρον ἄλλας αἰτίας, οἷον νοῦν ἢ φιλίαν· καὶ

---

[a] [In this paragraph the meaning of ἀρχή shifts. The 'unlimited' of Anaxagoras considered as the *initial state* of matter (ἀρχή) is an unlimited confused mass, actually existing as such before the moment of time when Mind makes a *beginning* (ἀρχή) of the 'separating-out.' This is contrasted with Democritus's multitude of immutably distinct atoms; but Aristotle suggests that 'body', of which these atoms all consist, may be regarded as a sort of indeterminate and 'unlimited' material *principle* (ἀρχή), though it never actually exists by itself. So both Anaxagoras and Democritus recognize an 'unlimited' (or 'indeterminate') *principle* (ἀρχή).—C.]

[b] [The archaic form of this traditional argument recalls Plato's proof of the immortality of the self-moving soul: *Phaedrus* 245 D ἀρχὴ δὲ ἀγένητον· ἐξ ἀρχῆς γὰρ ἀνάγκη πᾶν τὸ

And this principle is one, and is what he designates as 'Intelligence'; which Intelligence must have made some beginning of its action as such; so there must have been a time when all things were indistinguishably one, and a point of time at which the stirring of segregative movement began. Democritus, on the other hand, will not have it that any of his ultimate atoms can proceed out of any other; but nevertheless the matter itself that is distinguished in the different kinds of atom by size and shape, is a common principle that underlies them all.[a]

Clearly then the study of the undetermined or unlimited is germane to that of Nature. And again, all those who accept it are quite right in regarding it as a 'principle'; for, if it exists, it must affect things somehow, and it cannot affect them except as a principle; for everything is either determined by some principle or is a principle itself, and the undetermined cannot be determined at all, and so cannot depend upon anything else as its principle. And further, being a principle, it can have no beginning or end of existence; for whatever comes into being must come to an end, and there must be a term to any process of perishing.[b] So the 'unlimited' cannot be derived from any other principle, but is itself regarded as the principle of the other things, 'embracing and governing all,' as it is said to do by such as accept it, unless indeed they accept other principles alongside of it, such as 'Intelligence' or 'Amity.' This unlimited, then,

γιγνόμενον γίγνεσθαι, αὐτὴν δὲ μήδ' ἐξ ἑνός. εἰ γὰρ ἔκ του ἀρχὴ γίγνοιτο, οὐκ ἂν ἔτι ἀρχὴ γίγνοιτο. ἐπειδὴ δὲ ἀγένητόν ἐστιν, καὶ ἀδιάφθορον . . . Rostagni (*Verbo di Pitagora*, 137), suggests that Plato was echoing Alcmaeon, but Anaximander may have argued similarly.—C.]

203 b τοῦτ᾽ εἶναι τὸ θεῖον· ἀθάνατον γὰρ καὶ ἀνώλεθρον, ὥς φησιν ὁ Ἀναξίμανδρος καὶ οἱ πλεῖστοι τῶν φυσιολόγων.

Τοῦ δ᾽ εἶναί τι ἄπειρον ἡ πίστις ἐκ πέντε μάλιστ᾽ ἂν συμβαίνοι σκοποῦσιν· ἔκ τε τοῦ χρόνου (οὗτος γὰρ ἄπειρος)· καὶ ἐκ τῆς ἐν τοῖς μεγέθεσι διαιρέσεως (χρῶνται γὰρ καὶ οἱ μαθηματικοὶ τῷ ἀπείρῳ)· ἔτι τῷ οὕτως ἂν μόνως μὴ ὑπολείπειν γένεσιν καὶ φθοράν, εἰ ἄπειρον εἴη ὅθεν ἀφαιρεῖται τὸ γιγνόμενον· ἔτι τῷ τὸ πεπερασμένον ἀεὶ πρός τι περαίνειν, ὥστε ἀνάγκη μηδὲν εἶναι πέρας, εἰ ἀεὶ περαίνειν ἀνάγκη ἕτερον πρὸς ἕτερον· μάλιστα δὲ καὶ κυριώτατον, ὃ τὴν κοινὴν ποιεῖ ἀπορίαν πᾶσιν· διὰ γὰρ τὸ ἐν τῇ νοήσει μὴ ὑπολείπειν, καὶ ὁ ἀριθμὸς δοκεῖ ἄπειρος εἶναι καὶ τὰ μαθηματικὰ μεγέθη καὶ τὸ ἔξω τοῦ οὐρανοῦ. ἀπείρου δ᾽ ὄντος τοῦ ἔξω, καὶ σῶμα ἄπειρον εἶναι δοκεῖ καὶ κόσμοι· τί γὰρ μᾶλλον τοῦ κενοῦ ἐνταῦθα ἢ ἐνταῦθα; ὥστ᾽ εἴπερ μοναχοῦ, καὶ πανταχοῦ εἶναι τὸν ὄγκον. ἅμα δ᾽ εἰ καὶ ἔστι κενόν, καὶ τόπος ἄπειρος, καὶ σῶμα ἄπειρον εἶναι

---

[a] [Diels, *Vors.* Anaximandros 15. The φυσικοί who have not distinct moving causes (Anaxagoras' νοῦς, Empedocles' φιλία) regard the primary stuff of things as self-moving, *i.e.* animate with a divine (immortal) life, and 'governing' the

would be the divinity itself, being 'immortal and indestructible,' as Anaximander and most of the physicists declare it to be.[a]

Now the belief in the existence of something 'unlimited' appears to rest, in the main, upon the consideration of five things: (1) Time (which is regarded as having no limit); (2) the divisibility of magnitudes (which is treated by mathematicians as possible without limit); (3) the idea that from the unfailing persistence of genesis and its opposite it follows that the things which come into being are drawn from an unlimited store; or again (4) the argument that, since whatever is limited reaches its limit by coming up to something else, there can be no absolute limit, for nothing can be limited except by something else beyond its limit. But the most effective and central consideration, the weight of which all thinkers alike have felt, is that (5) the imagination can always conceive a 'beyond' reaching out from any limit, so that the series of numerals seems to have no limit, nor mathematical magnitudes, nor the 'beyond the heavens.' Moreover, it seems to follow from the 'beyond' being unlimited, that 'body' must be unlimited, and that there must be unlimited worlds. For why should there be more vacancy in one place than in another? So, if mass[b] is anywhere, must it not be everywhere? Or even if you assume vacancy everywhere, does it not follow that there is 'place' everywhere, and 'body'

movement of the finite things it 'embraces' (συγκρατεῖ, περιέχει Anaximenes, *frag.* 2: ἐκυβέρνησε Heracleitus, *frag.* 41, etc.).—C.]

[b] [ὄγκος was used as synonym of 'atom' by Democritus (Diog. Laert. ix. 44). Aristotle refers to the unlimited worlds of the Atomists.—C.]

203 b ἀναγκαῖον· ἐνδέχεσθαι γὰρ ἢ εἶναι οὐδὲν διαφέρει ἐν τοῖς ἀϊδίοις.

Ἔχει δ' ἀπορίαν ἡ περὶ τοῦ ἀπείρου θεωρία· καὶ γὰρ μὴ εἶναι τιθεμένοις πόλλ' ἀδύνατα συμβαίνει, καὶ εἶναι· ἔτι δὲ ποτέρως ἔστι—πότερον ὡς οὐσία, ἢ ὡς συμβεβηκὸς καθ' αὑτὸ φύσει τινί, ἢ οὐδετέρως, ἀλλ' οὐδὲν ἧττόν ἐστιν ἄπειρον ἢ
204 a ἄπειρα τῷ πλήθει. μάλιστα δὲ φυσικοῦ ἐπισκέψασθαι[1] εἰ ἔστι μέγεθος αἰσθητὸν ἄπειρον.

Πρῶτον οὖν διοριστέον ποσαχῶς λέγεται τὸ ἄπειρον. ἕνα μὲν δὴ τρόπον τὸ ἀδύνατον διελθεῖν τῷ μὴ πεφυκέναι διιέναι (ὥσπερ ἡ φωνὴ ἀόρατος)· ἄλλως δὲ τὸ διέξοδον ἔχον ἀτελεύτητον, ἢ ὃ μόλις, ἢ ὃ πεφυκὸς ἔχειν μὴ ἔχει διέξοδον ἢ πέρας.

Ἔτι ἄπειρον ἅπαν ἢ κατὰ πρόσθεσιν, ἢ κατὰ διαίρεσιν, ἢ ἀμφοτέρως.

[1] [ἐπισκέψασθαι Simplic. 469. 14 (lemma), Themist. 82. 27: ἐστὶ σκέψασθαι codd. Philop. 408. 22 (lemma).—C.]

---

[a] [*Or*, 'And even if it (*i.e.* τὸ ἔξω τοῦ οὐρανοῦ, ἄπειρον ὄν) is (not occupied by other worlds, but) vacant, "place" too will then be unlimited, and there must be no limit to "body"; for ("place" is that in which there is or may be "body" and) in things eternal there is no distinction between possibility and actuality.' Eudemus (Simpl. 467. 26) quotes an argument of Archytas for infinite extension of place and body on these lines. *Cf.* Heath, *Greek Maths.* i. 214.—C.]

[b] [*Or* (*reading* ἔστιν) 'there nevertheless exists an 'infinity' in the singular or the plural sense.'—C.]

[c] [The classification is this: 'infinite' is used to mean (*a*) 'what cannot be traversed' (διελθεῖν) at all, because it is not the sort of thing there can be any question of traversing, or (*b*) what can be traversed, but 'there is no getting to

occupying it everywhere, for in things eternal there is no distinction between possibility and actuality ? [a]

But the consideration of this ' unlimited ' presents us with a genuine problem ; for not only by asserting but also by denying its reality we seem to be landed in a number of untenable positions. And then we have to ask whether it is a substantively existing ' something ' ; or an attribute, necessarily pertaining to some substantively existent thing in nature ; or neither, though none the less unlimited, whether in its unity or its multiplicity.[b] But the physicist's chief concern is to inquire whether there is such a thing as an unlimited ' sensible ' magnitude.

Let us begin, then, by inquiring in how many senses we use the term. If by saying that a thing has ' no limit ' you mean ' no boundary,' then you mean that its nature is such that it would be nonsense to speak of ' passing through it from side to side,' just in the same way that a sound is ' invisible.' But you might also mean that, though it is of such nature that you can traverse it, it does not admit (whether you are speaking absolutely or practically) of your getting ' through ' it so as to come out beyond it. Or again you may mean that its nature would allow it to have a boundary such as to make it passable-through, but that in this case it has not.[c]

Again, the ' absence of limit ' may mean capacity for being multiplied indefinitely, or for being divided indefinitely, or both.

the end of it ' (τὸ διέξοδον ἔχον ἀτελεύτητον), either (1) in the sense that this is hardly possible or practically impossible, or (2) in the sense that it is absolutely impossible, because there is no actual end, though it is the sort of thing you might ' get to the end of ' if it had one.—C.] The distinction is between (*a*) ' eternal,' *e.g.* ratio of diagonal to side, and (*b*) ' endless,' *e.g.* time or number.—R. W.

CHAPTER V

ARGUMENT

[*The abstract* ' Infinity itself,' *which the Pythagoreans and Plato regarded as an element, cannot actually exist apart from sensible things* (204 a 8-9). (1) *If a substance, it would be indivisible, but they regard it as divisible* (a 9-14). *If an attribute, it cannot be an element* (a 14-17). (2) *There can be no ' infinite ' which is neither a number nor a magnitude* (a 17-20). (3) *Whether this alleged Infinite is divisible or indivisible, contradictions result* (a 20-34).

Physics, however, is concerned rather with the question whether an infinite material body *actually exists or not* (a 34-b 4).

Abstract arguments might be urged against its existence (b 4-10). *From natural facts it appears that an unlimited*

204 a 8 Χωριστὸν μὲν οὖν εἶναι τὸ ἄπειρον τῶν αἰσθητῶν, αὐτό τι ὂν ἄπειρον, οὐχ οἷόν τε.

Εἰ γὰρ μήτε μέγεθός ἐστι μήτε πλῆθος, ἀλλ' οὐσία αὐτό ἐστι τὸ ἄπειρον καὶ μὴ συμβεβηκός, ἀδιαίρετον ἔσται· τὸ γὰρ διαιρετὸν ἢ μέγεθος ἔσται ἢ πλῆθος. εἰ δὲ ἀδιαίρετον, οὐκ ἄπειρον, εἰ μὴ ὡς ἡ φωνὴ ἀόρατος· ἀλλ' οὐχ οὕτως οὔτε φασὶν εἶναι οἱ φάσκοντες εἶναι τὸ ἄπειρον· οὔτε ἡμεῖς ζητοῦμεν, ἀλλ' ὡς ἀδιέξοδον. εἰ δὲ κατὰ συμβεβηκός ἐστι τὸ ἄπειρον, οὐκ ἂν εἴη στοιχεῖον

[a] [The view ascribed above (203 a 4) to Plato and the Pythagoreans, who made ' Infinity itself ' an element (στοιχεῖον) in things.—C.]

CHAPTER V

ARGUMENT (*continued*)

body could not be either (1) *a compound of a limited number of elements* (b 10-22), *or* (2) *a simple and uniform element, or like Anaximander's unlimited stuff. If one element were unlimited, it would devour the rest ; and Anaximander's stuff does not exist* (b 22-205 a 8). *A further proof, based on the doctrine that each element has a 'natural place'* (a 8-b 1). *Anaxagoras gives the wrong reason for his 'Unlimited' remaining at rest* (b 1-23).

An unlimited body is inconsistent with the existence of natural places (b 23-31), *and of the distinctions between 'up' and 'down,' etc., in place* (b 31-35). *A body must be in a place ; but a place must be limited, and so no place can contain an unlimited body* (b 35-206 a 7).

Thus an unlimited body cannot exist in actuality (a 7-8). —*C.*]

Now, it is impossible that there should exist an 'unlimited' sejunct from objects of sense, and constituting a self-existing 'infinity.'

For (1) if the 'unlimited' is neither magnitude nor number, but infinity itself is its very being and not an attribute,[a] it must be indivisible ; for what is divisible must be either extended or numerous. But if it is 'indivisible,' it cannot be 'unlimited,' except in the sense in which speech is invisible ; but that is neither what those who assert the reality of the 'unlimited' mean by the term, nor what we are interested in investigating ; for what we all mean is that which it is not possible to get through to the end of. If the unlimited exists at all, therefore, it must be attributively ; but, if so, *qua* infinite it can no more be a constituent element of substantively

204 a τῶν ὄντων, ᾗ ἄπειρον, ὥσπερ οὐδὲ τὸ ἀόρατον τῆς διαλέκτου, καίτοι ἡ φωνή ἐστιν ἀόρατος.

Ἔτι πῶς ἐνδέχεται εἶναί τι αὐτὸ ἄπειρον, εἴπερ μὴ καὶ ἀριθμὸν καὶ μέγεθος, ὧν ἐστι καθ' αὑτὸ πάθος τι τὸ ἄπειρον; ἔτι γὰρ ἧττον ἀνάγκη ἢ τὸν ἀριθμὸν ἢ τὸ μέγεθος.

Φανερὸν δὲ καὶ ὅτι οὐκ ἐνδέχεται εἶναι τὸ ἄπειρον ὡς ἐνεργείᾳ ὂν καὶ ὡς οὐσίαν καὶ ἀρχήν. ἔσται γὰρ ὁτιοῦν αὐτοῦ ἄπειρον τὸ λαμβανόμενον, εἰ μεριστόν—τὸ γὰρ ἀπείρῳ εἶναι καὶ ἄπειρον τὸ αὐτό, εἴπερ οὐσία τὸ ἄπειρον καὶ μὴ καθ' ὑποκειμένου—ὥστ' ἢ ἀδιαίρετον ἢ εἰς ἄπειρα διαιρετόν. πολλὰ δ' ἄπειρα τὸ αὐτὸ εἶναι ἀδύνατον· ἀλλὰ μὴν ὥσπερ ἀέρος ἀὴρ μέρος, οὕτω καὶ ἄπειρον ἀπείρου, εἴ γε οὐσία ἐστὶ καὶ ἀρχή. ἀμέριστον ἄρα καὶ ἀδιαίρετον. ἀλλ' ἀδύνατον τὸ ἐντελεχείᾳ ὂν ἄπειρον (ποσὸν γάρ τι εἶναι ἀναγκαῖον). κατὰ συμβεβηκὸς ἄρα ὑπάρχει τὸ ἄπειρον. ἀλλ' εἰ οὕτως, εἴρηται ὅτι οὐκ ἐνδέχεται αὐτὸ λέγειν ἀρχήν, ἀλλ' ἐκεῖνο ᾧ συμβέβηκεν—τὸν ἀέρα ἢ τὸ ἄρτιον. ὥστε ἀτόπως ἂν ἀποφαίνοιντο οἱ λέγοντες οὕτως ὥσπερ οἱ Πυθαγόρειοί φασιν· ἅμα γὰρ οὐσίαν ποιοῦσι τὸ ἄπειρον καὶ μερίζουσιν.

Ἀλλ' ἴσως αὕτη μέν ἐστι καθόλου ἡ ζήτησις

---

[a] [*Literally,* '"to be infinite" (infiniteness) and "the infinite" are the same thing.' Its infiniteness is *ex hypothesi* the whole of its nature; so, if divided, its parts would be essentially and by nature infinite.—C.]

[b] [*Literally,* 'But an actually existing infinite could not (be indivisible into parts), for it would have to be a quantity.'—C.]

existent things than 'invisibility' can be a substantive component of 'language,' although sound is invisible.

(2) Again, how can the unlimited possibly have a substantive existence if magnitude and number themselves, of which it is, by its very nature, an affection or qualification, have no such existence? The substantive existence of the unlimited must be even more impossible than that of magnitude and number.

(3) It is further manifest that infinity cannot exist as an actualized entity and as substance or principle. For in that case, if it were divisible, then any several portion of it that you might take would itself be 'infinity'—for, if 'infinity' is a substance and not an attribute, then any portion of that which is infinity's self must by definition be infinite [a]—so that infinity must be either (*a*) altogether indivisible or (*b*) divisible into infinites. But (*b*) for one same thing to be many infinites is impossible; and yet this would follow, if infinity had a substantive existence and were an element, just as a portion of air is itself air. (*a*) If a substance, then, it must be indivisible and must not contain parts homogeneous with itself. But in that case it cannot be a realized entity at all; for if so, being essentially a quantum, it would be divisible.[b] Infinity, then, must exist, if at all, attributively. But in that case it has been shown that it cannot be infinity's self that is an element, but the infinite thing of which infinity is an attribute, as some have predicated of 'air' or of 'the even number.' This shows the futility of the Pythagorean contention, which gives infinity itself a substantive existence, and yet makes it divisible.

But, after all, the general question of the possibility

204 a 3 μᾶλλον, εἰ ἐνδέχεται τὸ ἄπειρον καὶ ἐν τοῖς μαθη-
204 b ματικοῖς εἶναι καὶ ἐν τοῖς νοητοῖς καὶ μηδὲν ἔχουσι μέγεθος· ἡμεῖς δ' ἐπισκοποῦμεν περὶ τῶν αἰσθητῶν καὶ περὶ ὧν ποιούμεθα τὴν μέθοδον, ἆρ' ἔστιν ἐν αὐτοῖς ἢ οὐκ ἔστι σῶμα ἄπειρον ἐπὶ τὴν αὔξησιν.

Λογικῶς μὲν οὖν σκοπουμένοις ἐκ τῶνδε δόξειεν ἂν οὐκ εἶναι. εἰ γάρ ἐστι σώματος λόγος τὸ ἐπιπέδῳ ὡρισμένον, οὐκ ἂν εἴη σῶμα ἄπειρον, οὔτε νοητὸν οὔτε αἰσθητόν. ἀλλὰ μὴν οὐδ' ἀριθμὸς οὕτως ὡς κεχωρισμένος καὶ ἄπειρος· ἀριθμητὸν γὰρ ἀριθμὸς ἢ τὸ ἔχον ἀριθμόν· εἰ οὖν τὸ ἀριθμητὸν ἐνδέχεται ἀριθμῆσαι, καὶ διεξελθεῖν ἂν εἴη δυνατὸν τὸ ἄπειρον.

Φυσικῶς δὲ μᾶλλον θεωροῦσιν ἐκ τῶνδε· οὔτε γὰρ σύνθετον οἷόν τε εἶναι οὔτε ἁπλοῦν.

Σύνθετον μὲν οὖν οὐκ ἔσται τὸ ἄπειρον σῶμα, εἰ πεπερασμένα τῷ πλήθει τὰ στοιχεῖα. ἀνάγκη γὰρ πλείω εἶναι, καὶ ἰσάζειν ἀεὶ τἀναντία, καὶ μὴ εἶναι ἓν αὐτῶν ἄπειρον· εἰ γὰρ ὁποσῳοῦν λείπεται ἡ ἐν ἑνὶ σώματι δύναμις θατέρου, οἷον εἰ τὸ πῦρ πεπέρανται ὁ δ' ἀὴρ ἄπειρος, ἔστι δὲ τὸ ἴσον πῦρ τοῦ ἴσου ἀέρος τῇ δυνάμει ὁποσαπλασιονοῦν, μόνον δὲ ἀριθμόν τινα ἔχον, ὅμως φανερὸν ὅτι τὸ

---

[a] [ἐπὶ τὴν αὔξησιν, as opposed to infinite *divisibility*.—C.]

[b] [λογικῶς, as here opposed to φυσικῶς, means arguing from abstract principles (λόγοι) rather than from the concrete facts (ἔργα) of our science.—C.]

[c] See Gen. Introd. pp. lxxxiii f.

[d] [*Literally*, 'Nor yet (can there exist) a number conceived as existing separately (from numbered things) and as infinite.' 'The infinite' cannot exist by itself in the form of an infinite number, any more than as an infinite body.—C.]

[e] [Otherwise the body would not be a compound.—C.]

of the unlimited in mathematics and, in the conceptual sphere, in things that have no magnitude, lies beyond our present scope; for we are engaged in the study of things cognizable by the senses, and what we are concerned to know is whether there exists any actual physical body or substance that is unlimited in expansion.[a]

If we argue on abstract grounds,[b] the following considerations would appear to be conclusive against the existence of any such body. If 'body' is defined as 'that, the limit or boundary of which is surface,'[c] an unlimited body is impossible, either conceptually or sensibly. Nor can 'a number,' even if abstract, be infinite;[d] for both numbers themselves and the things that are numbered are 'numerable,' *i.e.*, can be counted. If, then, 'a number' is by definition countable, then, if it were also unlimited, the unlimited would be passed through to the end or limit, which is a contradiction.

If, on the other hand, we argue from the natural properties of things, we may add these further considerations: the supposed infinite body can be neither (1) compound nor (2) simple.

(1) It cannot be compound, if the elemental substances are limited in number. For these elements must be more than one,[e] and they must permanently balance each other's contrasted qualities, so that no one of them must be infinite. For however relatively feeble the power of one element (say, air) might be with respect to that of another (say, fire), yet, if fire were limited, the air unlimited in quantity, and if there were *any* ratio of equivalence whatever between the assimilative power of a given volume of air and that of an equal volume of fire, then obviously

204 b ἄπειρον ὑπερβαλεῖ καὶ φθερεῖ τὸ πεπερασμένον. ἕκαστον δ' ἄπειρον εἶναι ἀδύνατον· σῶμα μὲν γάρ ἐστι τὸ πάντῃ ἔχον διάστασιν, ἄπειρον δὲ τὸ ἀπεράντως διεστηκός, ὥστε τὸ ἄπειρον σῶμα πανταχῇ ἔσται διεστηκὸς εἰς ἄπειρον.

Ἀλλὰ μὴν οὐδὲ ἓν καὶ ἁπλοῦν εἶναι ἐνδέχεται τὸ ἄπειρον σῶμα, οὔτε ὡς λέγουσί τινες τὸ παρὰ τὰ στοιχεῖα, ἐξ οὗ ταῦτα γεννῶσιν, οὔθ' ἁπλῶς. εἰσὶ γάρ τινες οἳ τοῦτο ποιοῦσι τὸ ἄπειρον ἀλλ' οὐκ ἀέρα ἢ ὕδωρ, ὡς μὴ τἆλλα φθείρηται ὑπὸ τοῦ ἀπείρου αὐτῶν· ἔχουσι γὰρ πρὸς ἄλληλα ἐναντίωσιν—οἷον ὁ μὲν ἀὴρ ψυχρός, τὸ δ' ὕδωρ ὑγρόν, τὸ δὲ πῦρ θερμόν—ὧν εἰ ἦν ἓν ἄπειρον, ἔφθαρτο ἂν ἤδη τἆλλα· νῦν δ' ἕτερον εἶναί φασιν ἐξ οὗ ταῦτα. ἀδύνατον δ' εἶναι τοιοῦτον, οὐχ ὅτι ἄπειρον (περὶ τούτου μὲν γὰρ κοινόν τι λεκτέον ἐπὶ παντὸς ὁμοίως, καὶ ἀέρος καὶ ὕδατος καὶ ὁτουοῦν), ἀλλ' ὅτι οὐκ ἔστι τοιοῦτον σῶμα αἰσθητὸν παρὰ τὰ στοιχεῖα καλούμενα· ἅπαντα γὰρ ἐξ οὗ ἐστι, καὶ διαλύεται εἰς τοῦτο, ὥστε ἦν ἂν ἐνταῦθα παρὰ ἀέρα καὶ πῦρ καὶ γῆν καὶ ὕδωρ· φαίνεται δ' οὐδέν.
205 a οὐδὲ δὴ πῦρ οὐδ' ἄλλο τι τῶν στοιχείων οὐδὲν ἄπειρον ἐνδέχεται εἶναι. ὅλως γὰρ καὶ χωρὶς τοῦ

---

[a] [Anaximander. *Cf.* 187 a 20.—C.]

[b] [Anaximander held that his infinite stuff existed in its undifferentiated state *outside* our world, which it envelopes. He also explicitly asserted ἐξ ὧν ἡ γένεσίς ἐστι τοῖς οὖσι, καὶ τὴν φθορὰν εἰς ταῦτα γίγνεσθαι, Diels, *Vors.* 2. 9.—C.]

the unlimited volume of air must vanquish and destroy the limited volume of fire. Nor could every element severally be unlimited ; for ' body ' is that which has dimension in every direction, and the unlimited that which extends without limit ; so an unlimited body must extend without limit in every direction.

(2) But if the infinite body cannot be compound, neither can it be simple and uniform, either (*a*) as itself the universal element, or (*b*) (as some [a] have supposed) as an unlimited something in addition to the elements, the matrix out of which they come. (*b*) The reason for supposing this additional something, rather than air or water for instance, to be the unlimited, was that it seemed to evade the dilemma set out above. For since air is cold, water moist, and fire hot, and these properties are mutually destructive, an infinity of one of them would mean that the others would have perished by this time ; but this, they say, would not apply to an undifferentiated something out of which they all came. But no such thing can exist ; not because of its infiniteness (as to which a more general consideration, equally applicable to air or water or any other substance whatever, will be urged in the immediate sequel), but because the senses reveal no such body (by hypothesis ' sensible') in addition to the elements as commonly enumerated ; for whatever things come out of when they come into existence, that they return into when they pass out of existence, so that if the elements came out of something else, that something would exist in our world [b] over and above air, fire, earth, and water ; but no such thing is apparent to observation. (*a*) Nor can fire or any other of the elements possibly be unlimited. For—

205 a ἄπειρον εἶναί τι αὐτῶν, ἀδύνατον τὸ πᾶν, κἂν ᾖ πεπερασμένον, ἢ εἶναι ἢ γίγνεσθαι ἕν τι αὐτῶν, ὥσπερ Ἡράκλειτός φησιν ἅπαντα γίνεσθαί ποτε πῦρ. ὁ δ' αὐτὸς λόγος καὶ ἐπὶ τοῦ ἑνός, οἷον ποιοῦσι παρὰ τὰ στοιχεῖα οἱ φυσικοί· πάντα γὰρ μεταβάλλει ἐξ ἐναντίου εἰς ἐναντίον, οἷον ἐκ θερμοῦ εἰς ψυχρόν. δεῖ δὲ περὶ παντὸς καὶ ἐκ τῶνδε[1] σκοπεῖν, εἰ ἐνδέχεται ἢ οὐκ ἐνδέχεται εἶναι ἄπειρον.[2]

Ὅτι δ' ὅλως ἀδύνατον εἶναι σῶμα ἄπειρον αἰσθητόν, ἐκ τῶνδε δῆλον.

Πέφυκε γὰρ πᾶν τὸ αἰσθητὸν πού εἶναι, καὶ ἔστι τόπος τις ἑκάστου, καὶ ὁ αὐτὸς τοῦ μορίου καὶ παντός, οἷον ὅλης τε τῆς γῆς καὶ βώλου μιᾶς, καὶ πυρὸς καὶ σπινθῆρος. ὥστε εἰ μὲν ὁμοειδές, ἀκίνητον ἔσται ἢ ἀεὶ οἰσθήσεται. καίτοι ἀδύνατον· τί γὰρ μᾶλλον κάτω ἢ ἄνω ἢ ὁπουοῦν[3]; λέγω δ'

[1] [To avoid making this sentence a mere duplicate of the next, I have followed Philoponus (436. 1). περὶ (I, corr. F, κατὰ E) παντὸς καὶ (FI ; E omits καὶ) ἐκ τῶνδε Simplic. 482. 1 (lemma), Philop. 436. 1 (lemma).—C.]

[2] [εἶναι ἄπειρον Simplic. 482.6 (paraphr.): εἶναι ἀρχήν (ἄπειρον t) Philop. 436. 6 (paraphr.): εἶναι E: εἶναι σῶμα (σῶμα εἶναι I) ἄπειρον αἰσθητόν cett.—C.]

[3] [ὁπουοῦν *Met.* 1067 a 10, Bonitz: ὁποιονοῦν F: ποῦ cett.—C.]

[a] [See *frag.* 63-66. It is disputed whether Heracleitus really meant a general conflagration (which seems inconsistent with his system), or that every part of the universe *at some time* passes through the fiery phase in the cycle of transformation.—C.]

[b] [And if the universe were one element or Anaximander's undifferentiated unity, there would be no opposite, and hence change would be impossible.—C.]

[c] [The question set aside above (*l.* 2).—C.]

[d] [Aristotle assumes his own doctrine that each element has a natural region (the centre for earth, the circumference

more generally and quite apart from the question of its being limited or unlimited—it is impossible that all the universe, even if it be limited, should be or should become any single one of the elements, as Heracleitus [a] supposed, when he said that all things sometime become fire. Against this the same argument can be urged as applies to the undifferentiated unity or matrix such as some physicists have assumed in addition to the elements; for things always change from one term of an opposition to the other, from hot to cold, for instance.[b] But the question whether an element can or cannot be unlimited [c] should be considered in the case of every element (not only of fire) in the light of the following considerations also.

So far on the supposition that the elemental bodies are limited in number; but as to the more general question whether it is possible in any case that a sensible body should be without limit, the following considerations are conclusive against any such possibility.

A sensible body or substance must be situated somewhere, and any given body has some proper site, which is the same for the part as for the whole; for instance, for earth as a whole and for a single clod of earth, or for fire as a whole and for a single spark.[d] Consequently (1) if the whole infinite body [e] were of uniform substance, it would either be motionless or always in motion. But the latter is impossible; for why more down than up, or to anywhere distinguishable from anywhere else? I

for fire, etc.), to which it is of its very nature to move, or in which it is of its nature to rest, *De caelo*, i. 2.—C.]

[e] [*Sc.* τὸ πᾶν (as in l. 20), the 'whole' opposed to the 'part' (as in *l.* 11).—C.]

οἷον εἰ βῶλος εἴη, ποῦ αὕτη κινηθήσεται ἢ ποῦ μενεῖ; ὁ γὰρ τόπος ἄπειρος τοῦ συγγενοῦς αὐτῇ σώματος. πότερον οὖν καθέξει τὸν ὅλον τόπον; καὶ πῶς; τίς οὖν ἢ ποῦ ἡ μονὴ καὶ ἡ κίνησις αὐτῆς; ἢ πανταχοῦ μενεῖ (οὐ κινηθήσεται ἄρα) ἢ πανταχοῦ κινηθήσεται (οὐκ ἄρα στήσεται). εἰ δ' ἀνόμοιον τὸ πᾶν, ἀνόμοιοι καὶ οἱ τόποι· καὶ πρῶτον μὲν οὐχ ἓν τὸ σῶμα τοῦ παντὸς ἀλλ' ἢ τῷ ἅπτεσθαι· ἔπειτα ἤτοι πεπερασμένα ταῦτ' ἔσται ἢ ἄπειρα τῷ εἴδει. πεπερασμένα μὲν οὖν οὐχ οἷόν τε· ἔσται γὰρ τὰ μὲν ἄπειρα τὰ δ' οὔ (εἰ τὸ πᾶν ἄπειρον), οἷον τὸ πῦρ ἢ τὸ ὕδωρ, φθορὰ δὲ τὸ τοιοῦτον τοῖς ἐναντίοις, καθάπερ εἴρηται πρότερον. (καὶ διὰ τοῦτ' οὐθεὶς τὸ ἓν καὶ ἄπειρον πῦρ ἐποίησεν οὐδὲ γῆν τῶν φυσιολόγων, ἀλλ' ἢ ὕδωρ ἢ ἀέρα ἢ τὸ μέσον αὐτῶν, ὅτι τόπος ἑκατέρου δῆλος ἦν διωρισμένος, ταῦτα δ' ἐπαμφοτερίζει τῷ ἄνω καὶ κάτω.) εἰ δ' ἄπειρα καὶ ἁπλᾶ, καὶ οἱ τόποι ἄπειροι καὶ ἔσται ἄπειρα τὰ στοιχεῖα. εἰ δὲ τοῦτ' ἀδύνατον καὶ πεπερασμένοι οἱ τόποι, καὶ

---

[a] [204 b 28.—C.]

mean: Suppose there is a clod. Whither is it to move, or where to rest, through the infinite region of substance indistinguishable from itself? Will it take possession of it all? But how can it? What then and where is its rest, or what and whither its movement? Either it will be at rest everywhere, in which case it will never move; or it will everywhere be in motion, and then it will never come to rest. If (2), on the other hand, the universe is not all homogeneous, then the sites proper to its unlike constituents are distinguishable; and in the first place the universe will have no unity save that of continuous contact of part with part; and in the second place the differences in *kind* between the parts will be either (*a*) limited or (*b*) unlimited in number. (*a*) They cannot be limited; for in that case, if the universe itself were unlimited in quantity, some constituents of it would have to be unlimited in quantity, others limited (say, fire limited and water unlimited), and the unlimited element would have destroyed its opposite and reduced the universe to uniformity, as we said above.[a] (This is why no one of the physicists has taken fire or earth as his unlimited unity, but water or air or something between them has always been selected; fire and earth have each a distinct region about which there could be no mistake, whereas water and air appear to be susceptible to movement either up or down.) (*b*) If, on the contrary, there were no limit to the differences of kind between the parts, and each part were a primary body, we should have an unlimited number of elements, and there would be no limit to the number of sites natural to them. But if there cannot be an unlimited number of elements and the sites are limited in number, the

205 a τὸ ὅλον πεπεράνθαι ἀναγκαῖον. ἀδύνατον γὰρ μὴ ἀπαρτίζειν τὸν τόπον καὶ τὸ σῶμα· οὔτε γὰρ ὁ τόπος ὁ πᾶς μείζων ἢ ὅσον ἐνδέχεται τὸ σῶμα ἅμα εἶναι (ἅμα δ' οὐδ' ἄπειρον ἔσται τὸ σῶμα),
205 b οὔτε τὸ σῶμα μεῖζον ἢ ὁ τόπος· ἢ γὰρ κενὸν ἔσται τι, ἢ σῶμα οὐδαμοῦ πεφυκὸς εἶναι.

Ἀναξαγόρας δ' ἀτόπως λέγει περὶ τῆς τοῦ ἀπείρου μονῆς· στηρίζειν γὰρ αὐτὸ αὑτό φησι τὸ ἄπειρον· τοῦτο δέ, ὅτι ἐν αὑτῷ (ἄλλο γὰρ οὐδὲν περιέχει), ὡς ὅπου ἄν τι ᾖ, πεφυκὸς ἐνταῦθα εἶναι. τοῦτο δ' οὐκ ἀληθές· εἴη γὰρ ἄν τί που βίᾳ καὶ οὐχ οὗ πέφυκεν. εἰ οὖν ὅτι μάλιστα μὴ κινεῖται τὸ ὅλον (τὸ γὰρ αὑτῷ στηριζόμενον καὶ ἐν αὑτῷ ὂν ἀκίνητον εἶναι ἀνάγκη), ἀλλὰ διὰ τί οὐ πέφυκε κινεῖσθαι, λεκτέον· οὐ γὰρ ἱκανὸν τὸ οὕτως εἰπόντα ἀπηλλάχθαι. εἴη γὰρ ἂν καὶ ὁτιοῦν ἄλλο οὐ κινούμενον, ἀλλὰ πεφυκέναι οὐδὲν κωλύει. ἐπεὶ καὶ ἡ γῆ οὐ φέρεται, οὐδ' εἰ ἄπειρος ἦν, εἰργμένη μέντοι ὑπὸ[1] τοῦ μέσου· ἀλλ' οὐχ ὅτι οὐκ ἔστιν ἄλλο οὗ ἐνεχθήσεται μείνειεν ἂν ἐπὶ τοῦ μέσου, ἀλλ' ὅτι πέφυκεν οὕτως· καίτοι ἐξείη ἂν λέγειν ὅτι στηρίζει αὑτήν. εἰ οὖν μηδ' ἐπὶ τῆς γῆς τοῦτο αἴτιον ἀπείρου οὔσης, ἀλλ' ὅτι βάρος ἔχει, τὸ δὲ βαρὺ μένει ἐπὶ τοῦ μέσου, ἡ δὲ γῆ

[1] [ὑπὸ Simpl. 486. 10 (paraphrase), Bonitz: ἀπὸ Philop. 451. 20, 22, codd.—C.]

[a] [That there cannot be an unlimited number of elements was proved in Bk. I. chap. vi., and it follows that the number of natural sites must be limited, and so their sum will be limited.—C.]

whole must also be limited.[a] For we cannot suppose that 'locality' in general and 'body' in general[b] do not fit; since neither can the totality of 'place' have a greater magnitude than the totality of 'body' can have at the same time—and besides, if this were so, body would not be infinite[c]—nor can 'body' be in excess of 'place'; otherwise, in the one case there would be vacant place, and in the other body of such a nature as to have no locality.

The remarks of Anaxagoras as to the Unlimited being stationary are futile; for he says that the Unlimited stabilizes itself because it is self-contained (since nothing embraces it)—as if it followed that, wherever a thing is, that must be its natural site. But this is not true, for a thing might be forced out of its natural site. However true it may be, therefore, that the universe does not move (for that which is self-stabilized and contained in itself must be motionless), it remains to be shown why it is its nature so to be; it will not do to let Anaxagoras off on his mere assertion. For any other body might be in fact motionless, and yet be naturally capable of motion. The earth is in fact motionless, and would be, if limitless, provided it were still held by the centre; but the reason why it would stay at the centre is not that it would have nowhere else to go, but that such is its nature; yet you might just as well say that the earth 'stabilizes itself' as that the Unlimited does so. If, then, in the case of the earth (supposed to be unlimited) this would not be why it was at the centre, but because it is heavy and what is heavy remains

[b] [*τὸν τόπον καὶ τὸ σῶμα* could also mean 'the place (of a body) and the body (occupying that place).'—C.]

[c] [Because place would be larger, and there cannot be a magnitude larger than an infinite magnitude.—C.]

205 b ἐπὶ τοῦ μέσου, ὁμοίως ἂν καὶ τὸ ἄπειρον μένοι ἐν αὑτῷ διά τιν' ἄλλην αἰτίαν καὶ οὐχ ὅτι ἄπειρον καὶ στηρίζει αὐτὸ αὑτό. ἅμα δὲ δῆλον ὅτι κἂν ὁτιοῦν μέρος δέοι μένειν· ὡς γὰρ τὸ ἄπειρον ἐν ἑαυτῷ μένει στηρίζον, οὕτω κἂν ὁτιοῦν ληφθῇ μέρος ἐν ἑαυτῷ μενεῖ. τοῦ γὰρ ὅλου καὶ τοῦ μέρους ὁμοειδεῖς οἱ τόποι (οἷον ὅλης γῆς καὶ βώλου κάτω, καὶ παντὸς πυρὸς καὶ σπινθῆρος ἄνω)· ὥστε εἰ τοῦ ἀπείρου τόπος τὸ ἐν αὑτῷ, καὶ τοῦ μέρους ὁ αὐτός· μενεῖ ἄρα ἐν ἑαυτῷ.

Ὅλως δὲ φανερὸν ὅτι ἀδύνατον ἅμα ἄπειρον λέγειν σῶμα καὶ τόπον τινὰ εἶναι τοῖς σώμασιν, εἰ πᾶν σῶμα αἰσθητὸν ἢ βάρος ἔχει ἢ κουφότητα, καὶ εἰ μὲν βαρύ, ἐπὶ τὸ μέσον ἔχει τὴν φορὰν φύσει, εἰ δὲ κοῦφον, ἄνω· ἀνάγκη γὰρ καὶ τὸ ἄπειρον, ἀδύνατον δὲ ἢ ἅπαν ὁποτερονοῦν ἢ τὸ ἥμισυ ἑκάτερον πεπονθέναι· πῶς γὰρ διελεῖς; ἢ πῶς τοῦ ἀπείρου ἔσται τὸ μὲν ἄνω τὸ δὲ κάτω, ἢ ἔσχατον ἢ μέσον;

Ἔτι πᾶν σῶμα αἰσθητὸν ἐν τόπῳ, τόπου δὲ εἴδη καὶ διαφοραὶ τἄνω καὶ κάτω καὶ ἔμπροσθεν καὶ ὄπισθεν καὶ δεξιὸν καὶ ἀριστερόν· καὶ ταῦτα οὐ μόνον πρὸς ἡμᾶς καὶ θέσει, ἀλλὰ καὶ ἐν αὐτῷ τῷ ὅλῳ διώρισται. ἀδύνατον δ' ἐν τῷ ἀπείρῳ εἶναι ταῦτα.

206 a Ἁπλῶς δ' εἰ ἀδύνατον τόπον ἄπειρον εἶναι, ἐν τόπῳ δὲ πᾶν σῶμα, ἀδύνατον ἄπειρόν τι εἶναι σῶμα. ἀλλὰ μὴν τό γε ποὺ ἐν τόπῳ, καὶ τὸ ἐν

---

[a] [And consequently there will be no motion at all; but Anaxagoras did not hold that there was no motion.—C.]

[b] See Gen. Introd. p. lxii.

at the centre, and the earth does so, similarly, if the Unlimited abides 'in itself,' it may be for some other reason than that it is unlimited and self-stabilized. And, moreover, if it is the nature of the unlimited element, *qua* unlimited, to abide where it is, every particle of it ought so to abide, and therefore every particle to abide in itself. For the localities of the whole and the part are homogeneous (as 'below' for earth at large and any one clod, or 'above' for fire at large or any one spark); hence, if the locality of the unlimited is 'in itself,' so likewise must the locality of any part of it be. It will therefore remain 'in itself.'[a]

But in truth it is antecedently impossible that there should be an unlimited body and yet that the different substances should each have a proper locality, if we grant that every sensible body has either weight or lightness, and, if heavy, tends downwards and, if light, upwards; for the unlimited body, too, must conform to this condition, and it is impossible that it should be either heavy or light in its entirety, or half one and half the other. For how could you determine the division? Or how could there be an up or down or an inner and outer in the unlimited?

Yet again, every sensible body has locality, and the distinctions between localities are 'above and below,' 'before and behind,' and 'to right and to left'; and these distinctions are not merely relative to us or conventional, but are defined in the universe itself.[b] Yet such distinctions cannot exist in the unlimited.

In a word, if there can be no such thing as an unbounded locality, and if every body has its locality, it is impossible that a body should be unlimited. Now whatever is 'somewhere' is in a place; and

206 a τόπῳ πού. εἰ οὖν μηδὲ ποσὸν οἷόν τ' εἶναι τὸ ἄπειρον—ποσὸν γάρ τι ἔσται, οἷον δίπηχυ ἢ τρίπηχυ· ταῦτα γὰρ σημαίνει τὸ ποσόν—οὕτω καὶ τὸ ἐν[1] τόπῳ, ὅτι πού· τοῦτο δὲ ἢ ἄνω ἢ κάτω ἢ ἐν ἄλλῃ τινὶ διαστάσει τῶν ἕξ· τούτων δ' ἕκαστον πέρας τί ἐστιν.

Ὅτι μὲν οὖν ἐνεργείᾳ οὐκ ἔστι σῶμα ἄπειρον, φανερὸν ἐκ τούτων.

CHAPTER VI

ARGUMENT

[*But infinity must somehow exist; otherwise time would have a beginning and end, there would be indivisible magnitudes, and counting would come to an end* (206 a 9-14).

The divisibility of spatial magnitude is potentially unlimited; but this potentiality will never be realized by actual division into an unlimited number of parts. The infinite only exists in the same sense that a day exists, by continuously coming into being; but in spatial magnitude each part that is taken persists, whereas in time and the succession of generations it is superseded by the next (a 14-b 3).

In a sense the unlimited by way of addition is the same as the unlimited by way of division. The subtraction of equal

Ὅτι δ', εἰ μὴ ἔστιν ἄπειρον ἁπλῶς, πολλὰ

[1] [E omits τὸ ἐν. Bonitz conjectured οὕτως οὐδὲ ἐν τόπῳ.—C.]

[a] [I have understood οὕτω καὶ τὸ ἐν τόπῳ (σημαίνει), ὅτι πού, leaving the construction of the whole sentence rather irregular. *Cf. Met.* 1067 a 31 τὸ γὰρ ἐν τόπῳ πού, τοῦτο δὲ σημαίνει ἢ ἄνω κτλ.—C.]

whatever is in a place is 'somewhere.' So if the unlimited cannot be a quantum at all, because, if it were, it must be of some definite quantity, such as two or three cubits length, these definite quantities being what Quantity denotes; in the same way being in a place means being 'somewhere,'[a] and that means either above or below or in some other of the six directions; but each of these is a limit.[b]

From all these considerations it is evident that an unlimited body cannot exist in accomplished fact.

CHAPTER VI

ARGUMENT (*continued*)

parts from a finite whole will finally exhaust the whole; but the subtraction of parts diminishing in a constant ratio will not. Conversely the addition of parts diminishing in a constant ratio will never construct a finite whole. In this sense a limited magnitude may be the actual seat of an illimitable process (b 3-33).

The infinite is beyond any definable limit; its essence is incompleteness; 'that beyond which there is nothing' is the 'whole.' The unlimited is analogous to the 'material' factor from which the 'whole' is constituted by the 'formal' (limiting) factor, and so cannot 'embrace' anything (b 33-207 a 32).—C.]

YET, if we frankly deny that anything at all can be without limit, we commit ourselves to many state-

[b] [Hence, if an unlimited body were in a place, it would be limited (which is a contradiction); if not in a place, it cannot be a body.—C.]

206 a ἀδύνατα συμβαίνει, δῆλον· τοῦ τε γὰρ χρόνου ἔσται τις ἀρχὴ καὶ τελευτή, καὶ τὰ μεγέθη οὐ διαιρετὰ εἰς μεγέθη, καὶ ἀριθμὸς οὐκ ἔσται ἄπειρος. ὅταν δὲ διωρισμένων οὕτως μηδετέρως φαίνηται ἐνδέχεσθαι, διαιτητοῦ δεῖ, καὶ δῆλον ὅτι πὼς μὲν ἔστι πὼς δ᾿ οὔ.

Λέγεται δὴ τὸ εἶναι τὸ μὲν δυνάμει τὸ δὲ ἐντελεχείᾳ, καὶ τὸ ἄπειρον ἔστι μὲν προσθέσει ἔστι δὲ καὶ ἀφαιρέσει. τὸ δὲ μέγεθος ὅτι μὲν κατ᾿ ἐνέργειαν οὐκ ἔστιν ἄπειρον, εἴρηται· διαιρέσει δ᾿ ἐστίν (οὐ γὰρ χαλεπὸν ἀνελεῖν τὰς ἀτόμους γραμμάς)· λείπεται οὖν δυνάμει εἶναι τὸ ἄπειρον. οὐ δεῖ δὲ τὸ δυνάμει ὂν λαμβάνειν ὥσπερ εἰ δυνατὸν τοῦτ᾿ ἀνδριάντα εἶναι, ὡς καὶ ἔσται τοῦτ᾿ ἀνδριάς, οὕτω καὶ ἄπειρόν τι ὃ ἔσται ἐνεργείᾳ· ἀλλ᾿ ἐπεὶ πολλαχῶς τὸ εἶναι, ὥσπερ ἡ ἡμέρα ἔστι καὶ ὁ ἀγὼν τῷ ἀεὶ ἄλλο καὶ ἄλλο γίνεσθαι, οὕτω καὶ τὸ ἄπειρον. καὶ γὰρ ἐπὶ τούτων ἐστὶ καὶ δυνάμει

---

[a] [The doctrine of indivisible lines, ascribed to Plato and his successor, Xenocrates, is discussed in Book VI. chaps. i.-ii. See also Ross on *Met.* 992 a 20.—C.]

ments that are obviously false ; for we should have to say that time had a beginning and will have an end, and that there are magnitudes that cannot be divided into magnitudes, and that numeration has a limit. So if we conceive the alternative to lie between the existence of some unlimited substance cognizable by the senses and there being nothing unlimited at all, we shall be landed in impossibilities either way ; so we must appeal to an umpire, who will obviously have to explain that in one sense the unlimited exists and in another sense not.

Now things are said to exist as potentialities or as actualities ; and there is no limit to the addition (or subtraction) of terms in a convergent series ; and though we have seen that a magnitude cannot actually be increased beyond limit by multiplication, it may be divided into something smaller yet than any parvitude you choose to mention—for there is no difficulty in refuting the doctrine that there are such things as atomic lines.[a] It results that the unlimited potentiality exists.

But how are we to understand 'potentiality' here ? Not in the sense in which we say that the potentiality of the statue exists in the bronze ; for that implies that the whole of the bronze may actually become the statue, whereas it is not so with an illimitable potentiality, since it can never become an unlimited actuality. As to this we must not be misled by the ambiguity of the word 'is,' for the only sense in which the unlimited is actualized at all is the sense in which we say that it 'actually is' such and such a day of the month, or that the games 'actually are' on ; for in these cases, too, the period of time or the succession of events in question is

206 a καὶ ἐνεργείᾳ· Ὀλύμπια γὰρ ἔστι καὶ τῷ δύνασθαι τὸν ἀγῶνα γίνεσθαι καὶ τῷ γίνεσθαι.

Ἄλλως δ' ἔν τε τῷ χρόνῳ δῆλον τὸ ἄπειρον καὶ ἐπὶ τῶν ἀνθρώπων καὶ ἐπὶ τῆς διαιρέσεως τῶν μεγεθῶν. ὅλως μὲν γὰρ οὕτως ἐστὶ τὸ ἄπειρον, τῷ ἀεὶ ἄλλο καὶ ἄλλο λαμβάνεσθαι, καὶ τὸ λαμβανόμενον μὲν ἀεὶ εἶναι πεπερασμένον, ἀλλ' ἀεί γε ἕτερον καὶ ἕτερον· [ἔτι τὸ εἶναι πλεοναχῶς λέγεται,[1] ὥστε τὸ ἄπειρον οὐ δεῖ λαμβάνειν ὡς τόδε τι, οἷον ἄνθρωπον ἢ οἰκίαν, ἀλλ' ὡς ἡ ἡμέρα λέγεται καὶ ὁ ἀγών, οἷς τὸ εἶναι οὐχ ὡς οὐσία τις γέγονεν, ἀλλ' ἀεὶ ἐν γενέσει ἢ φθορᾷ, εἰ καὶ πεπερασμένον, ἀλλ' ἀεί γε ἕτερον καὶ ἕτερον] ἀλλ'
206 b ἐν μὲν τοῖς μεγέθεσιν ὑπομένοντος τοῦ ληφθέντος τοῦτο συμβαίνει, ἐπὶ δὲ τοῦ χρόνου καὶ τῶν ἀνθρώπων φθειρομένων οὕτως ὥστε μὴ ἐπιλείπειν.

Τὸ δὲ κατὰ πρόσθεσιν τὸ αὐτό ἐστί πως καὶ τὸ κατὰ διαίρεσιν· ἐν γὰρ τῷ πεπερασμένῳ κατὰ πρόσθεσιν γίνεται ἀντεστραμμένως· ᾗ γὰρ διαιρούμενον ὁρᾶται εἰς ἄπειρον, ταύτῃ προστιθέμενον φανεῖται πρὸς τὸ ὡρισμένον. ἐν γὰρ τῷ πεπερασμένῳ μεγέθει ἂν λαβών τις ὡρισμένον προσλαμβάνῃ τῷ αὐτῷ λόγῳ, μὴ τὸ αὐτό τι τοῦ ὅλου μέγεθος περιλαμβάνων, οὐ διέξεισι τὸ πεπερασμένον· ἐὰν δ' οὕτως αὔξῃ τὸν λόγον ὥστε ἀεί τι τὸ αὐτὸ περιλαμβάνειν μέγεθος, διέξεισι, διὰ τὸ

[1] [ἔτι τὸ εἶναι πλεοναχῶς λέγεται Simpl. 495. 6, Philop. 468. 8 (lemma): ὅτι τὸ εἶναι πλεοναχῶς λέγεται E : om. cett. Philop. excised this sentence as not contained 'in the most accurate copies.' Diels (*Zur Textgesch.* p. 32), held it to be a duplicate of ἀλλ' ἐπεὶ πολλαχῶς . . . ἕτερον (ll. 21-29), from another recension. *Cf.* Simplic. *loc. cit.*—C.]

not (like the statue-potentialities of the bronze) all actualized at once, but is in course of transit as long as it lasts. The Olympic games, *as-a-whole*, are a potentiality only, even when they are in process of actualization.

Moreover, ' having no limit ' is not quite the same thing as applied to time or to the human race [a] and as applied to the possibility of continuously dividing a ' magnitude,' as it decreases. In all these cases, the ' absence of limit ' may be regarded as the open ' possibility of more,' the ' more ' that is actually taken being always limited, but always different; but when this occurs in the case of magnitudes what is once taken remains; whereas in the case of time or of the human race the parts taken are constantly perishing in such a way that the succession never fails.[b]

There is a certain process of endless addition that can be identified by reciprocity with endless division; for as we see the finite magnitude in process of divison *ad infinitum*, so we shall find the process of addition tending towards a definite limit. For if (1) one should take a definite piece away from a limited magnitude and then go on to take away the same *proportion of what is left* (not the same fraction of the original whole), and so on and so on, he will never work through to the end of the original magnitude; whereas, if (2) he increases the proportion of the remainder which he takes away each time, so as to make the actual magnitude taken away always the same, then he will get

[a] [The human race may never perish, but the individuals (or generations) of which it consists are always perishing.—C.]

[b] [μὴ ἐπιλείπειν. *Cf.* 208 a 8 ἵνα ἡ γένεσις μὴ ἐπιλείπῃ.—C.]

206 b πᾶν τὸ πεπερασμένον ἀναιρεῖσθαι ὁτῳοῦν ὡρισμένῳ. ἄλλως μὲν οὖν οὐκ ἔστιν, οὕτως δ' ἔστι τὸ ἄπειρον, δυνάμει τε καὶ ἐπὶ καθαιρέσει. καὶ ἐντελεχείᾳ δέ ἐστιν ὡς τὴν ἡμέραν εἶναι λέγομεν καὶ τὸν ἀγῶνα, καὶ δυνάμει οὕτως ὡς ἡ ὕλη, καὶ οὐ καθ' αὑτὸ ὡς τὸ πεπερασμένον. καὶ κατὰ πρόσθεσιν δὴ οὕτως ἄπειρον δυνάμει ἐστίν, ὃ ταὐτὸ λέγομεν τρόπον τινὰ εἶναι τῷ κατὰ διαίρεσιν· ἀεὶ μὲν γάρ τι αὐτοῦ ἔξω ἔσται λαμβάνειν, οὐ μέντοι ὑπερβαλεῖ παντὸς ὡρισμένου μεγέθους, ὥσπερ ἐπὶ τὴν διαίρεσιν ὑπερβάλλει παντὸς ὡρισμένου καὶ ἀεὶ ἔσται ἔλαττον. ὥστε δὲ παντὸς ὑπερβάλλειν κατὰ τὴν πρόσθεσιν, οὐδὲ δυνάμει οἷόν τε εἶναι, εἴπερ μὴ ἔστι κατὰ συμβεβηκὸς ἐντελεχείᾳ ἄπειρον, ὥσπερ φασὶν οἱ φυσιολόγοι τὸ ἔξω σῶμα τοῦ κόσμου, οὗ ἡ οὐσία ἢ ἀὴρ ἢ ἄλλο τι τοιοῦτον, ἄπειρον εἶναι. ἀλλ' εἰ μὴ οἷόν τε εἶναι ἄπειρον ἐντελεχείᾳ σῶμα αἰσθητὸν οὕτω, φανερὸν ὅτι οὐδὲ δυνάμει ἂν εἴη κατὰ πρόσθεσιν, ἀλλ' ἢ

[a] For instance, let AO be a finite line divided at B, C, D, etc., so that

AB = $\frac{1}{3}$AO, BC = $\frac{1}{3}$BO, CD = $\frac{1}{3}$CO, and so on and so on.

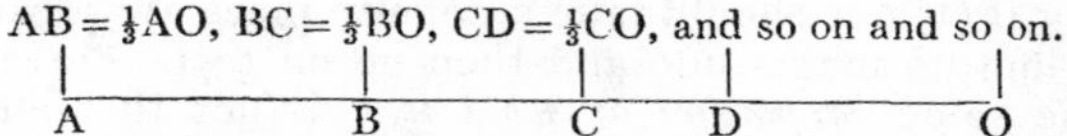

Then the subtraction of AB, BC, . . . etc., from AO can go on for ever and there will always be some of AO left. And the addition of AB, BC, etc., can go on for ever without completing AO. But if the points are so placed that

AB = BC = CD

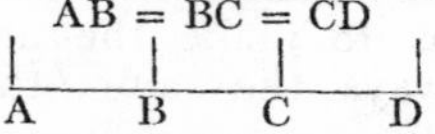

then the subtraction of AB, BC, and CD will exhaust AO and come to an end. And the addition of AB, BC, and CD will come to an end by completing AO.

through to the end; for successive withdrawals of any constant magnitude, however small, will exhaust any limited magnitude whatever.[a] The illimitable, then, exists only the way just described—as an endless potentiality of approximation by reduction of intervals. Illimitability is never actual except in the sense in which we can say 'the day' or 'the games' are actual, whereas as potentiality it is analogous to formless matter; it never exists as a *thing*, as a determined quantum does. In this sense, then, there is also illimitable potentiality of addition, which in a way is the same as what we describe as illimitable in respect of division, for in addition it will always be possible to find something beyond the total for the time being (in a convergent series) though the total will never exceed every assigned magnitude in the way that, in the direction of division, the result does pass every assigned magnitude and will always become still smaller. But, in the sense of exceeding *every* finite magnitude as the result of addition, the unlimited cannot exist even potentially, unless we accept the hypothesis of the physicists who suppose some such actual substance as air or the like to have bodily existence outside the universe and to be unlimited. In that case indeed 'infinity' would incidentally have an actual existence (though not itself substantial). But if (as has been shown) it is impossible that there should exist any such sensible body, as an accomplished actuality, it follows that there is no potentiality of a sum of additions extending beyond all assignable magnitude. The only potentiality of unlimited additions, then, is the additions that are the obverse

206 b ὥσπερ εἴρηται ἀντεστραμμένως τῇ διαιρέσει· ἐπεὶ καὶ Πλάτων διὰ τοῦτο δύο τὰ ἄπειρα ἐποίησεν, ὅτι καὶ ἐπὶ τὴν αὔξην δοκεῖ ὑπερβάλλειν καὶ εἰς ἄπειρον ἰέναι καὶ ἐπὶ τὴν καθαίρεσιν. ποιήσας μέντοι δύο οὐ χρῆται· οὔτε γὰρ ἐν τοῖς ἀριθμοῖς τὸ ἐπὶ τὴν καθαίρεσιν ἄπειρον ὑπάρχει (ἡ γὰρ μονὰς ἐλάχιστον) οὔτε ἐπὶ τὴν αὔξην (μέχρι γὰρ δεκάδος ποιεῖ τὸν ἀριθμόν).

Συμβαίνει δὲ τοὐναντίον ἄπειρον εἶναι ἢ ὡς 207 a λέγουσιν· οὐ γὰρ οὗ μηδὲν ἔξω, ἀλλ' οὗ ἀεί τι ἔξω ἔστι, τοῦτο ἄπειρόν ἐστιν. σημεῖον δέ· καὶ γὰρ τοὺς δακτυλίους ἀπείρους λέγουσι τοὺς μὴ ἔχοντας σφενδόνην, ὅτι ἀεί τι ἔξω ἔστι λαμβάνειν, καθ' ὁμοιότητα μέν τινα λέγοντες, οὐ μέντοι κυρίως· δεῖ γὰρ τοῦτό τε ὑπάρχειν καὶ μηδέποτε τὸ αὐτὸ λαμβάνεσθαι, ἐν δὲ τῷ κύκλῳ οὐ γίνεται οὕτως ἀλλ' ἀεὶ τὸ ἐφεξῆς μόνον ἕτερον. ἄπειρον μὲν οὖν ἐστιν οὗ κατὰ ποσὸν λαμβάνουσιν ἀεί τι λαβεῖν ἔστιν ἔξω· οὗ δὲ μηδὲν ἔξω, τοῦτ' ἐστὶ τέλειον καὶ ὅλον. οὕτω γὰρ ὁριζόμεθα τὸ ὅλον, οὗ μηθὲν ἄπεστιν, οἷον ἄνθρωπον ὅλον ἢ κιβωτόν· ὥσπερ δὲ τὸ καθ' ἕκαστον, οὕτω καὶ τὸ κυρίως,

---

[a] [See note on 201 b 20.—C.]

[b] [The Platonists adopted from the Pythagoreans the doctrine that all numbers are reducible to the numbers from one to ten, because after ten we go back to 1, 2, 3, etc. Theon Smyrn. p. 162 (Dupuis), *Met.* 1073 a 20, 1084 a 12. —C.]

[c] *i.e.* beyond *any* definable limit.

[d] On circular movement see Book VIII. chap. ix.

of the subtractions regulated by successive divisions as already explained. And this is why Plato himself distinguishes between two 'infinities,' thinking that he must have an infinite that could exceed all expansion as well as all reduction.[a] But although he postulates such an infinity, he never makes any use of it; for in numbers he does not admit either endless reduction (since the monad is the irreducible minimum), or increase without limit, for the series of numbers stops at the decad.[b]

The fact is that the unlimited is really the exact opposite of its usual description; for it is not that 'beyond which there is nothing,' but 'what is always beyond.'[c] And this is really implied in rings that have no gem-sockets being called 'endless,' because wherever you are you go on to more. But the analogy is not complete, for if anything is really to be 'endless' you must be able to go on, not only to 'more,' but to what you had never gone over or done before; and this is not so with the ring, for you keep on covering the same part of it once more and once more, and it is only the next point that is always different from the one before it.[d] The unlimited, then, is the open possibility of taking more, however much you have already taken; that of which there is nothing more to take is not unlimited, but whole or completed. For we define a whole precisely as that from which nothing is absent, for example, a 'whole man' or a 'whole chest.' And as with particular wholes, so when the word is used in the strict sense:[e] The Whole is that outside

[e] [Aristotle refers to the use of τὸ ὅλον and τὸ πᾶν to mean 'the universe'—*the* whole 'in the strict sense,' because it is not also a part of any larger whole.—C.]

207 a οἷον τὸ ὅλον οὗ μηδέν ἐστιν ἔξω· οὗ δ' ἐστὶν ἀπουσία ἔξω, οὐ πᾶν, ὅ τι ἂν ἀπῇ. ὅλον δὲ καὶ τέλειον ἢ τὸ αὐτὸ πάμπαν ἢ σύνεγγυς τὴν φύσιν ἐστίν, τέλειον δ' οὐδὲν μὴ ἔχον τέλος· τὸ δὲ τέλος πέρας.

Διὸ βέλτιον οἰητέον Παρμενίδην Μελίσσου εἰρηκέναι· ὁ μὲν γὰρ ἄπειρον τὸ ὅλον φησίν, ὁ δὲ τὸ ὅλον πεπεράνθαι ' μεσσοθεν ἰσοπαλές.' οὐ γὰρ λίνον λίνῳ συνάπτειν ἐστὶ τῷ ἅπαντι καὶ ὅλῳ τὸ ἄπειρον· ἐπεὶ ἐντεῦθέν γε λαμβάνουσι τὴν σεμνότητα κατὰ τοῦ ἀπείρου—τὸ ' πάντα περιέχειν ' καὶ τὸ ' πᾶν ἐν ἑαυτῷ ἔχειν '—διὰ τὸ ἔχειν τινὰ ὁμοιότητα τῷ ὅλῳ. ἔστι γὰρ τὸ ἄπειρον τῆς τοῦ μεγέθους τελειότητος ὕλη καὶ τὸ δυνάμει ὅλον, ἐντελεχείᾳ δ' οὔ, διαιρετὸν δ' ἐπί τε τὴν καθαίρεσιν καὶ τὴν ἀντεστραμμένην πρόσθεσιν· ὅλον δὲ καὶ πεπερασμένον οὐ καθ' αὑτὸ ἀλλὰ κατ' ἄλλο. καὶ οὐ περιέχει ἀλλὰ περιέχεται, ᾗ ἄπειρον. διὸ καὶ ἄγνωστον ᾗ ἄπειρον· εἶδος γὰρ οὐκ ἔχει ἡ ὕλη. ὥστε φανερὸν ὅτι μᾶλλον ἐν μορίου λόγῳ τὸ ἄπειρον ἢ ἐν ὅλου· μόριον γὰρ ἡ ὕλη τοῦ ὅλου ὥσπερ ὁ χαλκὸς τοῦ χαλκοῦ ἀνδριάντος. ἐπεὶ εἴ γε περιέχει ἐν τοῖς αἰσθητοῖς, καὶ ἐν τοῖς νοητοῖς τὸ μέγα καὶ τὸ μικρὸν ἔδει περιέχειν τὰ νοητά·

---

[a] [Parm. 8. 42 αὐτὰρ ἐπεὶ πεῖρας πύματον τετελεσμένον ἐστὶ | πάντοθεν εὐκύκλου σφαίρης ἐναλίγκιον ὄγκῳ, | μεσσόθεν ἰσοπαλὲς πάντῃ.—C.]

which there is nothing whatsoever ; whereas that from which something, no matter what, is missing and left outside is not 'All.' And 'whole' and 'complete,' if not absolutely the same, are very closely akin, and nothing is complete (*teleios*) unless it has an end (*telos*) ; but an end is a limit.

So Parmenides was nearer the mark than Melissos ; for Melissos speaks of 'the Whole' as 'unlimited,' whereas Parmenides sets boundaries to his 'whole,' that is 'equipoised on the centre.'[a] For 'whole' or 'all' and 'unlimited' are terms that cannot run in double harness. What led them to give to the Unlimited the impressive attributes of 'all-embracing' and 'all in itself containing' was the fact that it has a certain resemblance to the whole ; for the 'unlimited' is really the 'material' from which a magnitude is completed, and is the potential, though not the realized, whole. It is 'divisibility without limit' in the direction of reduction or converse expansion, and is not any determined whole in itself, but only as the unlimited and 'material' factor of the whole which is constituted as such by the limiting and 'formal' factor. As 'unlimited,' then, it is embraced and not embracing. Therefore *qua* unlimited it is unknowable, since 'material,' as such, is formless. So the unlimited were evidently better defined as a part than as the whole, in the sense in which 'part' means the 'material constituent,' as bronze is a part or constituent of the bronze statue. For, if in things of sense the undetermined 'great or small' were the continent and not the contained, analogy would demand that in the noetic world also the unintelligible should embrace the Ideas which are the norm of intelligibility ; but it is contradictory

207 a ἄτοπον δὲ καὶ ἀδύνατον τὸ ἄγνωστον καὶ τὸ ἀόριστον περιέχειν καὶ ὁρίζειν.

CHAPTER VII

ARGUMENT

[*Why number has a lower limit* (‘ *one* ’), *but no upper limit ; whereas there is no limit to the divisibility of magnitude but it cannot exceed every finite size either factually or conceptually* (207 a 33-b 21).—*R. W.*

Infinity of time is derived from infinity of motion, and this from the continuity (*infinite divisibility*) *of magnitude* (b 21-27).

Κατὰ λόγον δὲ συμβαίνει καὶ τὸ κατὰ πρόσθεσιν μὲν μὴ εἶναι δοκεῖν ἄπειρον οὕτως ὥστε παντὸς ὑπερβάλλειν μεγέθους, ἐπὶ τὴν διαίρεσιν δὲ εἶναι·
207 b περιέχεται γὰρ ὡς ἡ ὕλη ἐντὸς καὶ τὸ ἄπειρον, περιέχει δὲ τὸ εἶδος.

Εὐλόγως δὲ καὶ τὸ ἐν μὲν τῷ ἀριθμῷ εἶναι ἐπὶ τὸ ἐλάχιστον πέρας ἐπὶ δὲ τὸ πλεῖον ἀεὶ παντὸς ὑπερβάλλειν πλήθους, ἐπὶ δὲ τῶν μεγεθῶν τοὐναντίον ἐπὶ μὲν τὸ ἔλαττον παντὸς ὑπερβάλλειν μεγέθους ἐπὶ δὲ τὸ μεῖζον μὴ εἶναι μέγεθος ἄπειρον. αἴτιον δ᾽ ὅτι τὸ ἕν ἐστιν ἀδιαίρετον, ὅ τι περ ἂν ἓν ᾖ, οἷον ἄνθρωπος εἷς ἄνθρωπος καὶ οὐ πολλοί· ὁ δ᾽ ἀριθμός ἐστιν ἕνα πλείω καὶ πόσ᾽ ἄττα· ὥστ᾽ ἀνάγκη στῆναι ἐπὶ τὸ ἀδιαίρετον (τὰ γὰρ δύο καὶ τρία παρώνυμα ὀνόματά ἐστιν,

and impossible that the unknowable and undefined should embrace and define anything.

CHAPTER VII

ARGUMENT (*continued*)

Mathematicians do not require an actual infinite magnitude (b 27-34).

The infinite is ' material cause ' ; its essence, shortage ; its subject, the sensible-continuous. Hence it is that which is embraced or contained rather than the continent (b 34-208 a 4).—*C.*]

It follows also from the rationale of infinity that it cannot exceed all magnitude, but depends on the principle of division ; for since it is analogous to the ' material ' it is contained, whereas it is the ' form ' that is the continent.

It is also quite as it should be that in number there should be an inferior limit, whereas it is always possible to transcend any given number, but that in magnitudes, on the other hand, it should always be possible to make the small smaller, but there can be no magnitude of unlimited greatness. The reason is that unity, as unity, is atomic, the human unit, for instance, being one man and not more than one ;[a] whereas number is of more units than one, and specifically of ' so many,' so that you cannot go further back than the indivisible (for ' two ' and ' three,' that is, two ones and three ones, are both numbers, *qua* more than one, but different

207 b ὁμοίως δὲ καὶ τῶν ἄλλων ἀριθμῶν ἕκαστος). ἐπὶ δὲ τὸ πλεῖον ἀεὶ ἔστι νοῆσαι· ἄπειροι γὰρ αἱ διχοτομίαι τοῦ μεγέθους· ὥστε δυνάμει μὲν ἔστιν, ἐνεργείᾳ δ' οὔ, ἀλλ' ἀεὶ ὑπερβάλλει τὸ λαμβανόμενον παντὸς ὡρισμένου πλήθους. ἀλλ' οὐ χωριστὸς ὁ ἀριθμὸς οὗτος τῆς διχοτομίας, οὐδὲ μένει ἡ ἀπειρία ἀλλὰ γίνεται, ὥσπερ καὶ ὁ χρόνος καὶ ὁ ἀριθμὸς τοῦ χρόνου. ἐπὶ δὲ τῶν μεγεθῶν τοὐναντίον ἐστίν· διαιρεῖται μὲν γὰρ εἰς ἄπειρα τὸ συνεχές, ἐπὶ δὲ τὸ μεῖζον οὐκ ἔστιν ἄπειρον. ὅσον γὰρ ἐνδέχεται δυνάμει εἶναι, καὶ ἐνεργείᾳ ἐνδέχεται τοσοῦτον εἶναι· ὥστε, ἐπεὶ ἄπειρον οὐδέν ἐστι μέγεθος αἰσθητόν, οὐκ ἐνδέχεται παντὸς ὑπερβολὴν εἶναι ὡρισμένου μεγέθους· εἴη γὰρ ἄν τι τοῦ οὐρανοῦ μεῖζον.

Τὸ δ' ἄπειρον οὐ ταὐτὸν ἐν μεγέθει καὶ κινήσει καὶ χρόνῳ, ὡς μία τις φύσις, ἀλλὰ τὸ ὕστερον λέγεται κατὰ τὸ πρότερον, οἷον κίνησις μὲν ὅτι τὸ μέγεθος ἐφ' οὗ κινεῖται ἢ ἀλλοιοῦται ἢ αὐξά-

[a] [*παρώνυμα* are defined as things which *derive their name* from another name, with a difference of termination, *e.g.* 'grammarian' from 'grammar' (*Cat.* 1 a 11). 'Two' and 'Three' (etc.) might be regarded as the dual and plural (grammatical 'numbers') of 'one.' Simplic. 505. 18 explains that every number *derives its name* from the number of ones it contains: 'two' means 'that which consists of two ones,' and so on.—C.]

numbers, *qua* two or three respectively, and so with the rest).[a] But since you can always make another division of a magnitude into two, however many divisions you have already made to get it, you can always conceive a higher number of divisions than any given number however great; consequently the 'possibility of more' is inexhaustible and incapable of completion, but can be carried on through a greater than any assignable number of steps. This inexhaustible 'number,' however, is not separable from the dichotomy, and its 'illimitability' is not an accomplished *thing* like the magnitude itself that is the subject of the dichotomies, but is the accompaniment of the process of dichotomy, always in the making and never made; just like time, and the numerical register of time. So number cannot be reduced below unity, but can be increased indefinitely; but the reverse is true of magnitude; for a continuous magnitude can be divided beyond any given smallness, but cannot be increased above every assignable greatness. For any magnitude that can exist potentially can exist actually; so since, as we have seen, nothing sense-perceived can be unlimited, a magnitude in excess of every definite magnitude is an impossibility; it would have to transcend the universe.

Again 'illimitability' does not stand for the same thing in magnitude, movement, and time, but differs in accordance with their several natures in a determined order of priority. For continuity, in which the unlimited potentialities inhere, has a stable existence in magnitude, but movement (including modification and growth) is continuous because the magnitude over which it focuses is so, and time

νεται, ὁ χρόνος δὲ διὰ τὴν κίνησιν. νῦν μὲν οὖν χρώμεθα τούτοις, ὕστερον δὲ πειρασόμεθα λέγειν καὶ τί ἐστιν ἕκαστον καὶ διότι πᾶν μέγεθος εἰς μεγέθη διαιρετόν.

Οὐκ ἀφαιρεῖται δ' ὁ λόγος οὐδὲ τοὺς μαθηματικοὺς τὴν θεωρίαν, ἀναιρῶν οὕτως εἶναι τὸ ἄπειρον ὥστε ἐνεργείᾳ εἶναι ἐπὶ τὴν αὔξην ὡς ἀδιεξίτητον· οὐδὲ γὰρ νῦν δέονται τοῦ ἀπείρου οὐδὲ χρῶνται, ἀλλὰ μόνον εἶναι ὅσην ἂν βούλωνται τὴν πεπερασμένην· τῷ δὲ μεγίστῳ μεγέθει τὸν αὐτὸν ἔστι τετμῆσθαι λόγον ὁπηλικονοῦν μέγεθος ἕτερον. ὥστε πρὸς μὲν τὸ δεῖξαι ἐκείνοις οὐδὲν διοίσει, τὸ δ' εἶναι ἐν τοῖς οὖσιν ἔσται μεγέθεσιν.

Ἐπεὶ δὲ τὰ αἴτια διῄρηται τετραχῶς, φανερὸν ὅτι ὡς ὕλη τὸ ἄπειρόν ἐστιν αἴτιον, καὶ ὅτι τὸ μὲν εἶναι αὐτῷ στέρησις, τὸ δὲ καθ' αὑτὸ ὑποκείμενον τὸ συνεχὲς καὶ αἰσθητόν. φαίνονται δὲ πάντες καὶ οἱ ἄλλοι ὡς ὕλῃ χρώμενοι τῷ ἀπείρῳ· διὸ καὶ ἄτοπον τὸ περιέχον ποιεῖν αὐτὸ ἀλλὰ μὴ τὸ περιεχόμενον.

CHAPTER VIII

ARGUMENT

[*A final reply to the considerations urged* (*chap. iv.*) *in favour of the existence of something illimitable.—C.*]

Λοιπὸν δ' ἐπελθεῖν καθ' οὓς λόγους τὸ ἄπειρον

---

[a] [*Cf.* Bk. IV. chap. xi. 219 a 12 (p. 385). *Literally*, 'viz. a change (is called infinite) because the magnitude over which the change—of place or quality or size—takes place (is called infinite); and time (is called infinite) because the change (it measures is so).'—C.] The term 'infinite' here refers to the inexhaustible capacity for dichotomy.

is so because it is the comparative register of movements.[a] For the present, however, we shall deal with them all as we want them, though we shall not forget presently to attempt to give an account of what each of them is, and why every magnitude is divisible into magnitudes.[b]

Nor does this account of infinity rob the mathematicians of their study; for all that it denies is the actual existence of anything so great that you can never get to the end of it. And as a matter of fact, mathematicians never ask for or introduce an infinite magnitude; they only claim that the finite line shall be of any length they please; and it is possible to divide any magnitude whatsoever in the same proportion as the greatest magnitude.[c] So that the question under discussion does not affect their demonstrations; whereas actual dimensional existence can only be found in actually existent magnitudes.

As to the so-called 'four causes' or determinants, it is obviously the 'material determinant' to which the unlimited must be referred, and its essence is 'shortage,' while the subject in which it properly inheres will be the sensible-continuous. All other thinkers, too, agree in working with the unlimited as the 'material' (rather than formal) determinant; hence it is absurd to regard it as that which contains rather than as that which is contained.

CHAPTER VIII

It remains to disarm the considerations urged in

[b] See Book IV. chaps. xi.-xiii.

[c] [So they can obtain a line as *small* as they want.—C.]

208 a εἶναι δοκεῖ οὐ μόνον δυνάμει ἀλλ' ὡς ἀφωρισμένον· τὰ μὲν γάρ ἐστιν αὐτῶν οὐκ ἀναγκαῖα, τὰ δ' ἔχει τινὰς ἑτέρας ἀληθεῖς ἀπαντήσεις.

Οὔτε γὰρ ἵνα ἡ γένεσις μὴ ἐπιλείπῃ, ἀναγκαῖον ἐνεργείᾳ ἄπειρον εἶναι σῶμα αἰσθητόν· ἐνδέχεται γὰρ τὴν θατέρου φθορὰν θατέρου εἶναι γένεσιν, πεπερασμένου ὄντος τοῦ παντός.

Ἔτι τὸ ἅπτεσθαι καὶ τὸ πεπεράνθαι ἕτερον. τὸ μὲν γὰρ πρός τι καὶ τινός (ἅπτεται γὰρ πᾶν τινός) καὶ τῶν πεπερασμένων τινὶ συμβέβηκεν· τὸ δὲ πεπερασμένον οὐ πρός τι. οὐδ' ἅψασθαι τῷ τυχόντι τοῦ τυχόντος ἔστιν.

Τὸ δὲ τῇ νοήσει πιστεύειν ἄτοπον· οὐ γὰρ ἐπὶ τοῦ πράγματος ἡ ὑπεροχὴ καὶ ἡ ἔλλειψις, ἀλλ' ἐπὶ τῆς νοήσεως. ἕκαστον γὰρ ἡμῶν νοήσειέν ἄν τις πολλαπλάσιον ἑαυτοῦ αὔξων εἰς ἄπειρον· ἀλλ' οὐ διὰ τοῦτο ἔξω τοῦ ἄστεός τίς ἐστιν ἢ τοῦ τηλικοῦδε μεγέθους ὃ ἔχομεν, ὅτι νοεῖ τις, ἀλλ' ὅτι ἔστιν· τοῦτο δὲ συμβέβηκεν. ὁ δὲ χρόνος καὶ

---

[a] [See 203 b 18. The argument that there must be an inexhaustible reservoir (ἀρχή) from which things may come into existence is ascribed to Anaximander. (*Cf.* Diels, *Vors.* 2. 14.)—C.]

[b] [*Cf.* 203 b 20, the argument: 'What is limited is always limited by *coming up to something* (πρός τι περαίνειν), so there can be no (absolute) limit, if one thing must always come up to another.'—C.]

[c] [*Literally*, 'nor is contact possible between any two things you choose to mention.' This seems to be a distinct point in which 'being in contact' differs from 'being limited.'—C.]

[d] [*Cf.* 203 b 22, 'Number is thought to be infinite, because we can always conceive something beyond any limit.'—C.]

support of the existence of the unlimited not only as a potentiality but as actually compassed. Some of them do not follow as alleged from the admitted premises; and the rest can be met along some other line of sound reasoning.

(1) Admitting that things never cease to come into being,[a] it does not follow that there actually exists some sense-perceptible body unlimited in quantity; for though the sum of things be limited, things may come out of and pass into each other without end.

(2) Again, being in contact and being limited are different things.[b] Contact is a relation with something else, for there must be something to touch the touched; and this may happen to something limited incidentally; but 'being limited' is not a relation. Also a limited thing need not be touched by a thing homogeneous with itself and cannot be touched by any other.[c]

(3) It is futile to trust to what we can conceive as a guide to what is or can be;[d] for the excess or defect in such a case lies not in the thing but in the conceiving. One might conceive any one of us to be many times as big as we are, without limit; but if there does exist a man too big for the city to hold,[e] for instance, or even bigger than the men we know of, that is not because we have conceived him to exist, but because he does; and whether we have or have not conceived him to exist is a mere incident.

Time and movement are indeed unlimited, but

[e] [ἔξω τοῦ ἄστεος should mean 'outside the city'—a distinct illustration. Philoponus, 495. 6, says the most accurate copies did not contain τοῦ ἄστεος but only ἔξω τοῦ τηλικούτου μεγέθους (τουτέστιν οὗ ἔχομεν οἱ ἄνθρωποι). See Diels, *Zur Textgesch.* p. 39.—C.]

208 a ἡ κίνησις ἄπειρά ἐστι, καὶ ἡ νόησις, οὐχ ὑπομένοντος τοῦ λαμβανομένου. μέγεθος δὲ οὔτε τῇ καθαιρέσει οὔτε τῇ νοητικῇ αὐξήσει ἐστὶν ἄπειρον. Ἀλλὰ περὶ μὲν τοῦ ἀπείρου, πῶς ἔστι καὶ πῶς οὐκ ἔστι καὶ τί ἐστιν, εἴρηται.

---

[a] [*Or*, 'Time and movement are infinite only as processes in which any part you take does not subsist, and it is the same with our conception of them (*i.e.*, our conception can always reach out to *further* time and *further* change, but never embrace infinite time or infinite change as a completed whole): while magnitude (though infinitely divisible) is not

only as processes, and we cannot even suppose their successive stretches to exist.

All parts of a given magnitude do indeed exist all at once, but illimitable division does not—(either in reduction or expansion).[a]

So much as to the unlimited, the sense in which it exists and does not exist, and what it is.

reduced to the infinitely small by subtraction nor increased to the infinitely great by a process of augmentation that we conceive.' These may be taken as further arguments that our unlimited power of conception cannot establish the existence of infinites other than those Aristotle has recognized.—C.]

BOOK IV

INTRODUCTION

In this and the following books, the meaning of 'Motion' and 'Position' is examined and their connexion discussed.

These terms represent *Relations* of things to one another, not things or parts of things themselves.

Relations do not change, but the things which are related are capable of moving and changing, and in so doing they come into different relationships with one another.

When things are moving they are said to be 'in motion', when they are not moving they are said to be 'in position'; but nothing can be either 'in motion' or 'in position' except with relation to something else.

Being 'in motion' and being 'in position' are contrasted relationships, neither of which can be expressed in terms of the other: the first is dynamic, the second is static.[a] They are contradictory alternatives, with no intermediate between them: everything capable of being in motion must be actually in *either* one *or* the other of these relations; it cannot be in neither, nor can it be partially in both: for that which is already in position (with respect to certain other things) cannot move without

[a] The obvious fact that motion may be quicker or slower does not affect the question, for anything which is moving at all (whether quickly or slowly) is in motion and motion—whether slow or rapid—is a dynamic relation and as such is contrasted with position, the static relation.

quitting its position, therefore to assert that a thing which was formerly ' in position ' is now ' in motion ' is just another way of saying that it has quitted its position; the same reasoning applies (*mutatis mutandis*) to a thing which was formerly in motion and is now in position. Now anything which is capable of quitting its position is capable of ' being in motion ' and vice versa and therefore while actually ' in position ' it has the potentiality of ' being in motion ' and vice versa.

The Greek *tŏpos*, to the discussion of which the first section (chaps. i.-v.) of this Book is devoted, may mean either ' place ' (Latin *locus*) or ' space ' (Latin *spatium*); but what Aristotle is directly concerned with here is only ' place,' implying ' position,' and not abstract or absolute ' space ' at all.

Failure to understand this has led to grotesque misconceptions of Aristotle's teaching and depreciation of his intelligence.

When, in answer to the question ' where is it ? ' we mention the ' place ' that a thing is in, what exactly do we mean by ' place ' ?

The question had been confused by Zeno's attempt to show that the conception of things moving, or changing their places, or being in ' places ' at all, must involve us in hopeless contradictions ; and Aristotle, who, unlike Zeno, believes in the reality both of time and of place, begins his investigation, as usual, by asking how we currently use the word, and what are the perplexities in which it seems to land us. But from the very first he keeps firm hold of the principle that would now be expressed as ' the relativity of position.' To speak of the ' absolute ' position—either of a point or of the universe—would have been unmeaning to him, because ' position ' only exists in relation to some intrinsic ' frame of reference.' [a]

[a] Compare Clerk Maxwell's dictum (under the heading of ' Absolute space ') that anyone ' who will try to imagine the state of a mind conscious of knowing the absolute position of a point will ever after be content with our relative knowledge ' (*Matter and Motion*, London, 1876, p. 20).

It is impossible, then, to assign any place to the universe as a whole, and since Aristotle's universe was bounded by the outmost heavenly sphere it was impossible to assign to that sphere (regarded as a single whole) any place at all.

But Aristotle's universe has determinate dimensions, although it has no place. A thing's place, then, is not constituted by its own dimensions, nor can anything be its own place.

We turn now to the direct examination of Aristotle's doctrine of 'place,' which term applies to material bodies only. To the question 'Where is the wine ?' the answer may be 'In the flask' or 'In the chest' or 'In the house,' and so forth, till we come ultimately to 'Under heaven.' Now of these answers it is only the first that tells us of the wine's 'proper' place, which excludes other bodies. The other answers refer us to places 'common' to the wine and to other bodies. But they all assign some kind of physical *limit*, within which the wine is contained—either wine or together with certain other bodies.

In developing the implications of this all-important distinction between a thing's proper place and a place which is common to it and other things, it may be helpful to vary Aristotle's illustrations.

Suppose a stone to be sinking through a body of water (or a bubble rising through it). If at any moment its motion is arrested what will its place be ? Aristotle would answer 'The aqueous surface which at that moment constitutes its immediate envelope.' This is vital to his conception of a place-proper. The place-proper of the stone, at any moment, is not the whole body of water but the aqueous surface immediately enveloping it at that moment, and so too the proper place of the wine, as conceived by Aristotle, is not the bottle as a whole, but the inner surface of the bottle. It is therefore (like any other surface, or length, or extension) a reality, and even a quantitive reality, for things really are long, and just *so* long, and really have surfaces of definite dimensionality and (in the case of a spherical surface, for example) capacity. But an *actual* surface cannot exist apart from some physical body whose surface it is, in this case the whole undifferentiated mass of water in the flask.

Hence, for a body to be in a place it must be regarded as itself continuous, but as differentiated from its body-continent. A component portion of the water in the flask has no ' proper place ' unless it can be somehow differentiated from other portions. The component portions collectively compose the whole body of water, and the *proper* place of the water-as-a-whole is the ' collective place ' of its undifferentiated component portions. But if a definitely limited vortex could arise within the mass, it would be differentiated by its motion from its stationary aqueous envelope, and a ' proper place ' could be assigned to it : viz. the inner surface of the stationary envelope in contact with it. Conceptually a continuous mass can be subdivided indefinitely, but since surfaces cannot constitute a mass any more than points can constitute a line, there must always be some continuous mass between any two ' places.'[a] Thus a ' place ' is never ' in another place ' in the sense in which a content is in its continent.

A body, then, is in a place constituted by a surface of

[a] Aristotle, not being an atomist, undoubtedly believed in the continuity of the ultimate matter, so that, for instance, however thin the concentric laminae into which we suppose a sphere to be divided, there must always be undivided matter between the inner and the outer surfaces of any one of them. This, if the surfaces of lamina B divide it from lamina C within it and from lamina A outside it, the inner surface of B (coinciding with the outer surface of C) will constitute C's proper place, and also the place ' common ' to C and to all that is within it. Whereas the outer surface of B will coincide with the inner surface of A, which constitutes its (B's) proper place, and also the place ' common ' to it (B) and all that is within it.

Adjustments to less simple and symmetrical cases involve no new principle. The special case of *rotation* is treated in the Introduction to chap. v. p. 317.

Each lamina constitutes a ' vessel ' the inner surface of which is the ' place ' of its content, and this vessel-continent has its own place, viz. the inner surface of *its* vessel-continent.

its body-continent; and this surface, though not itself a physical body, must pertain to some physical body, of which it is the surface.[a] And that physical body (except in the case of the supreme heaven) must itself have a place. So, as we travel outwards, reaching, at every step, a place which is the proper place of the body-continent we have last reached (and the common place of it and of everything within it), we come ultimately to the revolving heaven, not contained by anything and therefore unplaced, but containing everything. Its inner surface is the universal place. Compare Gen. Introd. p. lxviii.

This inner celestial surface, as a whole, is stationary as to its position relatively to the Earth, and it determines its own centre geometrically. Together with the surface of the Earth (which Aristotle regards as fixed, Introduction to chap. v. p. 318), this heavenly sphere and its centre form a universal frame of reference, in relation to which all things can have geometrically determined positions assigned to them. These cosmic sites, so determined, are absolutely fixed and immovable, *with respect to the universal frame of reference*, but they are not physical places, as defined by Aristotle. Thus, if a vessel is moved about, with its content in it, the content remains in the same physical place; while that physical place itself (incidentally to the movement of the vessel whose inner surface it is) is changing its cosmically determined site.

It is obvious that dire confusion will arise if physically determined places and geometrically determined sites are confounded. Yet they are both ultimately referred to the same frame of reference, and can both be naturally called places (*tŏpoi*). It is Aristotle's concern to bring order into the chaos that these confusions have created.

The universe is a body, or mass, that has no place, because there is nothing else that it is in. Can there also —actually or conceivably—be a ' place ' that ' has nothing in it ' ?

[a] Existing in the body not as a material part but as an essential attribute.

It is obvious that this question may raise speculations as to 'void spaces' between, or within, the material constituents of the universe, just as the 'unplaced' universe itself suggests speculations as to an 'unmeasurable void' outside it. Thus a discussion of the 'void' which is the subject of the second section of this Book (chaps. vi.-ix.) naturally dovetails into the discussion of 'place' in the first section. Thus it happens that suggestions, at least, of 'space' or 'spaces' are, so to speak, always there in the offing, though Aristotle only indirectly concerns himself with them here. We may take it that κενόν [=void] comes much nearer to our 'space' than τόπος [=place].

Returning now to the question 'Can there be a place with nothing in it?' we see at once that Aristotle would have had no hestitation in declaring that at any rate there *is* no such place; for he regarded the universe as a *plenum*, with no inter-atomic or inter-corporeal void or voids within it. And moreover he thought he could prove that Democritus and other believers in the 'void,' while defining it as 'nothingness,' actually assigned physical attributes and activities to it which made it a 'something.'

Inside a place, therefore, Aristotle believed there was always a physical substance with dimensional extension of its own; its 'place,' being the inner surface of some other body, was coincident with its own outer surface.

This insistence that the 'place' of a thing embraces it and is not something intrapenetrant with it (the 'space' it occupies, as we might say), is vital to Aristotle's statement of his case. If it were no more than a matter of definition, it would, in Aristotle's opinion, be highly serviceable in clearing away confusions. But to him—with his disbelief in the 'void' and his conviction that to identify 'place' with it was to make it share its veiled materiality—it was more than a matter of definition. Compare the section on the 'void,' pp. 328 ff.

But in warring against the conception of a 'place' as a something inside the body-continent in addition to the content, but coinciding with and interpenetrating that content throughout its volume, Aristotle of course did not mean to deny the obvious fact that there are measurable *intramural* distances (so to express it) from point to point of the surface-continent itself, as measured across it. Every such surface-continent, if rigid, has a definite dimensional ' capacity,' and such capacity is exactly equal to the dimensional ' extension ' of the content. In the case of a bottle the surface-continent determines the form, or shape, of the fluid content. In the case of the sinking stone, on the other hand, it is the surface of the contained body as it passes through the fluid that determines the (constant) form of the changing surface-continent. But in either case the *capacity* of the ' place ' and the *extension* of the ' placed ' remain dimensionally equal to each other. Aristotle repeatedly and expressly insists on this dimensional equality, and his use of the word *chōra* (which may conveniently be translated by ' room ') as a synonym of *tŏpos* (place) shows that he never lost sight of it.

Aristotle, then, denies that there *is* such a thing as a really empty ' place with nothing in it.' But does his teaching preclude the conceptual *possibility* of such a thing ?

It appears from Themistius (4th century) that Galen (2nd century) pushed the hypothesis of a ' vessel with nothing in it,' and drew theoretical conclusions from it hostile to Aristotle's philosophy. Themistius himself can only answer that the hypothesis is absurd, and that if the content of a brazen vessel, for instance, *could* be taken out of it without the entry of something else, the vessel, however rigid, would collapse, like a bag or bladder.

But let us suppose the air-pump, or some analogous invention, to have suggested to Aristotle the possibility of a really ' exhausted ' receiver. It might no doubt have disturbed his system in many ways, but as to the matter now in hand I cannot think it would have disturbed him at all.

He would have said : The whole actual universe has no place, but it might *conceivably* have had one, if there had been something more outside it (though it would not then have been the whole universe), and therefore, though not an actual content (there being nothing to contain it), it is—so far as its own composition is concerned—a potential content. Just so, the inner surface of a rigid receiver, or vessel, if it had really nothing in it, would not actually be a ' place '; because it would not be the place of anything, and that which is not the place of anything is not a place at all. But it would be a potential place, and would have an inherent ' capacity ' of definite dimensional magnitude.

Finally (to return to the ' unplaced ' universe), does Aristotle (as has been alleged) believe that ' space ' is limited, and that there ' is no more of it outside the heavens ' ? Far from it. He does indeed hold that there is nothing ' outside everything, yet embracing everything, including itself.' But he expressly says that, conceptually, we can project a line beyond the limit of the universe, which implies, of course, that the universe itself might conceivably have been dimensionally greater or smaller than it is. There is no limit to its conceptual dimensions, then, but, whatever its dimensions might be, an actual universe must have an actual limit. For to speak of a universe at once actual and limitless would be to Aristotle a contradiction in terms ; since it would be speaking of ' the realized whole of that in which the possibility of more is still open and realizable.'

Thus Aristotle evades, if you like, or fails to recognize, the metaphysical problem of ' abstract-space.' But he does not talk the nonsense about it with which he has sometimes been discredited.

Δ

CHAPTER I

ARGUMENT

What account are we to give of the place *' in ' (' on,' or ' at ') which we say a thing is, and from which it may pass to another ' place ' ?*

All things in Nature are capable of passing from this place to that, but the heavenly bodies are susceptible of no other change, such as change of quality or size. Thus ' local mobility ' is the most universal phenomenon that the natural philosopher has to consider ; and the very language in which we describe every other mode of ' passing from this to that ' (every other trans-ition-*al change, that is) is derived from*
185 a 12 sq. *terms that primarily apply to local movement.*

Clearly, then, the natural philosopher must give some account of the ' place ' from and to which things move [a] (208 a 27-32).

[a] Note that in Greek there is a single word (*kinēsis*) the primary meaning of which is ' movement,' in the literal and local sense, but which can receive, without violence, a much wider derivative and extended sense than it is possible to give to ' movement ' in English. Hence *kinēsis* may always stand for the *typical representative* of every kind of transition, and may often be applied as an inclusive *appellation* of all transitions,—the passage from one colour to another, for instance, or the growth from a sapling into a tree. The translator must meet the difficulty thus encountered as best he can, and in different ways according to the context.

When Aristotle wants an exclusive term for ' local move-

BOOK IV

CHAPTER I

ARGUMENT (*continued*)

But it is no easy matter to meet this reasonable demand; for there turn out to be many assertions about the place of things which we should be inclined to accept without misgiving as points of departure, but which lead us to mutually contradictory conclusions.

To begin with, the phenomenon of 'displacement' seems to establish the 'place-continent' as something distinct from its 'content.'

And since the centre of the universe seems to have an influence on the movement of heavy bodies and the periphery on that of buoyant ones, cosmic 'locality' seems not only to be a reality distinct from its contents but to have a certain power of influencing things to come to it (8 a 32-b 16).

ment' he employs the word *phora*. This includes both movement in a straight line (the *translation* of modern text-books of mechanics) and movement round a fixed point (*rotation*). Aristotle was well aware of the principle that "any displacement of a body may be effected by a translation combined with a rotation" (Anthony and Brackett, *Text-book of Physics*, New York and London, 1908, p. 40). But he had no technical term for 'translation.' In many connexions *phora* is best translated not by 'local movement' but by the more 'active' or 'passive' terms (as the case may be) of 'trend' or 'vection.'

ARGUMENT (*continued*)

The primary and stable character of cosmic directions and positions vindicated and illustrated (b 16–25).

Real existence of place assumed in the definition of 'vacancy' and accepted by Hesiod—but seems very amazing. For what can it be? It has dimensions, but how can it be a body? If as 'continent' it is distinct from its 'content,' is every point in it (for in that which has dimensions there must be points) a 'point-continent' distinct from the 'point contained' in it? Is it compounded of elements? Hardly; for if its elements were conceptual it could have no concrete dimensions, which it has. And if its elements were physical it would be palpable to the senses (and most of all to the

208 a 27 Ὁμοίως δ' ἀνάγκη καὶ περὶ τόπου τὸν φυσικὸν ὥσπερ καὶ περὶ ἀπείρου γνωρίζειν, εἰ ἔστιν ἢ μή, καὶ πῶς ἐστι, καὶ τί ἐστιν.

Τά τε γὰρ ὄντα πάντες ὑπολαμβάνουσι εἶναί που (τὸ γὰρ μὴ ὂν οὐδαμοῦ εἶναι· ποῦ γάρ ἐστι τραγέλαφος ἢ σφίγξ;) καὶ τῆς κινήσεως ἡ κοινὴ μάλιστα καὶ κυριωτάτη κατὰ τόπον ἐστίν, ἣν καλοῦμεν φοράν.

Ἔχει δὲ πολλὰς ἀπορίας τί ποτ' ἔστιν ὁ τόπος· οὐ γὰρ ταὐτὸν φαίνεται θεωροῦσιν ἐξ ἁπάντων τῶν ὑπαρχόντων. ἔτι δ' οὐδ' ἔχομεν οὐδὲν παρὰ τῶν ἄλλων οὔτε προηπορημένον οὔτε προευπορη-
208 b μένον περὶ αὐτοῦ.

Ὅτι μὲν οὖν ἔστιν ὁ τόπος, δοκεῖ δῆλον εἶναι

[a] Aristotle himself does not accept this "general opinion.' Only bodies have locality, and other things than bodies 'exist.' Cf. Introduction, pp. 268 ff., also Chap. iii. of this Book, on the ambiguity of 'in.'

ARGUMENT (*continued*)

primary sense of touch), which it is not. Is it then itself an element? No; for nothing is 'made of it.' It is not the material of anything, therefore, nor can we imagine it to be any of the other three causal or essential determinants.

And, if it exists, 'where' does it exist? And if it exactly fits its content does it grow when its content grows? How can it? 194 b 16 sqq.

These are the perplexities to which 'common assumptions' lead us. We must start again, therefore, and assume nothing on the ground that it is 'commonly' assumed, not even that such a thing as 'place' exists at all (8 b 25–9 a 30).

The Natural Philosopher has to ask the same questions about 'place' as about the 'unlimited'; namely, whether such a thing exists at all, and (if so) after what fashion it exists, and how we are to define it.

It is generally assumed that whatever exists, exists 'somewhere'[a] (that is to say, 'in some place'), in contrast to things which 'are nowhere' because they are non-existent,—so that the obvious answer to the question "where is the goat-stag or the sphinx?" is "nowhere." And again the primary and most general case of 'passage' or transitional change from 'this' to 'that' is the case of local change from this to that 'place.'

But we encounter many difficulties when we attempt to say what exactly the 'place' of a thing is. For according to the data from which we start we seem to reach different and inconsistent conclusions. Nor have my precursors laid anything down, or even formulated any problems, on this subject.

To begin with, then, the phenomenon of 'replace-

208 b ἐκ τῆς ἀντιμεταστάσεως· ὅπου γάρ ἐστι νῦν ὕδωρ, ἐνταῦθα ἐξελθόντος ὥσπερ ἐξ ἀγγείου πάλιν ἀὴρ ἔνεστιν, ὅτε δὲ τὸν αὐτὸν τόπον τοῦτον ἄλλο τι τῶν σωμάτων κατέχει·[1] τοῦτο δὴ τῶν ἐγγινομένων καὶ μεταβαλλόντων ἕτερον πάντων εἶναι δοκεῖ· ἐν ᾧ γὰρ ἀήρ ἐστι νῦν, ὕδωρ ἐν τούτῳ πρότερον ἦν, ὥστε δῆλον ὡς ἦν ὁ τόπος τι καὶ ἡ χώρα ἕτερον ἀμφοῖν, εἰς ἣν καὶ ἐξ ἧς μετέβαλον.

Ἔτι δὲ αἱ φοραὶ τῶν φυσικῶν σωμάτων καὶ ἁπλῶν, οἷον πυρὸς καὶ γῆς καὶ τῶν τοιούτων, οὐ μόνον δηλοῦσιν ὅτι ἔστι τι ὁ τόπος, ἀλλ' ὅτι καὶ ἔχει τινὰ δύναμιν. φέρεται γὰρ ἕκαστον εἰς τὸν αὑτοῦ τόπον μὴ κωλυόμενον, τὸ μὲν ἄνω τὸ δὲ κάτω· ταῦτα δ' ἐστὶ τόπου μέρη καὶ εἴδη, τό τε ἄνω καὶ τὸ κάτω καὶ αἱ λοιπαὶ τῶν ἓξ διαστάσεων.

Ἔστι δὲ τὰ τοιαῦτα οὐ μόνον πρὸς ἡμᾶς, τὸ ἄνω καὶ κάτω καὶ δεξιὸν καὶ ἀριστερόν· ἡμῖν μὲν γὰρ οὐκ ἀεὶ τὸ αὐτό, ἀλλὰ κατὰ τὴν θέσιν, ὅπως ἂν στραφῶμεν, γίνεται, διὸ καὶ ταὐτὸ πολλάκις δεξιὸν καὶ ἀριστερόν ἐστι καὶ ἄνω καὶ κάτω καὶ πρόσθεν καὶ ὄπισθεν· ἐν δὲ τῇ φύσει διώρισται χωρὶς ἕκαστον. οὐ γὰρ ὅ τι ἔτυχέν ἐστι τὸ ἄνω ἀλλ' ὅπου φέρεται τὸ πῦρ καὶ τὸ κοῦφον· ὁμοίως δὲ καὶ τὸ κάτω οὐχ ὅ τι ἔτυχεν ἀλλ' ὅπου τὰ

[1] [I have repunctuated, taking ὅτε δέ to mean 'and at another time.' *Cf.* 209 b 26.—C.]

ment' seems at once to prove the independent existence of the 'place' from which—as if from a vessel—water, for instance, has gone out, and into which air has come, and which some other body yet may occupy in its turn; for the place itself is thus revealed as something different from each and all of its changing contents. For 'that wherein' air *is*, is identical with 'that wherein' water *was*; so that the 'place' or 'room' into which each substance came, or out of which it went, must all the time have been distinct from both of the substances alike.

Moreover the trends of the physical elements (fire, earth, and the rest) show not only that locality or position is a reality but also that it exerts an active influence; for fire and earth are borne, the one upwards and the other downwards, if unimpeded, each towards its own 'position,' and these terms—'up' and 'down' I mean, and the rest of the six dimensional directions—indicate subdivisions or distinct classes of positions or localities in general.

Now these terms—such as up and down and right and left, I mean—when thus applied to the trends of the elements are not merely relative to ourselves. For in this relative sense the terms have no constancy, but change their meaning according to our own position, as we turn this way or that; so that the same thing may be now to the right and now to the left, now above and now below, now in front and now behind; whereas in Nature each of these directions is distinct and stable independently of us. 'Up' or 'above' always indicates the 'whither' to which things buoyant tend; and so too 'down' or 'below' always indicates the 'whither' to which weighty

208 b ἔχοντα βάρος καὶ τὰ γεηρά, ὡς οὐ τῇ θέσει διαφέροντα μόνον ἀλλὰ καὶ τῇ δυνάμει. δηλοῖ δὲ καὶ τὰ μαθηματικά· οὐκ ὄντα γὰρ ἐν τόπῳ ὅμως κατὰ τὴν θέσιν τὴν πρὸς ἡμᾶς ἔχει δεξιὰ καὶ ἀριστερά, ὥστε μόνον αὐτῶν νοεῖσθαι τὴν θέσιν, ἀλλὰ μὴ ἔχειν φύσιν τούτων ἕκαστον.[1]

Ἔτι οἱ τὸ κενὸν φάσκοντες εἶναι τόπον λέγουσιν· τὸ γὰρ κενὸν τόπος ἂν εἴη ἐστερημένος σώματος.

Ὅτι μὲν οὖν ἔστι τι ὁ τόπος παρὰ τὰ σώματα, καὶ πᾶν σῶμα αἰσθητὸν ἐν τόπῳ, διὰ τούτων ἄν τις ὑπολάβοι· δόξειε δ' ἂν καὶ Ἡσίοδος ὀρθῶς λέγειν ποιήσας πρῶτον τὸ χάος. λέγει γοῦν "πάντων μὲν πρώτιστα χάος γένετ', αὐτὰρ ἔπειτα γαῖ' εὐρύστερνος," ὡς δέον πρῶτον ὑπάρξαι χώραν τοῖς οὖσι, διὰ τὸ νομίζειν, ὥσπερ οἱ πολλοί, πάντα εἶναί που καὶ ἐν τόπῳ.

Εἰ δ' ἐστὶ τοιοῦτο, θαυμαστή τις ἂν εἴη ἡ τοῦ τόπου δύναμις καὶ προτέρα πάντων· οὗ γὰρ ἄνευ
209 a τῶν ἄλλων οὐδὲν ἔστιν, ἐκεῖνο δ' ἄνευ τῶν ἄλλων, ἀνάγκη πρῶτον εἶναι· οὐ γὰρ ἀπόλλυται ὁ τόπος τῶν ἐν αὐτῷ φθειρομένων.

Οὐ μὴν ἀλλ' ἔχει γ' ἀπορίαν, εἰ ἔστι, τί ἐστι,

[1] [ὥστε μόνον αὐτῶν νοεῖσθαι τὴν θέσιν is Alexander's correction. Laas, Diels, and Prantl prefer what Simplicius (526. 5, 16) read: ὡς τὰ μόνον λεγόμενα διὰ θέσιν οὐκ ἔχοντα φύσει τούτων ἕκαστον. FGI adopt Alexander's correction, but retain οὐκ ἔχοντα φύσει τούτων ἕκαστον, which belongs to Simplicius's reading. E substitutes ἀλλὰ μὴ ἔχειν φύσιν τούτων ἕκαστον to fit Alexander's correction. We should follow either Alexander or Simplicius, but not both.—C.]

and earthy matters tend, and does not change with circumstance; and this shows that 'above' and 'below' not only indicate definite and distinct localities, directions, and positions, but also produce distinct effects. The comparison of mathematical figures illustrates the point. For such figures occupy no real positions of their own, but nevertheless acquire a right and left with reference to us, thus showing that their positions are merely such as we mentally assign to them and are not intrinsically distinguished by anything in Nature.

Further, the thinkers who assert the existence of the 'void' agree with all others in recognizing the reality of 'place,' for the 'void' is supposed to be 'place without anything in it.'

One might well conclude from all this that there must be such a thing as 'place' independent of all bodies, and that all bodies cognizable by the senses occupy their several distinct places. And this would justify Hesiod in giving primacy to *Chaos* [=the 'Gape'] where he says: "First of all things was Theog 116
Chaos, and next broad-bosomed Earth"; since before there could be anything else 'room' must be provided for it to occupy. For he accepted the general opinion that everything must be somewhere and must have a place.

And if such a thing should really exist well might we contemplate it with wonder—capable as it must be of existing without anything else, whereas nothing else could exist without it, since 'place' is not destroyed when its contents vanish.

But then, if we grant that such a thing exists, the question as to *how* it exists and what it really is must

209 a πότερον ὄγκος τις σώματος ἤ τις ἑτέρα φύσις· ζητητέον γὰρ τὸ γένος αὐτοῦ πρῶτον.

Διαστήματα μὲν οὖν ἔχει τρία, μῆκος καὶ πλάτος καὶ βάθος, οἷς ὁρίζεται σῶμα πᾶν. ἀδύνατον δὲ σῶμα εἶναι τὸν τόπον· ἐν ταὐτῷ γὰρ ἂν εἴη δύο σώματα.

Ἔτι εἴπερ ἔστι σώματος τόπος καὶ χώρα, δῆλον ὅτι καὶ ἐπιφανείας καὶ τῶν λοιπῶν περάτων· ὁ γὰρ αὐτὸς ἁρμόσει λόγος· ὅπου γὰρ ἦν πρότερον τὰ τοῦ ὕδατος ἐπίπεδα, ἔσται πάλιν τὰ τοῦ ἀέρος. ἀλλὰ μὴν οὐδεμίαν διαφορὰν ἔχομεν στιγμῆς καὶ τόπου στιγμῆς, ὥστ' εἰ μηδὲ ταύτης ἕτερόν ἐστιν ὁ τόπος, οὐδὲ τῶν ἄλλων οὐδενός, οὐδ' ἔστι τι παρ' ἕκαστον τούτων ὁ τόπος.

Τί γὰρ ἄν ποτε καὶ θείημεν εἶναι τὸν τόπον; οὔτε γὰρ στοιχεῖον οὔτ' ἐκ στοιχείων οἷόν τ' εἶναι τοιαύτην ἔχοντα φύσιν, οὔτε τῶν σωματικῶν οὔτε τῶν ἀσωμάτων· μέγεθος μὲν γὰρ ἔχει, σῶμα δ' οὐδέν· ἔστι δὲ τὰ μὲν τῶν αἰσθητῶν σωμάτων στοιχεῖα σώματα, ἐκ δὲ τῶν νοητῶν οὐδὲν γίνεται μέγεθος.

Ἔτι δὲ καὶ τίνος ἄν τις θείη τοῖς οὖσιν αἴτιον εἶναι τὸν τόπον; οὐδεμία γὰρ αὐτῷ ὑπάρχει αἰτία τῶν τεττάρων· οὔτε γὰρ ὡς ὕλη τῶν ὄντων (οὐδὲν γὰρ ἐξ αὐτοῦ συνέστηκεν) οὔτε ὡς εἶδος καὶ λόγος

---

[a] On the Categories *cf.* Gen. Introd. pp. l *sq.*

[b] ἐκ δὲ τῶν νοητῶν κτλ. develops μέγεθος . . . ἔχει, while ἔστι δὲ τὰ μὲν κτλ. develops σῶμα δ' οὐδέν.

give us pause. Is it some kind of corporeal bulk? Or has it some other mode of existence? For we must begin by determining to what category it belongs.[a]

Now a 'place,' as such, has the three dimensions of length, breadth, and depth, which determine the limits of all bodies; but it cannot itself be a body, for if a 'body' were in a 'place' and the place itself were a body, two bodies would coincide.

Another difficulty. If a body has 'place' and 'room,' the reasoning already employed would show that its surfaces and other limits must also have them. For 'where' the surfaces of the water were, 'there' the surfaces of the air now are. But we cannot distinguish between a point and its own position, and if there is no distinction here neither is there in any of the others, line, surface, bulk or capacity. So what would then become of the thesis that all bodies have places distinguishable from themselves?

What kind of thing then must we conceive 'a place' to be? Its properties forbid us to think of it either as being an element itself or as being compounded of elements—whether physical or conceptual. For it has size, which conceptual components could not give it; but it is not a body, which it would necessarily be if compounded of elements cognizable by sense.[b]

Again, how are we to suppose that place affects or determines things in any way? For it cannot be brought under any one of the four casual or essential determinants:—not as the 'material' of things, for nothing is composed of it; nor as their 'form' or

209 a τῶν πραγμάτων οὔθ' ὡς τέλος, οὔτε κινεῖ τὰ ὄντα.

Ἔτι δὲ καὶ αὐτὸς εἰ ἔστι τι τῶν ὄντων, ποῦ ἔσται; ἡ γὰρ Ζήνωνος ἀπορία ζητεῖ τινα λόγον· εἰ γὰρ πᾶν τὸ ὂν ἐν τόπῳ, δῆλον ὅτι καὶ τοῦ τόπου τόπος ἔσται, καὶ τοῦτο εἰς ἄπειρον πρόεισιν.

Ἔτι ὥσπερ ἅπαν σῶμα ἐν τόπῳ, οὕτω καὶ ἐν τόπῳ ἅπαντι σῶμα· πῶς οὖν ἐροῦμεν περὶ τῶν αὐξανομένων; ἀνάγκη γὰρ ἐκ τούτων συναύξεσθαι τὸν τόπον αὐτοῖς, εἰ μήτ' ἐλάττων μήτε μείζων ὁ τόπος ἑκάστου.

Διὰ μὲν οὖν τούτων οὐ μόνον τί ἐστιν, ἀλλὰ καὶ εἰ ἔστιν, ἀπορεῖν ἀναγκαῖον.

CHAPTER II

ARGUMENT

This chapter consists of two unequal and disparate portions.

In the first, Aristotle, following his usual method, puts aside provisionally the difficulties he has enumerated and opens his systematic investigation.

If (to vary Aristotle's own illustration) we ask "Where is such and such a book?" the answer may be "In the house" or "In the study" or "On such and such a shelf" or "Between such and such volumes"—(which volumes are on the shelf, which shelf is in the study which . . . is under the heavens).

Here 'between the two volumes' is the 'proper' place of

[a] Aristotle has already answered this by anticipation. The ultimate cosmic limits exercise an effective causation in actualizing the potentialities of buoyant and heavy bodies. *Cf.* 208 b 19.

[b] [Diels, *Vors.* 19 A 24.—C.]

constituent definition ; nor as their contemplated 'end'; nor as setting them in motion, or otherwise changing them.[a]

And yet again : If it has an existence of its own, 'where' does it exist ? For we cannot ignore Zeno's dilemma [b] : If everything that exists, exists in some 'place,' then if the place itself exists it too must have a place to exist in, and so on *ad infinitum*.

Further : If each body exactly occupies the place it is in, then reciprocally each place is exactly occupied by the body in it. But in that case what account are we to give of 'growing' things ? It would seem that their places must also grow, to keep company with them, since they can never be less than the places they occupy, nor the places they occupy be greater than they are.

So, after all, we are forced by these perplexities not only to ask what a 'place' is, but also to reopen the question that appeared to be closed, and ask whether there is such a thing as 'place' at all.

CHAPTER II

ARGUMENT (*continued*)

the book in question, from which it excludes all other bodies. All the other localities, being 'common' to it and to other bodies, are only called its place in virtue of the fact that they include its real and proper place.

Thus Aristotle introduces, almost incidentally, his own definition of 'place' (to be elaborated and rendered more precise subsequently) as that which immediately encompasses the 'place-occupying' body, or substance, in question (a 31–b 1).

In the second section, by a rather forced transition,

ARGUMENT (*continued*)

Aristotle returns (with special reference to Plato) to the thesis that ' place ' can be identified neither with ' form ' nor

209 a Ἐπεὶ δὲ τὸ μὲν καθ' αὑτὸ τὸ δὲ κατ' ἄλλο λέγεται, καὶ τόπος ὁ μὲν κοινός, ἐν ᾧ ἅπαντα τὰ σώματά ἐστιν, ὁ δ' ἴδιος, ἐν ᾧ πρώτῳ.

Λέγω δ' οἷον σὺ νῦν ἐν τῷ οὐρανῷ ὅτι ἐν τῷ ἀέρι, οὗτος δ' ἐν τῷ οὐρανῷ, καὶ ἐν τῷ ἀέρι δὲ ὅτι ἐν τῇ γῇ, ὁμοίως δὲ καὶ ἐν ταύτῃ ὅτι ἐν
209 b τῷδε τῷ τόπῳ, ὃς περιέχει οὐδὲν πλέον ἢ σέ.

Εἰ δή ἐστιν ὁ τόπος τὸ πρῶτον περιέχον τῶν σωμάτων ἕκαστον, πέρας τι ἂν εἴη, ὥστε δόξειεν ἂν τὸ εἶδος καὶ ἡ μορφὴ ἑκάστου ὁ τόπος εἶναι, ᾧ ὁρίζεται τὸ μέγεθος καὶ ἡ ὕλη ἡ τοῦ μεγέθους· τοῦτο γὰρ ἑκάστου πέρας. οὕτω μὲν οὖν σκοποῦσιν ὁ τόπος τὸ ἑκάστου εἶδός ἐστιν·

---

[a] The spot of Earth is obviously only a part (though a distinctive one) of the whole envelope. But Aristotle is not meticulously scrupulous as to this, either here or elsewhere. [ἐν τῷδε τῷ τόπῳ can mean 'in this (proper or primary) place,' which has a position on the earth's surface.—C.]

[b] A rigid continent, regarded as a mould, determines the ' shape ' and the ' appearance ' (so far at least as the visible contour is concerned) of the fluid poured into it (say, molten metal), which is itself indeterminate in shape.

Now the two original meanings of the Greek words *morphe* and *eidos* (which Aristotle most frequently uses to signify 'form,' in the philosophical sense of 'collectivity of distinguishing characteristics') are precisely 'shape' and 'appearance.' Moreover, in the case, so often used as an illustration, of a statue, the 'shape' does actually constitute the ' form ' in the philosophic sense, as distinct from the metallic or other ' material,' of the thing. Thus content and continent become symbols of undifferentiated material and differentiating form in general. And so, if 'place' be identified with continent it may come to be confused with form.

ARGUMENT (*continued*)

with 'matter.' This, together with a stray return to another point already dealt with, closes the chapter (9 b 1–10 a 13).

We have seen that attributions are made directly, in virtue of their immediate applicability, or mediately, because, though not immediately applicable themselves, they include, involve, or imply something that is immediately applicable. And so, too, a 'place' may be assigned to an object either primarily because it is its special and exclusive place, or mediately because it is 'common' to it and other things, or is the universal place that includes the proper places of *all* things. 195 a 33 *sqq.*

I mean, for instance, that you, at this moment, are in the universe because you are in the air, which air is in the universe; and in the air because on the earth; and in like manner on the earth because on the special place which 'contains and circumscribes you, and no other body.' [a]

But if what we mean by the 'place' of a body is its immediate envelope, then 'place' is a limiting determinant, which suggests that it is the specifying or moulding 'form' by which the concrete *quantum*, together with its component matter, is 'determined.' For it is just the office of a limit so to determine or mould something. From this point of view, then, we should identify 'place' with 'form.' [b]

Whereas if identified with intramural dimensionality it may come to be confused with material. Hence Aristotle's repeated insistence (9 a 20 f., the present passage, and 11 b 10-19, in Chap. iv.) on the point that place is neither form nor material. It would appear that Greek thinkers, hardly less than English students, were in perpetual danger of being dragged back from the full philosophical significance of *morphe* (= form) by the vividness of its visual suggestions,—sight being, then as now, 'the most despotic of the senses.'

209 b Ἡ δὲ δοκεῖ ὁ τόπος εἶναι τὸ διάστημα τοῦ μεγέθους, ἡ ὕλη· τοῦτο γὰρ ἕτερον τοῦ μεγέθους· τοῦτο δ' ἐστὶ τὸ περιεχόμενον ὑπὸ τοῦ εἴδους καὶ ὡρισμένον, οἷον ὑπὸ ἐπιπέδου καὶ πέρατος. ἔστι δὲ τοιοῦτον ἡ ὕλη καὶ τὸ ἀόριστον· ὅταν γὰρ ἀφαιρεθῇ τὸ πέρας καὶ τὰ πάθη τῆς σφαίρας, λείπεται οὐδὲν παρὰ τὴν ὕλην.

Διὸ καὶ Πλάτων τὴν ὕλην καὶ τὴν χώραν ταὐτό φησιν εἶναι ἐν τῷ Τιμαίῳ· τὸ γὰρ μεταληπτικὸν καὶ τὴν χώραν ἓν καὶ ταὐτόν. ἄλλον δὲ τρόπον ἐκεῖ τε λέγων τὸ μεταληπτικὸν καὶ ἐν τοῖς λεγομένοις ἀγράφοις δόγμασιν, ὅμως τὸν τόπον καὶ τὴν χώραν τὸ αὐτὸ ἀπεφήνατο. λέγουσι μὲν γὰρ πάντες εἶναί τι τὸν τόπον, τί δ' ἐστίν, οὗτος μόνος ἐπεχείρησεν εἰπεῖν.

Εἰκότως δ' ἐκ τούτων σκοπουμένοις δόξειεν ἂν εἶναι χαλεπὸν γνωρίσαι τί ἐστιν ὁ τόπος, εἴπερ τούτων ὁποτερονοῦν ἐστιν, εἴτε ἡ ὕλη εἴτε τὸ εἶδος· ἄλλως τε γὰρ τὴν ἀκροτάτην ἔχει θέαν, καὶ χωρὶς ἀλλήλων οὐ ῥᾴδιον γνωρίζειν.

Ἀλλὰ μὴν ὅτι γε ἀδύνατον ὁποτερονοῦν τούτων εἶναι τὸν τόπον, οὐ χαλεπὸν ἰδεῖν. τὸ μὲν γὰρ εἶδος καὶ ἡ ὕλη οὐ χωρίζεται τοῦ πράγματος, τὸν δὲ τόπον ἐνδέχεται· ἐν ᾧ γὰρ ἀὴρ ἦν, ἐν τούτῳ πάλιν ὕδωρ, ὡς ἔφαμεν, γίνεται, ἀντιμεθισταμένων ἀλλήλοις τοῦ τε ὕδατος καὶ τοῦ ἀέρος,

---

[a] [τοῦτο in *l.* 8 (as in *l.* 7) means τὸ διάστημα. 'This (dimensionality) **is** what is contained and defined by the shape (form), *e.g.* by a surface or limit. And that description applies to the matter or the indeterminate factor.'—C.]

[b] Aristotle too identifies 'place' and 'room' (*cf.* Introduction to this Book, p. 272), but Aristotle assimilates them

But if we think of a thing's place as its 'dimensionality' or 'room-occupancy' (to be distinguished from the thing itself, as a concrete *quantum*) we must then regard it as 'matter' rather than as 'form,' for matter[a] is the factor that is bounded and determined by the form, as a surface, or other limit, moulds and determines; for it is just that which is in itself undetermined, but capable of being determined, that we mean by matter. Thus, if a concrete sphere, *e.g.*, be stripped of its limit, as well as of its other determining characteristics, nothing but its matter is left.

This is why Plato, in the *Timaeus*, identifies 52 A
'matter' and 'room,' because 'room' and 'the receptive-of-determination' are one and the same thing. His account of the 'receptive' differs in the *Timaeus* and in what are known as his *Unwritten Teachings*, but he is consistent in asserting the identity of 'place' and 'room.' Thus, whereas everyone asserts the reality of 'place,' only Plato has so much as attempted to tell us what it is.[b]

It is no wonder that, when thus regarded,—either as matter or as form, I mean,—'place' should seem hard to grasp, especially as matter and form themselves stand at the very apex of speculative thought, and cannot well, either of them, be cognized as existing apart from the other.

But in truth it is easy enough to see that its place cannot possibly be either the matter or the form of a thing; for neither of these is separable from the thing itself, as its place undoubtedly is. For we have
already explained that 'where' the air was 'there' 8 b 6 *sqq.*
again the water is, when the water and air succeed

both to the surface-continent, and Plato to the intramural dimensionally. See Plato, *Tim.* 52 A and Archer Hind, *ad loc.*

καὶ τῶν ἄλλων σωμάτων ὁμοίως, ὥστε οὔτε μόριον οὔθ' ἕξις ἀλλὰ χωριστὸς ὁ τόπος ἑκάστου ἐστίν.

Καὶ γὰρ δοκεῖ τοιοῦτό τι εἶναι ὁ τόπος οἷον τὸ ἀγγεῖον· ἔστι γὰρ τὸ ἀγγεῖον τόπος μεταφορητός· τὸ δ' ἀγγεῖον οὐδὲν τοῦ πράγματός ἐστιν.

Ἧ μὲν οὖν χωριστός ἐστι τοῦ πράγματος, ταύτῃ μὲν οὐκ ἔστι τὸ εἶδος· ᾗ δὲ περιέχει, ταύτῃ δ' ἕτερος τῆς ὕλης. δοκεῖ δὲ ἀεὶ τὸ ὄν που αὐτό τε εἶναί τι καὶ ἕτερόν τι ἐκτὸς αὐτοῦ.

Πλάτωνι μέντοι λεκτέον, εἰ δεῖ παρεκβάντας εἰπεῖν, διὰ τί οὐκ ἐν τόπῳ τὰ εἴδη καὶ οἱ ἀριθμοί, εἴπερ τὸ μεθεκτικὸν ὁ τόπος, εἴτε τοῦ μεγάλου καὶ τοῦ μικροῦ ὄντος τοῦ μεθεκτικοῦ εἴτε τῆς ὕλης, ὥσπερ ἐν τῷ Τιμαίῳ γέγραφεν.

Ἔτι πῶς ἂν φέροιτο εἰς τὸν αὑτοῦ τόπον, εἰ ὁ τόπος ἦν ἡ ὕλη ἢ τὸ εἶδος; ἀδύνατον γὰρ οὗ μὴ κίνησις μηδὲ τὸ ἄνω ἢ κάτω ἐστί, τόπον εἶναι. ὥστε ζητητέος ἐν τοῖς τοιούτοις ὁ τόπος.

Εἰ δ' ἐν αὐτῷ[1] ὁ τόπος (δεῖ γάρ, εἴπερ ἢ μορφὴ ἢ ὕλη[2]), ἔσται ὁ τόπος ἐν τόπῳ· μεταβάλλει γὰρ ἅμα τῷ πράγματι καὶ κινεῖται καὶ τὸ εἶδος καὶ τὸ ἀόριστον,

[1] [αὐτῷ Simpl. 547. 33 (lemma) (*i.e.* ἐν αὐτῷ τῷ πράγματι, 548. 1), Philop. 525. 16 (lemma): αὐτῷ codd.—C.]

[2] [εἴπερ ἡ μορφὴ ἢ ἡ ὕλη G. *Cf.* Simpl. 548. 6.—C.]

[a] But consult p. 313, Chap. iv., 212 a 14 ff. and Argument.

[b] *Cf.* p. 40. [Syrianus on *Met.* 1092 a 17, says that the later

each other, and so too with any other substance ; and therefore its ' place ' can be neither a factor nor an intrinsic possession of the thing, but is something separable from it.

In fact a ' place ' seems to resemble a vessel, a ' vessel ' being ' a place that can itself be moved about.' [a] And just as the vessel is no part of its content, so the place is no part of that which is in it.

A place, then, is neither the form of its ' content ' (inasmuch as it is not integral to it) nor its matter (inasmuch as it is not the ' content ' but the ' continent '). It appears, then, that whatever is ' somewhere ' is itself a definite ' something ' and also has a definite ' something else ' outside it.

At which point we may remark parenthetically that Plato ought to tell us why the Ideas and Numbers have no locality or place, if ' place ' is indeed the ' receptive factor,'—and this whether the said receptive factor is ' the great and small ' or (as he writes in the *Timaeus*) ' matter.' [b]

Again, how could the elements severally make for their proper places if place were matter or form ? For there can be no place where there is neither motion nor ' above ' and ' below.' We must therefore look for place where such things are.

And if a thing's place is intrinsic to itself (as it must be if it is either its form or its matter), ' place ' must itself occupy ' place,' for both the defining (formal) and the intrinsically indefinite (material) factor move, and change place, together with the thing itself.

Platonism distinguished four kinds of place or space : *ἄλλος οὖν τόπος σωμάτων φυσικῶν* (physical space), *ἄλλος ἐνύλων εἰδῶν* (matter), *ἄλλος μαθηματικῶν σωμάτων* (imagination), *ἄλλος ἀύλων λόγων* (higher intellectual conception).—C.]

210 a οὐκ ἀεὶ ἐν τῷ αὐτῷ ἀλλ' οὗπερ καὶ τὸ πρᾶγμα· ὥστε τοῦ τόπου ἔσται τόπος.

Ἔτι ὅταν ἐξ ἀέρος ὕδωρ γένηται, ἀπόλωλεν ὁ τόπος· οὐ γὰρ ἐν τῷ αὐτῷ τόπῳ τὸ γενόμενον σῶμα· τίς οὖν ἡ φθορά;

Ἐξ ὧν μὲν τοίνυν ἀναγκαῖον εἶναί τι τὸν τόπον, καὶ πάλιν ἐξ ὧν ἀπορήσειεν ἄν τις αὐτοῦ περὶ τῆς οὐσίας, εἴρηται.

CHAPTER III

ARGUMENT

(This chapter, still digressing from the direct line of advance, is mainly occupied by a discussion, elaborated in substance, but highly elliptical in expression, whether a thing can be ' in itself,' so as to constitute its own place.

The importance of this question hardly reveals itself till, in a later part of the treatise (Book VIII. Chaps. iv. ff.), we come to examine the sense in which an animal can be said to ' move itself' ; which latter question is crucial to Aristotle's whole theory of the First Cause. Compare General Introduction, pp. xl, lxvii sqq.)

Different meanings of ' in ' (a 14–24).

Can a thing be ' in itself' ? Only if it is ambiguously designated after its parts. Thus (again to vary Aristotle's illustration) we might have : Question : *' Has the kettle*

Μετὰ δὲ ταῦτα ληπτέον ποσαχῶς ἄλλο ἐν ἄλλῳ λέγεται. ἕνα μὲν δὴ τρόπον ὡς ὁ δάκτυλος ἐν τῇ χειρὶ καὶ ὅλως τὸ μέρος ἐν τῷ ὅλῳ. ἄλλον

[a] [οὐκ ἀεὶ ἐν τῷ αὐτῷ can be construed with μεταβάλλει καὶ κινεῖται. Form and matter change at the same time as the

They do not stay where they were but are always where the thing itself is.[a] So the place that was already intrinsic in the thing has now come to occupy the place the thing itself has come into.

Further, when air becomes water the place (if intrinsic to the air) will have vanished, as the air has. For the nascent body (water) has not the same place as the vanished body (air) had. But how are we to conceive of a ' place ' vanishing ?

So much for the considerations, on the one hand, that compel us to suppose that ' place ' has an actual existence, and those, on the other hand, which perplex us as to the mode of its existence.

CHAPTER III

ARGUMENT (*continued*)

come back from the tinker yet ?' Answer : *' Yes, and I've put it on the fire, and it's just boiling.' Here the ' kettle ' means* 1° *the empty kettle,* 2° *the kettle with the water in it,* 3° *the water in the kettle. So kettle* 3° *is in kettle* 1° *and both are comprised in kettle* 2°. *Not otherwise can a thing be ' in itself'* (a 25–b 22).

The solution of Zeno's dilemma, already mentioned, and a summary of conclusions, close the chapter (b 22–31). 9 a 23 *sqq.*

At this point we must examine the different senses in which one thing is said to be ' in ' another. (i.) The finger may be said to be ' included in ' the hand, or, to put it more generally, the part ' in ' the

thing, *not in every case* in respect of such changes as can occur *in the same place* (*e.g.* rotation of a sphere, changes of quality) *but* moving *wherever the thing moves.*—C.]

210 a δὲ ὡς τὸ ὅλον ἐν τοῖς μέρεσιν· οὐ γάρ ἐστι παρὰ τὰ μέρη τὸ ὅλον. ἄλλον δὲ τρόπον ὡς ὁ ἄνθρωπος ἐν ζῴῳ καὶ ὅλως εἶδος ἐν γένει. ἄλλον δὲ ὡς τὸ γένος ἐν τῷ εἴδει καὶ ὅλως τὸ μέρος τοῦ εἴδους ἐν τῷ τοῦ εἴδους λόγῳ. ἔτι ὡς ἡ ὑγίεια ἐν θερμοῖς καὶ ψυχροῖς καὶ ὅλως τὸ εἶδος ἐν τῇ ὕλῃ. ἔτι ὡς ἐν βασιλεῖ τὰ τῶν Ἑλλήνων καὶ ὅλως ἐν τῷ πρώτῳ κινητικῷ. ἔτι ὡς ἐν τῷ ἀγαθῷ καὶ ὅλως ἐν τῷ τέλει· τοῦτο δ' ἐστὶ τὸ οὗ ἕνεκα. πάντων δὲ κυριώτατον τὸ ὡς ἐν ἀγγείῳ καὶ ὅλως ἐν τόπῳ.

Ἀπορήσειε δ' ἄν τις, ἆρα καὶ αὐτό τι ἐν ἑαυτῷ ἐνδέχεται εἶναι, ἢ οὐθέν, ἀλλὰ πᾶν ἢ οὐδαμοῦ ἢ ἐν ἄλλῳ.

Διχῶς δὲ τοῦτ' ἐστίν, ἤτοι καθ' αὑτὸ ἢ καθ' ἕτερον. ὅταν μὲν γὰρ ᾖ μόρια τοῦ ὅλου τὸ ἐν ᾧ καὶ τὸ ἐν τούτῳ, λεχθήσεται τὸ ὅλον ἐν ἑαυτῷ· λέγεται γὰρ καὶ κατὰ μέρη, οἷον λευκὸς ὅτι ἡ ἐπιφάνεια λευκή, καὶ ἐπιστήμων ὅτι τὸ λογιστικόν.

---

[a] According to Aristotle the heart warms and the brain cools. Health lies in the balance. *Cf.* General Introduction, p. lxvi.

[b] 'Surface,' here and on p. 297, means, of course, the physical surface, or skin, not the abstract surface of mathematics.

whole. (ii.) The whole is said to 'consist in' the full tale of its parts; for there is none of the whole 'outside' of the collective parts. (iii.) In one way 'man' is 'included in' the wider term 'animal,' or more generally the species 'in' the genus. (iv.) But yet in another way the *genus* as one factor (the *differentia* being the other) is 'included in' the definition of the *species*. Or (v.) health may be said to 'have its seat in' the balance of warm and cold [a]; or generally the 'form' to 'have its seat in' the material. Or (vi.) one may say "The affairs of Greece lie 'in' the King's hands"; or generally 'wherever' or 'in what place soever' the prime initiative (or efficient cause) resides. Or again (vii.) we may say that the motive to action 'is found in' the 'expected good,' or more generally 'in the contemplated end' (or final cause). But the primary sense, from which all these are derived, is that in which we say that a thing is 'in' a vessel, or more generally 'in a place.'

And here the question may arise whether a thing can be in itself or whether it must always be in something other than itself, if it be anywhere at all.

Now, a thing may be said to be 'in' this or that either primarily on its own account, or in virtue of some less direct relation. For instance, that which is properly applicable to the part may be predicated, by extension, of the whole. Thus we attribute pallor to a thing the *surface* [b] of which is pale, and knowledge to one whose *intellect* has ordered contents. If, then, a whole should chance to consist of two parts, one of which is the 'content' and the other the 'continent,' then that whole, in virtue of the contained part, may be said to be 'in' that same whole, in virtue of its

210 a ὁ μὲν οὖν ἀμφορεὺς οὐκ ἔσται ἐν ἑαυτῷ, οὐδ' ὁ οἶνος· ὁ δὲ τοῦ οἴνου ἀμφορεὺς ἔσται· ὅ τε γὰρ καὶ ἐν ᾧ, ἀμφότερα τοῦ αὐτοῦ μόρια.

Οὕτω μὲν οὖν ἐνδέχεται αὐτό τι ἐν ἑαυτῷ εἶναι, πρώτως δ' οὐκ ἐνδέχεται, οἷον τὸ λευκὸν ἐν σώματι.
210 b ἡ ἐπιφάνεια γὰρ ἐν σώματι, ἡ δ' ἐπιστήμη ἐν ψυχῇ. κατὰ ταῦτα δ' αἱ προσηγορίαι μέρη ὄντα, ὥς γε ἐν ἀνθρώπῳ. ὁ δ' ἀμφορεὺς καὶ ὁ οἶνος χωρὶς μὲν ὄντα οὐ μέρη, ἅμα δέ. διὸ ὅταν ᾖ μέρη, ἔσται αὐτὸ ἐν ἑαυτῷ, οἷον τὸ λευκὸν ἐν ἀνθρώπῳ ὅτι ἐν σώματι, καὶ ἐν τούτῳ ὅτι ἐν ἐπιφανείᾳ· ἐν δὲ ταύτῃ οὐκέτι κατ' ἄλλο.

Καὶ ἕτερά γε τῷ εἴδει ταῦτα, καὶ ἄλλην φύσιν ἔχει ἑκάτερον καὶ δύναμιν, ἥ τ' ἐπιφάνεια καὶ τὸ λευκόν.

Οὔτε δὴ ἐπακτικῶς σκοποῦσιν οὐδὲν ὁρῶμεν ἐν ἑαυτῷ κατ' οὐδένα τῶν διορισμῶν, τῷ τε λόγῳ δῆλον ὅτι ἀδύνατον· δεήσει γὰρ ἀμφότερα ἑκάτερον ὑπάρχειν, οἷον τὸν ἀμφορέα ἀγγεῖόν τε καὶ οἶνον εἶναι καὶ τὸν οἶνον οἶνόν τε καὶ ἀμφορέα, εἴπερ ἐνδέχεται αὐτό τι ἐν ἑαυτῷ εἶναι. ὥστε εἰ ὅτι μάλιστα ἐν ἀλλήλοις εἶεν, ὁ μὲν ἀμφορεὺς δέξεται τὸν οἶνον οὐχ ᾗ αὐτὸς οἶνος ἀλλ' ᾗ ἐκεῖνος, ὁ δ'

[a] I have changed the order of the clauses in this paragraph, for the sake of lucidity.

[b] I might 'drink a bottle of wine' or I might 'break a bottle of wine'; and the bottle of wine I might drink is in the bottle of wine I might break; but neither of them, severally, is in itself.

[c] One is a quality and the other is its immediate material substrate [*Met.* 1022 a 16 ἐν ᾧ πρώτῳ πέφυκε γίγνεσθαι, οἷον τὸ χρῶμα ἐν τῇ ἐπιφανείᾳ], and therefore neither of them can be

containing part.[a] For example, though neither the flask nor the wine can be said to be 'in itself,' yet the 'flask-of-wine' can. For the wine-content and the flask-continent are both of them parts of the same whole.[b]

In this sense, therefore, a thing can be in itself, but not in the primary and direct sense.

It is like this :—Pallor is in a man (that is in his body), because the pale 'surface' is ; and knowledge is in him, because the well-stored 'mind' is. But these attributions are extended to the man in virtue of the parts, *as existing in the man*, not otherwise. And in the same way it is only while the wine and the flask are taken together as a whole (not severally, and asunder) that they can be regarded as parts of a whole, to which their attributions can be extended.

And note that these two, the pallor and the surface, belong to different categories, and are unlike each other in their nature and in their capabilities.[c]

Nor, if we take all the meanings of 'in' successively, shall we find that a thing can be in itself according to any of them.

Nor can a thing be in itself, by virtue of definition.[d] For to make it so, each of the two parts would have to be defined as being both of them, the flask as being both flask and wine and the wine both wine and flask. So, however much they were 'in each other' by definition, the flask would continue to receive the wine not in the capacity of itself being wine, but in its flask-capacity ; and the wine would be in the

so identified with the other as to make one 'being in the other' the same thing as 'being in itself.'

[d] [τῷ λόγῳ, 'by process of reasoning,' as opposed to a review of all the instances (ἐπακτικῶς).—C.]

210 b οἶνος ἐνέσται ἐν τῷ ἀμφορεῖ οὐχ ᾗ αὐτὸς ἀμφορεὺς ἀλλ' ᾗ ἐκεῖνος. κατὰ μὲν οὖν τὸ εἶναι ὅτι ἕτερον, δῆλον· ἄλλος γὰρ ὁ λόγος τοῦ ἐν ᾧ καὶ τοῦ ἐν τούτῳ.

Ἀλλὰ μὴν οὐδὲ κατὰ συμβεβηκὸς ἐνδέχεται· ἅμα γὰρ δύο ἐν ταὐτῷ ἔσται· αὐτός τε γὰρ ἐν αὑτῷ ὁ ἀμφορεὺς ἔσται, εἰ οὗ ἡ φύσις δεκτική, τοῦτ' ἐνδέχεται ἐν ἑαυτῷ εἶναι καὶ ἔτι ἐκεῖνο οὗ δεκτικόν, οἷον, εἰ οἴνου, ὁ οἶνος.

Ὅτι μὲν οὖν ἀδύνατον ἐν ἑαυτῷ τι εἶναι πρώτως, δῆλον.

Ὃ δὲ Ζήνων ἠπόρει, ὅτι εἰ ἔστι τι ὁ τόπος, ἐν τίνι ἔσται, λύειν οὐ χαλεπόν· οὐδὲν γὰρ κωλύει ἐν ἄλλῳ μὲν εἶναι τὸν πρῶτον τόπον, μὴ μέντοι ὡς ἐν τόπῳ ἐκείνῳ, ἀλλ' ὥσπερ ἡ μὲν ὑγίεια ἐν τοῖς θερμοῖς ὡς ἕξις, τὸ δὲ θερμὸν ἐν σώματι ὡς πάθος. ὥστ' οὐκ ἀνάγκη εἰς ἄπειρον ἰέναι.

Ἐκεῖνο δὲ φανερόν, ὅτι ἐπεὶ οὐδὲν τὸ ἀγγεῖον τοῦ ἐν αὑτῷ[1] (ἕτερον γὰρ τὸ πρώτως ὅ τε καὶ ἐν ᾧ), οὐκ ἂν εἴη οὔτε ἡ ὕλη οὔτε τὸ εἶδος ὁ τόπος, ἀλλ' ἕτερον. ἐκείνου γάρ τι ταῦτα τοῦ ἐνόντος, καὶ ἡ ὕλη καὶ ἡ μορφή.

Ταῦτα μὲν οὖν ἔστω διηπορημένα.

[1] [αὑτῷ Simpl. 564. 27; Bonitz: αὐτῷ codd.—C.]

---

[a] Compare 195 a 35 ff. The man of general culture and the artificer may coincide in the person of Polycleitus; but this does not make room for two men in his body.

[b] The sting of an *ad infinitum* as a *reductio ad absurdum* consists in its being a series of ' steps that make no progress,' 12 b 27 sqq. each apparent advance landing you exactly where you were before. This is the point of the illustration from health and warmth. The ' in ' that you end with is *not* the same ' in ' that you began with. But we have not yet been told what a ' place ' really is ' in ' and how it is ' in it.' We have only been told

flask not in the capacity of itself being flask, but in its wine-capacity. Obviously then, as actually existing, they differ from each other ; for the rationale of the 'continent' is other than that of the 'content.'

Nor is it possible for a thing to be in itself by coincidence of subject, since, if it were, two bodies would have to coincide.[a] For the flask itself can only be in itself if that which has the receptive character can be in itself in addition to its own content, to wit the wine—or whatever it may be—that is already in it.

Thus the impossibility of anything being, in the primary sense, in itself, is clearly demonstrable.

We shall now have no difficulty in escaping Zeno's dilemma that either there is no such thing as a place, or it must itself occupy a place ; for there is no reason why the proper place of a thing should not itself be 'in' something else—only not in the same local sense as that in which its own content is 'in' it. Similarly health resides '*in*' the balance of warmth and coolness *as a habit* (and is a vital experience), whereas warmth and coolness, in their turn, 'exist *in*' the body *as physical modifications*. And this involves no *ad infinitum* process.[b]

So much, then, is clear : that since the vessel is no part of its own content (for the primary and proper 'thing in the place' is other than the 'place the thing is in'), a thing's place can be neither its matter nor its form. It must be something other than either of these, for both matter and form are intrinsic to the 'content,'[c] whereas its place is extrinsic to it.

Let this close the discussion.

that it need not be a 'place.' *Cf.* chap. iv. 211 b 1-5, and Introduction, p. 270.

[c] [τοῦ ἐνόντος, 'the thing that is in the place.' *Cf.* Simpl. 564. 29 τοῦ γὰρ ἐν τόπῳ ὄντος ἐστὶ ταῦτα.—C.]

CHAPTER IV

ARGUMENT

Returning to the directly constructive side of his argument (from the digression beginning at 9 b 1), *Aristotle proceeds to consider :*

The data for the inquiry and the conditions which a satisfactory solution should fulfil (210 b 32-211 a 11).

The problem of the nature *of* place *arises in connexion with the observed phenomenon of motion, which involves the* change *of* place. *Varieties of local motion. Shifting, rotation, expansion or contraction. Direct and dependently-involved motion* (a 12-23).

Place-proper, place-common, and place-universal. Cf. *chap. ii. and Argument* (a 23-29).

Place-proper implies a recognizable division, marking off the content from its embracing continent and allowing of motion of the content 'into,' 'out of' or 'within' the continent,—whether that continent itself is stationary or in motion. Understanding by 'undifferentiated' the absence *of any such division, we may speak of portions of an undifferentiated mass as being 'in' the whole mass 'as component parts'; but for a part to have a 'place' in that whole mass there must be both differentiation and contact where the said part, as content, meets its continent,—the continent being no part of the content, but dimensionally fitting it exactly.* Cf. *Introd. p.* 271 (a 29-b 1).

A part, or factor, of a whole, which is not 'in it as in a place,' may inhere in it, as vision does in the eye, or may be coherent with it, as the hand is with the body; but in neither case when the part moves 'with' the whole does it move because it is 'in' it, in the sense in which water is 'in' the cask, i.e. *as a content in a continent, or a placed in a place* (b 1-5).

This seems to establish our definition of place. But it will be instructive to confirm it by the method of exhaustion.

CHAPTER IV

ARGUMENT (*continued*)

Enumeration of the four conceivable alternatives (b 5-9).

Exclusion of ' form ' (b 9-14).

Exclusion of ' internal dimensionality.' If a place is conceived as something (other than the material content) inside the vessel-continent, extending from inner wall to inner wall in every direction, then the vessel will be able to carry that place about with it when it moves. But since (on this showing) the vessel itself also occupies a place, which place extends from its *outer wall to outer wall, and therefore includes the place occupied by its content, two places will be in the same place, one upon the other. And if the vessel itself changes its place, it takes the ' place that is in it ' out of ' the place it is itself in ' and puts it down in another place. Besides (on this showing) all the undifferentiated concentric parts of the content are in places in the whole place. But the whole place extends from wall to wall, and so the places of the parts (one within the other indefinitely) must all be superimposed upon each other. And when the whole place shifts they must be collectively superimposed upon another place*[a] (b 9-29).

Exclusion of ' matter ' (b 29-212 a 2).

Residuary hypothesis vindicated (a 2-7).

Causes of bewilderment as to ' place ' enumerated (a 7-14).

As a vessel has been called, by an extension of terms, a ' movable place ' (because the place, though not directly

[a] Remembering that Aristotle regarded the ' void ' of Democritus as demonstrably material, and equating the ' void ' with what we should call ' space,' we can see that he is trying to commit believers in the ' void ' to the paradox that ' pieces of space ' can be fenced off and carried about, or laid upon each other, so that they coincide with and interpenetrate each other.

ARGUMENT (*continued*)

movable, accompanies the moving vessel, and is involved in its movement), so a 'site,' i.e. *a cosmically determined place may, by a like extension, be called an 'immovable vessel,' because it really is 'immovable' and is like a 'vessel' in so far as it is a continent.*

The place-universal is such a fixed continent, or vessel, So are (potentially) all cosmically determined sites. Thus, if there is a physically determined place-common (like the bed and banks of a river) which is cosmically stable, and a fluid body (like the water of the river) is passing through it while a solid body (like a boat on the river) is changing its place-proper in the fluid, the mind tends to regard the water

210 b Τί δέ ποτέ ἐστιν ὁ τόπος, ὧδ' ἂν γένοιτο φανερόν. λάβωμεν δὲ περὶ αὐτοῦ ὅσα δοκεῖ ἀληθῶς καθ' αὐτὸ ὑπάρχειν αὐτῷ.

Ἀξιοῦμεν δὴ τὸν τόπον εἶναι πρῶτον μὲν περι-
211 a έχον ἐκεῖνο οὗ τόπος ἐστί, καὶ μηδὲν τοῦ πράγματος εἶναι· ἔτι τὸν πρῶτον τόπον μήτ' ἐλάττω μήτε μείζω· ἔτι ἀπολείπεσθαι ἑκάστου καὶ χωριστὸν εἶναι· πρὸς δὲ τούτοις πάντα τόπον ἔχειν τὸ ἄνω καὶ κάτω, καὶ φέρεσθαι φύσει καὶ μένειν ἐν τοῖς οἰκείοις τόποις ἕκαστον τῶν σωμάτων, τοῦτο δὲ ποιεῖν ἢ ἄνω ἢ κάτω.

Ὑποκειμένων δὲ τούτων τὰ λοιπὰ θεωρητέον. δεῖ δὲ πειρᾶσθαι τὴν σκέψιν οὕτω ποιεῖσθαι,[1] ὅπως

[1] [The comma after ποιεῖσθαι might be omitted, to indicate that ὅπως . . . ἀποδοθήσεται is not governed by οὕτω. 'We must try so to conduct our inquiry designed to give an account of its nature, as to' (satisfy *three* conditions which any such account should satisfy).—C.]

ARGUMENT (*continued*)

as ' carrying ' the boat (as a vessel carries something that is moving about inside it), and to locate the boat not by reference to its own shifting place-proper, but with reference to the banks which form a stable frame of reference both for it and the water that forms ' the whole river ' at the moment (12 a 14-21).

The cosmic place-universal is the ultimate frame of reference (12 a 21-29).

Epilogue. Cosmically determined sites do not grow, but the expansion of a body involves the expansion of the physically determined surface-continent, or place, that fits and environs it (12 a 29 sq.).

It should now be possible to reach a clear conception of what the place of a thing really is, by gathering up such properties of ' place ' as appear to have emerged securely as its characteristics, and proceeding as set forth in the sequel.

Well then, to begin with, we may safely assert—(i.) That the place of a thing is no part or factor of the thing itself, but is that which embraces it; (ii.) that the immediate or ' proper ' place of a thing is neither smaller nor greater than the thing itself; (iii.) that the place where the thing is can be quitted by it, and is therefore separable from it; and lastly, (iv) that any and every place implies and involves the correlatives of ' above ' and ' below,' and that all the elemental substances have a natural tendency to move towards their own special places, or to rest in them when there—such movement being ' upward ' or ' downward,' and such rest ' above ' or ' below.'

Taking these as our data we may now pursue our investigation, and we must try so to conduct it (1) that a full account shall be given of the meaning and

211 a τὸ τί ἐστιν ἀποδοθήσεται, ὥστε τά τε ἀπορούμενα λύεσθαι, καὶ τὰ δοκοῦντα ὑπάρχειν τῷ τόπῳ ὑπάρχοντα ἔσται, καὶ ἔτι τὸ τῆς δυσκολίας αἴτιον καὶ τῶν περὶ αὐτὸν ἀπορημάτων ἔσται φανερόν· οὕτω γὰρ ἂν κάλλιστα δεικνύοιτο ἕκαστον.

Πρῶτον μὲν οὖν δεῖ κατανοῆσαι ὅτι οὐκ ἂν ἐζητεῖτο ὁ τόπος, εἰ μὴ κίνησίς τις[1] ἦν ἡ κατὰ τόπον· διὰ γὰρ τοῦτο καὶ τὸν οὐρανὸν μάλιστ' οἰόμεθα ἐν τόπῳ, ὅτι ἀεὶ ἐν κινήσει.

Ταύτης δὲ τὸ μὲν φορά, τὸ δ' αὔξησις καὶ φθίσις· καὶ γὰρ ἐν τῇ αὐξήσει καὶ φθίσει μεταβάλλει, καὶ ὃ πρότερον ἦν ἐνταῦθα πάλιν μεθέστηκεν εἰς ἔλαττον ἢ μεῖζον.

Ἔστι δὲ κινούμενον τὸ μὲν καθ' αὑτὸ ἐνεργείᾳ, τὸ δὲ κατὰ συμβεβηκός· τοῦ δὲ κατὰ συμβεβηκὸς τὸ μὲν ἐνδεχόμενον κινεῖσθαι καθ' αὑτό, οἷον τὰ μόρια τοῦ σώματος καὶ ὁ ἐν τῷ πλοίῳ ἧλος, τὰ δ' οὐκ ἐνδεχόμενα ἀλλ' αἰεὶ κατὰ συμβεβηκός, οἷον ἡ λευκότης καὶ ἡ ἐπιστήμη· ταῦτα γὰρ οὕτω μεταβέβληκε τὸν τόπον, ὅτι ἐν ᾧ ὑπάρχουσι μεταβάλλει.

Ἐπεὶ δὲ λέγομεν εἶναι ὡς ἐν τόπῳ τῷ οὐρανῷ, διότι ἐν τῷ ἀέρι, οὗτος δ' ἐν τῷ οὐρανῷ· καὶ ἐν

[1] [τις om. F, Simpl. 603. 18, Philop. 543. 27 (lemma), Themist. 111. 18.—C.]

---

[a] We are to see later on that it is a fact that 'heaven' is always moving (with the motion of rotation), but that it is a mistake (however natural) to infer that it must therefore be always changing its place, and so must have a place to change. *Cf.* chap. v.

[b] [Prantl takes the clauses from ἐπεὶ (*l.* 24) to ἁπτομένων (*l.* 34) as a single sentence. The new point is in the apodosis, beginning ὅταν μὲν οὖν (*l.* 29).—C.]

nature of ' place ' ; (2) that the problems we have encountered on our way shall find their solution in it ; (3) that the characteristics of ' place ' which we have noted shall reveal themselves as integral to its nature as we have defined it ; and lastly (4) that the perplexities we have met shall be seen to rise naturally out of the facts, as explained. No more can be demanded of any solution than that it should satisfy all these conditions.

To begin with, then, we must recognize that no speculations as to place would ever have arisen had there been no such thing as movement, or *change* of place. Indeed, the chief reason of the persistent tendency to think of heaven itself as having a ' place ' is that it is always moving.[a]

Now change of place may occur either by way of translation, or by way of increase and decrease, for in this case too what was formerly in such and such a place is now in a larger or a smaller one.

Again, of things which are in motion some are moved by the actualizing of their own inherent potentialities, and others only by being involved in the movement of something else in which they inhere ; and of this latter class again some (being substantive) are *capable* of movement on their own account as well (such are limbs of the body, or a rivet in a ship), whereas others, like ' whiteness ' or ' wisdom ' (being attributive) are not so much as capable of moving otherwise than incidentally, for the only sense in which they move at all is that they inhere in something that moves.

Again,[b] when we say that a thing is ' in the universe,' as though that were its place, it is because it is in the air, which air is in the universe. And we

211 a τῷ ἀέρι δὲ οὐκ ἐν παντί, ἀλλὰ διὰ τὸ ἔσχατον αὐτοῦ καὶ περιέχον ἐν τῷ ἀέρι φαμὲν εἶναι· εἰ γὰρ πᾶς ὁ ἀὴρ τόπος, οὐκ ἂν ἴσος εἴη ἑκάστου ὁ τόπος καὶ ἕκαστον, δοκεῖ δέ γε ἴσος εἶναι.

Τοιοῦτος δ' ὁ πρῶτος ἐν ᾧ ἐστιν.

Ὅταν μὲν οὖν μὴ διῃρημένον ᾖ τὸ περιέχον ἀλλὰ συνεχές, οὐχ ὡς ἐν τόπῳ λέγεται εἶναι ἐν ἐκείνῳ, ἀλλ' ὡς μέρος ἐν ὅλῳ· ὅταν δὲ διῃρημένον ᾖ καὶ ἁπτόμενον, ἐν πρώτῳ ἐστὶ τῷ ἐσχάτῳ τοῦ περιέχοντος, ὃ οὔτ' ἐστὶ μέρος τοῦ ἐν αὐτῷ ὄντος οὔτε μεῖζον τοῦ διαστήματος ἀλλ' ἴσον· ἐν γὰρ τῷ αὐτῷ τὰ ἔσχατα τῶν ἁπτομένων.

Καὶ συνεχὲς μὲν ὂν οὐκ ἐν ἐκείνῳ κινεῖται ἀλλὰ μετ' ἐκείνου, διῃρημένον δὲ ἐν ἐκείνῳ. καὶ ἐάν
211 b τε κινῆται τὸ περιέχον ἐάν τε μή, οὐδὲν ἧττον. ἔτι[1] ὅταν μὴ διῃρημένον ᾖ, ὡς μέρος ἐν ὅλῳ λέγεται, οἷον ἐν τῷ ὀφθαλμῷ ἡ ὄψις ἢ ἐν τῷ σώματι ἡ χείρ, ὅταν δὲ διῃρημένον, οἷον ἐν τῷ κάδῳ τὸ ὕδωρ ἢ ἐν τῷ κεραμίῳ ὁ οἶνος· ἡ μὲν γὰρ χεὶρ μετὰ τοῦ σώματος κινεῖται, τὸ δὲ ὕδωρ ἐν τῷ κάδῳ.

Ἤδη τοίνυν φανερὸν ἐκ τούτων τί ἐστιν ὁ τόπος.

[1] [b 1 ἔτι . . . 5 κάδῳ was rejected as a doublet of the preceding sentences by Alexander and ignored by Aspasius (Simpl. 570. 28).—C.]

say it is in the air not because it is in all of it, but in virtue of its being embraced by a certain surface of the air, which fits round it; for if the whole of the air were its place, then the place a thing occupied would not be of the same size as the thing itself, which equality of size we have accepted as one of our data.

Such then—the inner surface of the envelope, namely—is the immediate place of a thing.

Now if a thing is not separated from its embracing environment, but is undifferentiated from it, it is indeed 'included in' it—not, however, as in its *place*, but only in the sense in which a *part* is said to be 'included in' its *whole*; whereas when the thing is differentiated from its immediate continent but is in direct contact with it, then it is 'in' the inner surface of this continent as its immediate and proper 'place,'—which inner surface is neither a part of the content nor dimensionally greater than it, but equal to it; for when things touch each other their surfaces coincide.

A constituent or part of a whole (visual faculty as a 'constituent' of a seeing eye, or the 'hand' as a 'part' of the body) cannot change 'the place it occupies in' the whole, but it 'shares with' the whole any change of place that whole may make; whereas a 'contained part,' such as water in a cask or wine in a flask, which is contiguous with but detached from the 'continent part,' can 'move about in it'—and that whether the said continent move or no. For the movement of the hand is 'involved' in that of the body, but the water is 'contained in' the cask.

From all this the answer to the question 'What

211 b σχεδὸν γὰρ τέτταρά ἐστιν ὧν ἀνάγκη τὸν τόπον ἕν τι εἶναι· ἢ γὰρ μορφή, ἢ ὕλη, ἢ διάστημά τι τὸ μεταξὺ τῶν ἐσχάτων, ἢ τὰ ἔσχατα εἰ μὴ ἔστι μηδὲν διάστημα παρὰ τὸ τοῦ ἐγγινομένου σώματος μέγεθος.

Τούτων δ' ὅτι οὐκ ἐνδέχεται τὰ τρία εἶναι, φανερόν· ἀλλὰ διὰ μὲν τὸ περιέχειν δοκεῖ ἡ μορφὴ εἶναι· ἐν ταὐτῷ γὰρ τὰ ἔσχατα τοῦ περιέχοντος καὶ τοῦ περιεχομένου. ἔστι μὲν οὖν ἄμφω πέρατα, ἀλλ' οὐ τοῦ αὐτοῦ, ἀλλὰ τὸ μὲν εἶδος τοῦ πράγματος, ὁ δὲ τόπος τοῦ περιέχοντος σώματος.

Διὰ δὲ τὸ μεταβάλλειν πολλάκις μένοντος τοῦ περιέχοντος τὸ περιεχόμενον καὶ διῃρημένον, οἷον ἐξ ἀγγείου ὕδωρ, τὸ μεταξὺ εἶναί τι δοκεῖ διάστημα, ὡς ὄν τι παρὰ τὸ σῶμα τὸ μεθιστάμενον.

Τὸ δ' οὐκ ἔστιν, ἀλλὰ τὸ τυχὸν ἐμπίπτει[1] σῶμα τῶν μεθισταμένων καὶ ἅπτεσθαι πεφυκότων. εἰ δ' ἦν τι [τὸ] διάστημα τὸ πεφυκὸς καὶ μένον ἐν τῷ αὐτῷ [τόπῳ], ἄπειροι ἂν ἦσαν τόποι· μεθισταμένου γὰρ τοῦ ὕδατος καὶ τοῦ ἀέρος ταὐτὸ ποιήσει τὰ μόρια πάντα ἐν τῷ ὅλῳ ὅπερ ἅπαν τὸ ὕδωρ

[1] [μετεμπίπτει Simpl. 572. 17 (lemma), Philop. 553. 9.—C.]

[a] [*Cf.* Simpl. 571. 22 τὸ διάστημα τὸ μεταξὺ τῶν ἐσχάτων τοῦ περιέχοντος τὸν τόπον, not a gap or interval between container and 'contained,' *e.g.* between the pail and the water it holds (as interpreted *e.g.* by A. E. Taylor, *Commentary on Timaeus*, p. 673).—C.]

[b] In rejecting τό (b 19) and τόπῳ (b 20) I follow the text (apparently) of all the Greek commentators. τόπῳ is omitted by F. [εἰ δ' ἦν τι διάστημα καθ' αὑτὸ πεφυκὸς εἶναι καὶ μένειν ἐν ἑαυτῷ (Them. 116. 12), Laas, Prantl, Carteron.—C.]

is place?' already emerges. For we may take it that it must be either (i.) the form or (ii.) the matter of the body itself, or (iii.) some kind of dimensional extension lying between the points of the containing surface,[a] or (iv.)—if there be no such 'intervenient,' apart from the bulk of the included body—the containing surface itself.

Now it is clear that no one of the three first alternatives is admissible; though 'that which embraces' (i.) may suggest the moulding 'form'; for the limiting surfaces of the embracing and the embraced coincide. It is true, then, that both the 'place' and the 'form' are limits, but not of the same thing, for the form determines the thing itself, but the place the body-continent.

But because the encircled content may be taken (iii.) out and changed again and again, while the encircling continent remains unchanged—as when water passes out of a vessel—the imagination pictures a kind of dimensional entity left there, distinct from the body that has shifted away.

But this is not so; for what really happens is that (instead of anything being *left*) some other body—it matters not what, so long as it is mobile and tangible—*succeeds* the vacating body without break and continuously. Whereas if that 'which abides where it was' (the place itself, to wit) were a dimensional entity,[b] there would be an unlimited number of 'places' there; for, when the water—or air—shifts, all the portions of the water will play the same part in the whole mass of water that that mass itself plays in the vessel; that is to say, each will leave behind it a dimensional deposit constituting a 'place'; and at the same time the whole place of the mass of

211 b ἐν τῷ ἀγγείῳ· ἅμα δὲ καὶ ὁ τόπος ἔσται μεταβάλλων. ὥστ' ἔσται τοῦ τόπου τ' ἄλλος τόπος, καὶ πολλοὶ τόποι ἅμα ἔσονται.

Οὐκ ἔστι δὲ ἄλλος τόπος ὁ τοῦ μορίου, ἐν ᾧ κινεῖται, ὅταν ὅλον τὸ ἀγγεῖον μεθίστηται, ἀλλ' ὁ αὐτός· ἐν ᾧ γάρ ἐστιν, ἀντιμεθίσταται ὁ ἀὴρ καὶ τὸ ὕδωρ ἢ τὰ μόρια τοῦ ὕδατος, ἀλλ' οὐκ ἐν ᾧ γίνονται τόπῳ, ὃς μέρος ἐστὶ τοῦ τόπου ὅς ἐστι τόπος ὅλου τοῦ οὐρανοῦ.

Καὶ ἡ ὕλη δὲ δόξειεν ἂν εἶναι τόπος, εἴ γε ἐν ἠρεμοῦντί τις σκοποίη καὶ μὴ κεχωρισμένῳ ἀλλὰ συνεχεῖ. ὥσπερ γὰρ εἰ ἀλλοιοῦται, ἔστι τι ὃ νῦν μὲν λευκὸν πάλαι δὲ μέλαν, καὶ νῦν μὲν σκληρὸν πάλαι δὲ μαλακόν (διό φαμεν εἶναί τι τὴν ὕλην), οὕτω καὶ ὁ τόπος διὰ τοιαύτης τινὸς εἶναι δοκεῖ φαντασίας, πλὴν ἐκεῖνο μὲν διότι ὃ ἦν ἀήρ, τοῦτο νῦν ὕδωρ, ὁ δὲ τόπος ὅτι οὗ ἦν ἀήρ, ἐνταῦθ' ἐστὶ νῦν ὕδωρ.

212 a Ἀλλ' ἡ μὲν ὕλη, ὥσπερ ἐλέχθη ἐν τοῖς πρότερον, οὔτε χωριστὴ τοῦ πράγματος οὔτε περιέχει, ὁ δὲ τόπος ἄμφω.

Εἰ τοίνυν μηδὲν τῶν τριῶν ὁ τόπος ἐστί, μήτε τὸ

---

[a] [*Literally*, 'place in which they come to be' (when the vessel with its contents is moved).—C.]

[b] For συνεχές = ὁμοιομερές *cf.* 212 b 5. For χωρίζεσθαι as applied to the differentiation of the homogeneous *cf. De gen. et corr.* 315 a 8. [ἐν ἠρεμοῦντι, 'in (or 'in the case of') a stationary body.'—C.]

[c] [*Or*, 'the only difference being that we imagine a subject matter underlying change because air is transformed into water, while we believe in the existence of the place because water has taken the place of air.' The objection follows in the next sentence.—C.]

water would change its place, if the vessel itself shifted, so that a place would come to occupy another place, itself the while being a nest of superimposed places.

But in truth when the vessel as a whole, contents and all, changes its place, it is not true that the contents, as a part of that whole, also change *their* place. On the contrary, they remain in the same place ; for when we speak of water and air ' replacing ' each other we refer to the ' proper ' place in which the one was and the other is, not to any common place,[a] ultimately determined by its relation to the whole cosmic place.

Again, if a man looks upon ' place ' as undisturbed (ii.) by changes that come to pass in it, and also as undifferentiated in itself, and as a *continuum*, he may liken it to matter.[b] For just as, when there is a qualitative modification, there is a something that is now white and before was black, or now hard that was formerly soft—which is what makes us assert the existence of matter,—so too when one thing goes out and another comes in, while the place abides undisturbed, that ' place ' may, by an analogous working of imagery upon our minds, seem to be an abiding ' matter.' But there is all the difference between ' what ' was air, ' that ' being now water, and ' where ' was air, ' there ' being now water.[c]

And then there is the conclusive consideration that the ' matter ' of a thing is not separable from it, which its place is, and does not embrace it, which its place does.

If then a thing's place is no one of these three—

212a εἶδος μήτε ἡ ὕλη μήτε διάστημά τι ἀεὶ ὑπάρχον ἕτερον παρὰ τὸ τοῦ πράγματος τοῦ μεθισταμένου, ἀνάγκη τὸν τόπον εἶναι τὸ λοιπὸν τῶν τεσσάρων, τὸ πέρας τοῦ περιέχοντος σώματος.[1] λέγω δὲ τὸ περιεχόμενον σῶμα τὸ κινητὸν κατὰ φοράν.

Δοκεῖ δὲ μέγα τι εἶναι καὶ χαλεπὸν ληφθῆναι ὁ τόπος διά τε τὸ παρεμφαίνεσθαι τὴν ὕλην καὶ τὴν μορφήν, καὶ διὰ τὸ ἐν ἠρεμοῦντι τῷ περιέχοντι γίνεσθαι τὴν μετάστασιν τοῦ φερομένου· ἐνδέχεσθαι γὰρ φαίνεται εἶναι διάστημα μεταξὺ ἄλλο τι τῶν κινουμένων μεγεθῶν. συμβάλλεται δέ τι καὶ ὁ ἀὴρ δοκῶν ἀσώματος εἶναι· φαίνεται γὰρ οὐ μόνον τὰ πέρατα τοῦ ἀγγείου εἶναι ὁ τόπος, ἀλλὰ καὶ τὸ μεταξὺ ὡς κενόν.

Ἔστι δ' ὥσπερ τὸ ἀγγεῖον τόπος μεταφορητός, οὕτω καὶ ὁ τόπος ἀγγεῖον ἀμετακίνητον.

Διὸ ὅταν μὲν ἐν κινουμένῳ τι κινῆται καὶ μεταβάλλῃ τὸ ἐντός, οἷον ἐν ποταμῷ πλοῖον, ὡς <ἐν>[2] ἀγγείῳ χρῆται μᾶλλον ἢ τόπῳ τῷ περιέχοντι. βούλεται δ' ἀκίνητος εἶναι ὁ τόπος· διὸ ὁ πᾶς μᾶλλον ποταμὸς τόπος, ὅτι ἀκίνητος ὁ πᾶς. ὥστε

[1] [After σώματος Simpl. (580. 3) and Themist. (118. 8) read καθ' ὃ συνάπτει τῷ περιεχομένῳ. (' For the continent might have some other limit as well ' Simpl.)—C.]

[2] [ὡς ἐν ἀγγείῳ I (according to Bekker). This may be a trace of an alternative reading; *cf.* Simplic. 583. 23 (paraphr.) τότε ὡς ἐν ἀγγείῳ ἐστὶν ἐν ἐκείνῳ τοῦ ὕδατος τῷ μέρει τῷ περιέχοντι αὐτὸ μᾶλλον ἢ ὡς ἐν τόπῳ. But χρῆται requires ὡς ἀγγείῳ (cett., Themist. 118. 29).—C.]

neither its form, nor its matter, nor a dimensional entity distinct from the dimensions of the entering or vacating body—it must needs be the fourth of the alternatives, namely, the limiting surface of the body continent—the content being a material substance susceptible of movement by transference.

So we see that what makes ' place ' appear so mysterious and hard to grasp is its illusive suggestion now of matter and now of form, and the fact that while the continent is at rest the transferable content may change, for this suggests that there may be a dimensional something that stays there other than the entering and vacating *quanta*,—air too contributing to this last illusion since it looks as if it were incorporeal,—so that the 'place,' instead of being recognized as constituted solely by the adjacent surface of the vessel, is held to be the dimensional interval within the surface, conceived as ' vacancy.'

But now note that we have as good a right to regard a *place* as ' an immovable vessel ' as we had to regard a *vessel* as a ' movable place.'

And, from this point of view, if one thing is moving about inside another, which other is also in motion, as when a boat moves through the flowing water of a river, the water is related to the boat as a vessel-continent rather than as a place-continent ; and if we look for stability in ' place,' then the river as a permanent and stable whole,[a] rather than the flowing water in it at the moment, will be the boat's site.

[a] Thus, when we speak of ' the River Thames ' or ' the Falls of Niagara ' we do not mean the mass of water which at the moment constitutes them, but a (comparatively at least) permanent and cosmically stable site or location.

212 a τὸ τοῦ περιέχοντος πέρας ἀκίνητον πρῶτον, τοῦτ' ἔστιν ὁ τόπος.

Καὶ διὰ τοῦτο τὸ μέσον τοῦ οὐρανοῦ καὶ τὸ ἔσχατον τὸ πρὸς ἡμᾶς τῆς κύκλῳ φορᾶς δοκεῖ εἶναι τὸ μὲν ἄνω τὸ δὲ κάτω μάλιστα πᾶσι κυρίως, ὅτι τὸ μὲν ἀεὶ μένει, τοῦ δὲ κύκλου[1] τὸ ἔσχατον ὡσαύτως ἔχον μένει.

Ὥστ' ἐπεὶ τὸ μὲν κοῦφον τὸ ἄνω φερόμενόν ἐστι φύσει, τὸ δὲ βαρὺ τὸ κάτω, τὸ μὲν πρὸς τὸ μέσον περιέχον πέρας κάτω ἐστί, καὶ αὐτὸ τὸ μέσον, τὸ δὲ πρὸς τὸ ἔσχατον ἄνω, καὶ αὐτὸ τὸ ἔσχατον· καὶ διὰ τοῦτο δοκεῖ ἐπίπεδόν τι εἶναι καὶ οἷον ἀγγεῖον ὁ τόπος καὶ περιέχον. ἔτι ἅμα τῷ πράγματι ὁ τόπος· ἅμα γὰρ τῷ πεπερασμένῳ τὰ πέρατα.

[1] [κύκλου E, Simplic. 607. 13, Philop. 591. 8 (lemma): κύκλω FGI, κύκλῳ Simpl. 585. 11, 603. 27.—C.]

[a] I take πρῶτον to mean 'basal' here. It might also mean 'the first stable surface-continent we come to' (moving outwards). This would amount to the same thing. But I cannot take it (with the Commentators) as meaning 'proper,' which would surely be a contradiction in terms. [Simplic. explains πρῶτον as meaning τὸ προσεχές (immediately contiguous) and equivalent to καθ' ὃ συνάπτει τῷ περιεχομένῳ (see 212 a 6 note). This sentence contains the final definition of place, rather than the conclusion of the preceding argument.—C.]

CHAPTER V

INTRODUCTORY NOTE

The noteworthy departures from Aristotle's linguistic
212 a 31-b 29 usage in the first section of this chapter make it difficult

Thus whatever fixed environing surface we take our reckoning from [a] will be the place.

So the centre of the universe and the inner surface of the revolving heavens constitute the supreme 'below' and the supreme 'above'; the former being absolutely stable, and the latter constant in its position as a whole.

And since what we mean by 'upwards' is the direction taken by what is buoyant, and 'downwards' that taken by what is heavy, the limiting surface of a body towards the centre and that centre itself are below it, and the limiting surface towards the inner periphery and that periphery itself above it. So it appears that place is a surface-continent that embraces its content after the fashion of a vessel; and further, that the place of the content coinciding with its outer surface keeps pace with it as it increases or diminishes dimensionally.[b]

[b] I take this clause to refer to datum (ii.) on p. 303, ἅμα carrying its usual meaning of 'keeping pace with.' Those who take it to mean 'coinciding with' have to apologize for the phrase as an inaccuracy. But it would be worse than an inaccuracy, for it would be a surrender of Aristotle's whole case, since according to him the 'thing contained' emphatically does *not* coincide throughout with its place. It is the error he is most anxious to refute. In any case this clause has the air of a 'boulder' that has detached itself from some other context and is a 'stray' here.

CHAPTER V

INTRODUCTORY NOTE (*continued*)

to believe that it came, as we have it, from the Master's own hand. But nevertheless it rests upon a very sound

INTRODUCTORY NOTE (*continued*)

Aristotelian basis and suggests valuable developments of his teaching. To understand and appreciate it, it is necessary to discuss in some detail a difficulty that every careful student of Aristotle's cosmography must have felt.

It must be borne in mind that Aristotle conceived ' heaven,' as a whole, to be a nest of concentric and hollow revolving spheres, in frictionless contact with each other, each of them containing, immovably fixed in its substance (between its outer and its inner superficies, that is), the heavenly body (or, in the case of the stellar sphere, the heavenly bodies) which it bore with it as it revolved.

These spheres, taken separately, are ' parts ' of heaven as a whole and may be called ' the heavens ' or, individually, the ' stellar heaven,' the ' heaven of Saturn,' and so forth. They are composed of a kind of ' matter ' which, unlike that of our elements and their compounds, is capable of no modification or change save that of rotatory movement.

This premised, we have to consider two unquestionable articles of Aristotle's belief: (1) The universe as a whole, and specifically ' heaven ' as a whole, and most specifically of all the stellar heaven individually, ' has no place.' (2) Heaven, as a whole, rotates; and so does each heaven in particular, including the stellar heaven.

Now each of these propositions carries its own implications with it, and Aristotle nowhere explicitly develops and harmonizes them. They are, however, perfectly consistent with each other, as I must now try to show, without travelling outside Aristotle's own data.

The contents of a stationary vessel can obviously change their relative positions, and therefore the places-proper which they mutually constitute for each other. But can they, as a whole, change their collective place-common? Clearly not; for a cosmic place cannot change at all, and a physically determined place can only move with the body or substance which determines it — in this instance, the vessel which we have supposed to be stationary.

But here two questions arise. (i.) *Question*: ' You say

INTRODUCTORY NOTE (*continued*)

that the contents of a stationary vessel cannot change their place as a whole. But if *every part* of what is contained in the vessel is changing its place with respect to the other parts, may not the *whole* be said to 'be moving'? *Answer*: 'Yes, but not *as* a whole.' It is 'all of it moving,' then, but it is not 'changing its collective place.'

(ii.) Suppose the several parts of the content to retain their mutually determined places and positions with respect to each other (as if they were rigidly connected and in that sense formed a continuous whole), but to be differentiated, say, by colour. If such a content should rotate within the vessel-continent, would the whole, as a whole, be moving with respect to its place-continent?

For this question to have any meaning we must suppose the vessel-continent (though rigid and stationary by hypothesis) to be in some way differentiated, no less than the content.

In such a case we may at least say that the change of angular distances between points on the outer surface of the content and points on the inner surface of the continent constitute a change of 'rotational position,' or a change in 'the kind of place' that content and continent constitute by rotation.

But in such a case what could we mean by saying that a given change in 'rotation-constituted place' was due entirely to a rotation of the content, the continent having remained stationary?

We could only mean that with respect to some chosen type of stability—say a more comprehensive vessel that contained them both—the content showed a rotatory change, and the continent none. But this system of reference will at last bring us to the outmost shell of the universe—according to Aristotle the stellar heaven. Here there is nothing outside the vessel-continent to refer to, and we seem to come to a stand.

Up to this point we have followed pretty closely (as will be seen by an actual examination of the text) the hints

INTRODUCTORY NOTE (*continued*)

and obvious implications contained in the section of chapter v. now under review—whether we ascribe these hints to an authentic Aristotelian basis or to the work of an intelligent and sympathetic redactor. But whoever the author or redactor may be he fails us just at this critical point.

But it is also just at this point that Aristotle himself, in a passage of unchallenged authenticity in the *De caelo*,
II. 14 takes up the tale; for he finds the standard of stability
(296 b 23 *sqq.*) that is wanted in the 'direction of gravitation.' He
assumes (and here it must be admitted that he really begs the question—though he begs it in a highly original and instructive manner) that if a body were projected vertically from the Earth (*i.e.* perpendicularly to the tangential plane at the point) it would fall back again upon the same spot from which it rose. If we accepted this assumption as true, it would follow that, during the ascent and descent of the missile, the heavens had turned round over the stable line of gravitation, while the Earth had remained stationary under it. Had the missile alighted at a different point from that at which it started it would have followed that the Earth had turned round under the direction of gravitation, during the period in question. This completes the answer to the second question.

It is startling to notice that (in spite of the complications introduced into the question by the consideration of the law of Inertia—unknown to Aristotle, though he was in grievous and almost conscious need of it) Galileo, with his experiment from the summit of the tower of Pisa, and Foucault, with his pendulum, had to disprove Aristotle's axiom (and show that the earth *does* turn round under the constant action of gravitation upon the missile or the pendulum) before they could disprove his deduction from it. They brought a more instructed plea before the court and obtained a contrary verdict. But it was the same court that they appealed to.

INTRODUCTORY NOTE (*continued*)

What the author or redactor of the section under review has done, therefore, is to show that all the heavens except the outmost of them have ' places,' though not of the same kind as those which bodies capable of translation have.

Further, Simplicius quotes a certain Maximus, presumably Maximus of Epirus (a reputed teacher of Julian the Apostate, and therefore contemporary with Themistius), in connexion with this very phrase ' the kind of place ' which the heavens make for each other. It is to the effect that when Aristotle, in the *De caelo*, says that rotation " to the left " is the natural rotation of heaven, he is applying local language to it (though it is not ' in a place '). II. 2 (285 b 14 *sqq.*)

And lastly, Themistius makes the naïve but suggestive remark that though the outmost heaven is not neighboured by anything on its outer surface, yet since it is neighboured by the heaven of Saturn on its inner surface, it may be considered as in a place " on the inside."

One can see that all these tentatives are feeling for the needed distinction between two senses in which bodies A and B can be said to have ' moved ' or to have ' undergone a change of place ' with respect to each other as wholes—the one sense involving translation and the other not involving translation but only rotation. This distinction was presumably perfectly clear to Aristotle, for though he nowhere expounds it explicitly with reference to the matter now in hand, yet he is never embarrassed, as his commentators are, by any perplexities of thought on the subject.

The section that follows must be spurious. It was evidently absent from the text of Themistius, and it causes much embarrassment to Simplicius and also, in his degree, to Philoponus. Whatever its origin it is an unintelligent attempt to supply what its author felt to be an omission in the demonstration that the account given of ' place ' fulfils the conditions set forth earlier in this Book. 212 b 29-13 a 10 (211 a 8)

ARGUMENT

The universe, as a whole, has no place—except the kind of place in which Rotation only, and not Translation, is possible. But since both the components of the universe, heavenly and elemental alike, are all susceptible of ' movement ' after their respective fashions, the universe itself may be regarded as ' moving,' in a certain sense, in virtue of the mobility of all its parts. Likewise everything, including the outmost sphere, *may be said to be ' in the universe,' in the sense in which ' the component parts are in the whole,' though not in the local sense of ' content in continent '* (212 a 31-b 18).

In the local sense of content in continent, each of the elements from earth up to fire finds its natural abode ' in the place ' constituted by the elemental or heavenly body (as the case may be), that is its ' next neighbour above.' But the heavenly body itself, as a whole, has not any place (b 18-22).

212 a ᾯ μὲν οὖν σώματι ἔστι τι ἐκτὸς σῶμα περιέχον
αὐτό, τοῦτό ἐστιν ἐν τόπῳ, ᾧ δὲ μή, οὔ. διὸ κἂν
ὕδωρ γένηται τοιοῦτο, τὰ μὲν μόρια κινήσεται
αὐτοῦ (περιέχεται γὰρ ὑπ' ἀλλήλων), τὸ δὲ πᾶν
ἔστι μὲν ὡς κινήσεται ἔστι δ' ὡς οὔ. ὡς μὲν
γὰρ ὅλον, ἅμα τὸν τόπον οὐ μεταβάλλει, κύκλῳ
212 b δὲ κινήσεται· τῶν μορίων γὰρ οὗτος ὁ τόπος·

Καὶ ἄνω μὲν καὶ κάτω οὔ, κύκλῳ δ', ἔνια. τὰ δὲ καὶ ἄνω καὶ κάτω, ὅσα ἔχει πύκνωσιν καὶ μάνωσιν.

---

[a] ' Instead of being the celestial substance which it actually is.' As to ' having no place ' the nature of the continent-uncontained would make no difference. The essential is that there should be nothing outside it. But this parenthetical supposition of a possible alternative is lost sight of at once, and the discourse proceeds on the assumption that we are dealing with a universe whose outer shell is not ' water or anything you like,' but the celestial substance

ARGUMENT (*continued*)

The nature of ' place ' being now understood, all the perplexities that were noted in chap. i. of this Book, as rising out of confused ideas on the subject, at once disappear. We have not to suppose that cosmic places grow, nor that a point has a position other than itself, for we have shown that a limit is in no case ' in ' that which it limits as content in continent, though it does belong to it and is ' involved in ' it. Neither need we suppose two bodies to be in the same place when the contents of a vessel change, for the vacating body leaves no materially dimensional entity behind it (b 22-29). 209 a 2 sqq.

[*The elements seek the relative positions proper to them because in these positions they have neighbours akin to them; and stay there because, in a sense, they form, at their margins, parts of homogeneous wholes with those neighbours* (b 29-213 a 10).]

Epilogue (a 10-11).

It follows that if a body is encompassed by another body, external to it, it is ' in a place ' ; but if not, not. As to such an ' unplaced ' body (be it water or anything else [a]) its parts may be in motion, for they embrace each other, but as a whole it can be said to ' move ' only in a special sense. For as a whole it cannot change its collective place. But it may have a motion of rotation, which motion is what constitutes the ' kind of place ' with respect to which the rotating *parts* move.

Now in the universe there are some parts which do not move up and down but do rotate ; and others (such as are susceptible to condensation and rarefication [b]) which can either rotate or move up and down.

of the concentric spheres. Simplicius notes the " great obscurity " caused by the linguistic violence of this passage.

[b] Aristotle's own phrase here would surely have been " such as have weight and buoyancy."

212 b Ὥσπερ δ' ἐλέχθη, τὰ μὲν ἔστιν ἐν τόπῳ κατὰ δύναμιν, τὰ δὲ κατ' ἐνέργειαν· διὸ ὅταν μὲν συνεχὲς ᾖ τὸ ὁμοιομερές, κατὰ δύναμιν ἐν τόπῳ τὰ μέρη, ὅταν δὲ χωρισθῇ μὲν ἅπτηται δ' ὥσπερ σωρός, κατ' ἐνέργειαν.

Καὶ τὰ μὲν καθ' αὑτά, οἷον πᾶν σῶμα ἢ κατὰ φορὰν ἢ κατ' αὔξησιν κινητὸν καθ' αὑτό που, ὁ δ' οὐρανός, ὥσπερ εἴρηται, οὔ που ὅλος οὐδ' ἔν τινι τόπῳ ἐστίν, εἴ γε μηδὲν αὐτὸν περιέχει σῶμα· ἐφ' ᾧ δὲ κινεῖται, ταύτῃ καὶ τόπος ἐστὶ τοῖς μορίοις· ἕτερον γὰρ ἑτέρου ἐχόμενον τῶν μορίων ἐστίν.

Τὰ δὲ κατὰ συμβεβηκός, οἷον ἡ ψυχή· καὶ ὁ οὐρανός, τὰ γὰρ μόρια ἐν τόπῳ πως πάντα, ἐπὶ τῷ κύκλῳ γὰρ περιέχει ἄλλο ἄλλο.

Διὸ κινεῖται μὲν κύκλῳ τὸ ἄνω, τὸ δὲ πᾶν οὔ που. τὸ γάρ που αὐτό τ' ἐστί τι, καὶ ἔτι ἄλλο τι δεῖ εἶναι παρὰ τοῦτο ἐν ᾧ ὃ περιέχει· παρὰ δὲ τὸ πᾶν καὶ ὅλον οὐδέν ἐστιν ἔξω τοῦ παντός.

Καὶ διὰ τοῦτο ἐν τῷ οὐρανῷ πάντα, ὁ γὰρ οὐρανὸς

---

[a] [Simplic., Philop. and Themist. understand ἐν τόπῳ ἐστί (not κινεῖται) with τὰ μὲν καθ' αὑτά (*l.* 7) and τὰ δὲ κατὰ συμβεβηκός (*l.* 12). 'Some things are in a place *per se*, *e.g.* every body capable of translation or growth is *per se* 'somewhere'; but the heaven as a whole is not 'somewhere' nor in any particular place . . .'—C.]

[b] The Commentators note that qualification is necessary in the case of the outmost sphere. *Cf.* Introduction to this chapter, p. 319.

[c] Themistius (120. 21-26) and Philoponus (603. 10) point out that κατ' ἄλλο, not κατὰ συμβεβηκός, would

Moreover, as already observed, some things are only potentially in places but others actually so. So that if a mass is continuous and homogeneous, its parts are only potentially in places-proper, but if they were so divided as to be in mutual contact (as if in a heap), they would have actualized places-proper.

And again, some things move[a] on their own account (all elemental *bodies*, to wit, are susceptible of change of place, whether by transference or growth); but the heavenly mass as has been said cannot change its place as a whole. Nor indeed has it a place to change, seeing that there is no body-continent embracing it. But after the fashion of its own motion it constitutes places for its own parts,[b] since one part embraces another.

Other things there are which cannot move on their own account, but whose motion is involved in that of something else. Such is the 'vitality,' 'aliveness' of a living organism. And such too is the heavenly mass, for all its parts are in places, after their own fashion, since they embrace each other spherically.[c]

Heaven therefore 'rotates,' but the universe has not a 'where,' for to have a 'where' a thing must not only exist itself but must be embraced by something other than itself; and there *is* nothing other than the universe-and-the-sum-of-things, outside that sum, and therefore nothing to embrace *it*.

It is true, then, to say—'all things are in heaven' (if you admit the doubtful usage of calling the uni-

be the proper Aristotelian qualification of the motion *τοῦ οὐρανοῦ*. The former refers to chap. iii. of this Book to illustrate the distinction. I think they are right, but perhaps 354 b 8-10 somewhat blunts the force of their objection.

212 b τὸ πᾶν ἴσως, ἔστι δ' ὁ τόπος οὐχ ὁ οὐρανός, ἀλλὰ τοῦ οὐρανοῦ τι τὸ ἔσχατον καὶ ἁπτόμενον τοῦ κινητοῦ σώματος πέρας ἠρεμοῦν.

Καὶ διὰ τοῦτο ἡ μὲν γῆ ἐν τῷ ὕδατι, τοῦτο δ' ἐν τῷ ἀέρι, οὗτος δ' ἐν τῷ αἰθέρι, ὁ δ' αἰθὴρ ἐν τῷ οὐρανῷ, ὁ δ' οὐρανὸς οὐκέτι ἐν ἄλλῳ.

Φανερὸν δ' ἐκ τούτων ὅτι καὶ αἱ ἀπορίαι πᾶσαι λύοιντ' ἂν οὕτω λεγομένου τοῦ τόπου. οὔτε γὰρ συναύξεσθαι ἀνάγκη τὸν τόπον, οὔτε στιγμῆς εἶναι τόπον, οὔτε δύο σώματα ἐν τῷ αὐτῷ τόπῳ, οὔτε διάστημά τι εἶναι σωματικόν· σῶμα γὰρ τὸ μεταξὺ τοῦ τόπου τὸ τυχόν, ἀλλ' οὐ διάστημα σώματος. καὶ ἔστιν ὁ τόπος καὶ πού, οὐχ ὡς ἐν τόπῳ δέ, ἀλλ' ὡς τὸ πέρας ἐν τῷ πεπερασμένῳ. οὐ γὰρ πᾶν τὸ ὂν ἐν τόπῳ, ἀλλὰ τὸ κινητὸν σῶμα.

Καὶ φέρεται δὴ εἰς τὸν αὑτοῦ τόπον ἕκαστον εὐλόγως· ὃ γὰρ ἐφεξῆς καὶ ἁπτόμενον μὴ βίᾳ, συγγενές· καὶ συμπεφυκότα μὲν ἀπαθῆ, ἁπτόμενα δὲ παθητικὰ καὶ ποιητικὰ ἀλλήλων.

---

[a] Aristotle notes this curious use of the word οὐρανός in the *De caelo* i. 9 (278 b 20 f.).

[b] With Themistius I understand κινητόν to be used in the sense of παθητικόν, which would be equivalent to 'elemental' or 'susceptible of up and down movement.'

[c] Here 'aether' is obviously a synonym for 'fire.' Aristotle frequently warns us that when Anaxagoras says 'aether' he means 'fire,' but he himself condemns this usage. On the other hand he accepts the current usage of his own time, by which the fifth or heavenly σῶμα (the substance of the celestial spheres) was called aether. But (so far as the indexes of Bonitz and Heintz reveal) Aristotle himself does not employ the word at all, apart from quotations and references. (It occurs in the spurious *De mundo*, 392 a 5.)

De caelo, 270 b 25, 302 b 4

verse 'heaven'[a]), but what constitutes the place-universal is not the whole rotating aetherial mass but the inner surface of that mass, which is at rest and is in contact with the world of modifiable substances.[b]

So earth is naturally surrounded and embraced by water, water by air, air by aether,[c] aether by heaven, and heaven itself not at all.

It is an easy matter now to show that our definition of place smoothes out all the perplexities that have arisen. For we are not compelled to think of 'place' growing, on its own account, to accompany a growing body; nor of a point occupying 'place'; nor of two bodies being in the same place; nor of 'intervals' having an independent material existence. For what lies between point and point of the continent is the body, whatever it may be, that forms its content, not the dimensions of that body, as existing independently of it. Also you can assign a 'something wherein it resides' to a place itself, but only in the sense in which the limit 'resides in' the body it limits, not as the content is 'in' the continent; for 'in' has local significance only when applied to materially movable bodies, and there are existences other than these.

[And that each element should make for its own place amongst the others is only what we should expect; for in the succession of elementary bodies the neighbours with which each is in unforced contact are kindred to it. And note that between the several parts of the homogeneous element itself there is no action and reaction; but at the margins where the different (but kindred) elements are in contact, they act and react upon each other.

212 b Καὶ μένει δὴ φύσει πᾶν ἐν τῷ οἰκείῳ τόπῳ ἕκαστον οὐκ ἀλόγως· καὶ γὰρ τὸ μέρος τόδε ἐν ὅλῳ τῷ τόπῳ ὡς διαιρετὸν μέρος πρὸς ὅλον
213 a ἐστίν, οἷον ὅταν ὕδατος κινήσῃ τις μόριον ἢ ἀέρος.

Οὕτω δὲ καὶ ἀὴρ ἔχει πρὸς ὕδωρ· οἷον ὕλη γάρ, τὸ δὲ εἶδος, τὸ μὲν ὕδωρ ὕλη ἀέρος, ὁ δ' ἀὴρ οἷον ἐνέργειά τις ἐκείνου· τὸ γὰρ ὕδωρ δυνάμει ἀήρ ἐστιν, ὁ δ' ἀὴρ δυνάμει ὕδωρ ἄλλον τρόπον. διοριστέον δὲ περὶ τούτων ὕστερον· ἀλλὰ διὰ τὸν καιρὸν ἀνάγκη μὲν εἰπεῖν, ἀσαφῶς δὲ νῦν ῥηθὲν τότ' ἔσται σαφέστερον. εἰ οὖν τὸ αὐτὸ ἡ ὕλη καὶ ἡ ἐντελέχεια (ὕδωρ γὰρ ἄμφω, ἀλλὰ τὸ μὲν δυνάμει τὸ δ' ἐντελεχείᾳ), ἔχοι ἂν ὡς μόριόν πως πρὸς ὅλον. διὸ καὶ τούτοις ἁφή ἐστιν· σύμφυσις δέ, ὅταν ἄμφω ἐνεργείᾳ ἓν γένωνται.

Καὶ περὶ μὲν τόπου, καὶ ὅτι ἔστι καὶ τί ἐστιν, εἴρηται.

---

[a] A vortex, so individualized, would not tend to rise or fall in its homogeneous environment.

[b] The general doctrine of the transition from water to air, etc., is expounded by Aristotle in the *De gen. et cor.* i. 5 and ii. 4, and the *De caelo* iii. 5. An account of it will be best reserved for treatment in connexion with the doctrine of Growth and the Void (p. 340 of this vol.). The present passage is a confused and unintelligent attempt to foreshadow it. Simplicius points out (and tries in vain to palliate) the fallacy of treating the material as 'part' of the

And it is equally natural for the elements severally to rest in the places that belong to them, for the relation of each element, in its own portion of the whole place-universal, to its neighbours is that of a separated part of a single element to the whole of that element—if one should make a swirl within a body of water or air, I mean.[a]

And this is how air is related to water; for air is the form and water the material, the air in a sense being the actualizing of water, since the water is potentially air, as indeed the air is potentially water, but in a different way. This will be gone into more closely later on.[b] It was, however, necessary to touch on it here, for the context demanded it; and what is vague in this exposition will become clearer under fuller treatment. If, then, the same thing is both matter and fulfilment (water, to wit, being both; but the one in potentiality and the other in actuality), each aspect of it would stand somewhat as a part to its whole. Water and air, then, when separated are merely in contact, but are homogeneous when both become one in actuality.]

This concludes the account of 'place,' and the demonstration of its real existence and of what it is.

form (before it has received it) instead of part of the *compositum* (after it has received the form).

The transition from water to air is from the lower to the higher form, and is therefore quasi-constructive; whereas the reverse transition is quasi-destructive. Hence the 'but in a different way.'

CHAPTER VI

ARGUMENT

Close connexion between the problem of ' place,' and that of ' void ' or ' empty space,' conceived as ' place with nothing in it ' (213 a 12–22).

Anaxagoras and others who denied the existence of the void missed the point, for they only showed that ' air is not nothing ' not ' that there are no interspaces of vacancy in the world ' (a 22–b 2).

The believers in the void, on the contrary, did at least argue to the point. They alleged as proofs : (1) *that nothing could*

213 a 12 Τὸν αὐτὸν δὲ τρόπον ὑποληπτέον εἶναι τοῦ φυσικοῦ θεωρῆσαι καὶ περὶ κενοῦ, εἰ ἔστιν ἢ μή, καὶ πῶς ἔστι καὶ τί ἐστιν, ὥσπερ καὶ περὶ τόπου· καὶ γὰρ παραπλησίαν ἔχει τήν τε ἀπιστίαν καὶ τὴν πίστιν διὰ τῶν ὑπολαμβανομένων· οἷον γὰρ τόπον τινὰ καὶ ἀγγεῖον τὸ κενὸν τιθέασιν οἱ λέγοντες, δοκεῖ δὲ πλῆρες μὲν εἶναι, ὅταν ἔχῃ τὸν ὄγκον οὗ δεκτικόν ἐστιν, ὅταν δὲ στερηθῇ, κενόν, ὡς τὸ αὐτὸ μὲν ὂν κενὸν καὶ πλῆρες καὶ τόπον, τὸ δ' εἶναι αὐτοῖς οὐ ταὐτὸ ὄν.

Ἄρξασθαι δὲ δεῖ τῆς σκέψεως λαβοῦσιν ἅ τε λέγουσιν οἱ φάσκοντες εἶναι καὶ πάλιν ἃ λέγουσιν οἱ μὴ φάσκοντες, καὶ τρίτον τὰς κοινὰς περὶ αὐτῶν δόξας.

Οἱ μὲν οὖν δεικνύναι πειρώμενοι ὅτι οὐκ ἔστιν οὐχ ὃ βούλονται λέγειν οἱ ἄνθρωποι κενόν, τοῦτ' ἐξελέγχουσιν, ἀλλ' ἁμαρτάνοντες λέγουσιν, ὥσπερ

CHAPTER VI

ARGUMENT (*continued*)

move unless there were empty places for it to move into (Melissus advanced this argument as sound, but treated it as a reductio ad absurdum, *and so concluded that ' nothing can move '*) (b 2–14) ; (2) *that bodies can contract into the void spaces contained in them* (b 14–18) ; *and* (3) *that without vacancy nothing could grow or expand* (b 18–22).

The Pythagoreans, too, had their doctrine of numerical and other void interspaces (b 22–29).

We must take it as part of the natural philosopher's task to reflect on the existence or non-existence, and the ' how ' and the ' what,' of ' vacancy ' no less than of ' place.' For indeed the conceptions formed of it lead to arguments for and against its very existence exactly analogous to those concerning place, inasmuch as the believers in its reality present it to us as if it were some kind of receptive place or vessel, which may be regarded as full when it contains the bulk of which it is capable, and empty when it does not. Thus ' place ' and the ' filled ' and the ' vacant ' would all be one identical entity under varying aspects or conditions of existence.

We must begin our inquiry by examining the grounds on which the existence of the void is maintained, those on which it is denied, and what we may assume to be universally admitted as fundamental to our conception of it.

We shall find that those who have attempted to refute the doctrine of the void have not grappled with that which its advocates are really driving at, but only with an incidental error in their presentation

213 a ᾿Αναξαγόρας καὶ οἱ τοῦτον τὸν τρόπον ἐλέγχοντες. ἐπιδεικνύουσι γὰρ ὅτι ἔστι τι ὁ ἀήρ, στρεβλοῦντες τοὺς ἀσκοὺς καὶ δεικνύντες ὡς ἰσχυρὸς ὁ ἀήρ, καὶ ἐναπολαμβάνοντες ἐν ταῖς κλεψύδραις. οἱ δ' ἄνθρωποι βούλονται κενὸν εἶναι διάστημα ἐν ᾧ μηδέν ἐστι σῶμα αἰσθητόν· οἰόμενοι δὲ τὸ ὂν ἅπαν εἶναι σῶμα φασίν, ἐν ᾧ ὅλως μηδέν ἐστι, τοῦτ' εἶναι κενόν· διὸ τὸ[1] πλῆρες ἀέρος κενὸν εἶναι. οὔκουν τοῦτο δεῖ δεικνύναι, ὅτι ἔστι τι ὁ ἀήρ, ἀλλ' ὅτι οὐκ ἔστι διάστημα ἕτερον τῶν σωμάτων, οὔτε χωριστὸν οὔτε ἐνεργείᾳ ὄν, ὃ διαλαμβάνει τὸ πᾶν σῶμα ὥστ' εἶναι μὴ συνεχές (καθάπερ
213 b λέγουσι Δημόκριτος καὶ Λεύκιππος καὶ ἕτεροι πολλοὶ τῶν φυσιολόγων), ἢ καὶ εἴ τι ἔξω τοῦ παντὸς σώματος ἔστιν ὄντος συνεχοῦς.

Οὗτοι μὲν οὖν οὐ κατὰ θύρας πρὸς τὸ πρόβλημα ἀπαντῶσιν, ἀλλ' οἱ φάσκοντες εἶναι μᾶλλον.

[1] [διὸ τὸ] *διότι* E: *οὐ δὴ τὸ* Prantl; *cf.* Simpl. 648. 3 *τὸ δὲ τὸν ἀέρα μὴ εἶναι σῶμα, καὶ διὰ τοῦτο μηδὲν εἶναι ἐν τῷ διαστήματι τῷ τὸν ἀέρα ἔχοντι, τοῦτο ἀπατώμενοι λέγουσι*, Themist. 123. 6 *οἰόμενοι δὲ τὸν ἀέρα μὴ εἶναι σῶμα αἰσθητὸν τὸ πλῆρες ἀέρος κενὸν νομίζουσι.*—C.]

---

[a] [*Or* ' Those who try to show that there is no such thing do not disprove what ordinary people mean by ' vacancy,' but their arguments miss the point.'—C.]

[b] [Diels, *Vors.* 46 A 68–69.—C.]

[c] To illustrate this take a small flower pot, put your finger over the hole at the bottom, then turn it upside down and force it into a bowl of water. The water will not rise in it unless you remove your finger. This shows that the pot was not empty but filled with air which escapes and makes room for the water to rise when you remove your finger. [*Cf. Problems* xvi. 8, 914 b 9 ff. ; Empedocles, *Frag.* 100.—C.]

of the case.[a] This applies to Anaxagoras[b] and all who have followed his line of argument: what they did was to demonstrate that air is a physical substance by inflating bladders and showing their strength of resistance to compression, and by intercepting the escape of the air from water-clocks.[c] But what those who believe in the void really mean by the word is 'dimensional interval without any substance perceptible to the senses occupying it'; and, because they regard sensible substances as the only things that exist at all, they express this by saying 'in which there is nothing at all.' And they regard things full of air as empty only because they had not recognized the sense-perceptibility of air and so did not know that 'anything' was there.[d] But what the deniers of the void have really to show is not that air is an actual substance, but that there is no such thing as a dimensional entity, other than that of material substances, existing in such detachment and actuality[e] as to intervene and break the continuity of a body. That there is such an entity is the contention of Democritus and Leucippus, and many other natural philosophers. Or it might perhaps be maintained that, though material existence is continuous throughout the cosmos, the void is a something existing outside it.

So they who deny the void are not fairly facing the question in dispute. In this the defenders of the

[d] The old commentators found no difficulty in filling up the ellipticity of this passage as I have done. But the Didot text cuts the knot by inserting a *μή* before *κενόν*, I think wrongly; for though it makes the construction and translation easier it is far from improving the argument. It may well be, however, that some words have fallen out here.

[e] The 'actuality' is important.

213 b λέγουσι δ' ἓν μὲν ὅτι κίνησις ἡ κατὰ τόπον οὐκ ἂν εἴη (αὕτη δ' ἐστὶ φορὰ καὶ αὔξησις)· οὐ γὰρ ἂν δοκεῖν εἶναι κίνησιν, εἰ μὴ εἴη κενόν· τὸ γὰρ πλῆρες ἀδύνατον εἶναι δέξασθαί τι. εἰ δὲ δέξεται καὶ ἔσται δύο ἐν ταὐτῷ, ἐνδέχοιτ' ἂν καὶ ὁποσαοῦν εἶναι ἅμα σώματα· τὴν γὰρ διαφοράν, δι' ἣν οὐκ ἂν εἴη τὸ λεχθέν, οὐκ ἔστιν εἰπεῖν. εἰ δὲ τοῦτο ἐνδέχεται, καὶ τὸ μικρότατον δέξεται τὸ μέγιστον· πολλὰ γὰρ μικρά ἐστι τὸ μέγα· ὥστ' εἰ πολλὰ ἴσα ἐνδέχεται ἐν ταὐτῷ εἶναι, καὶ πολλὰ ἄνισα. Μέλισσος μὲν οὖν καὶ δείκνυσιν ὅτι τὸ πᾶν ἀκίνητον ἐκ τούτων· εἰ γὰρ κινήσεται, ἀνάγκη εἶναι (φησί) κενόν, τὸ δὲ κενὸν οὐ τῶν ὄντων.

Ἕνα μὲν οὖν τρόπον ἐκ τούτων δεικνύουσιν ὅτι ἔστι τι κενόν, ἄλλον δ' ὅτι φαίνεται ἔνια συνιόντα καὶ πιλούμενα, οἷον καὶ τὸν οἶνόν φασι δέχεσθαι μετὰ τῶν ἀσκῶν τοὺς πίθους, ὡς εἰς τὰ ἐνόντα κενὰ συνιόντος τοῦ πυκνουμένου σώματος.

Ἔτι δὲ καὶ ἡ αὔξησις δοκεῖ πᾶσι γίγνεσθαι διὰ κενοῦ· τὴν μὲν γὰρ τροφὴν σῶμα εἶναι, δύο δὲ σώματα ἀδύνατον ἅμα εἶναι. μαρτύριον δὲ καὶ τὸ περὶ τῆς τέφρας ποιοῦνται, ἣ δέχεται ἴσον ὕδωρ ὅσον τὸ ἀγγεῖον τὸ κενόν.

Εἶναι δ' ἔφασαν καὶ οἱ Πυθαγόρειοι κενόν, καὶ ἐπεισιέναι αὐτῷ[1] τῷ οὐρανῷ ἐκ τοῦ ἀπείρου

[1] [αὐτῶ G, αὐτῷ Philop. 615. 19 (lemma), Bonitz, Diels (Vors. 45 B 30): αὐτὸ cett.—C.]

---

[a] [Diels, *Vors.* 20 B 7, § 7; *De gen. et corr.* 325 a 2.—C.]

[b] Aristotle evidently attaches little value to such allegations. He takes no serious notice of them. *Cf.* p. 340.

doctrine have the advantage over them ; for they maintain (1) that there could not be such a thing as local movement at all (whether of transference or of growth) without ' emptiness ' for it to move into, since the ' full ' could not allow anything to move into it. Because, if the full could receive anything into itself so that two bodies coincided, then any number of bodies might coincide ; for no reason can be given why not three and so forth, if two. And, if that were so, the smallest body could receive the greatest into itself ; for many smalls make a great, so that if many equals can coincide, so can many unequals. Melissos [a] indeed makes use of this very argument to prove that there is no movement in the universe ; for he accepts the proof that movement implies the existence of void, but denies the possibility of any such existence, since ' void ' is not an ' existent.'

Such then is one of the alleged proofs of the existence of the void. Another (2) is that certain substances seem to contract and to thicken. This is exemplified in the alleged [b] fact that casks can receive their full complement of wine and the wine-skins as well, which is attributed to the bodily substances drawing up into the vacant interspaces and so becoming denser.

And further (3) it is generally supposed that growth is made possible only by the void, ' food ' being a bodily substance and its coincidence with the ' fed ' being therefore impossible. In support of this, what is said about ashes—that they will absorb as much water as the vessel they are in would hold if it were empty—is appealed to.

The Pythagoreans too asserted the existence of the void and declared that it enters into the heavens out

213 b πνεύματος[1] ὡς ἀναπνέοντι καὶ τὸ κενόν, ὃ διορίζει τὰς φύσεις, ὡς ὄντος τοῦ κενοῦ χωρισμοῦ τινος τῶν ἐφεξῆς καὶ [τῆς][2] διορίσεως· καὶ τοῦτ' εἶναι πρῶτον ἐν τοῖς ἀριθμοῖς· τὸ γὰρ κενὸν διορίζειν τὴν φύσιν αὐτῶν.

Ἐξ ὧν μὲν οὖν οἱ μέν φασιν εἶναι οἱ δ' οὔ φασι, σχεδὸν τοιαῦτα καὶ τοσαῦτά ἐστιν.

[1] [πνεύματος codd.: πνεῦμα, Heidel: πνεῦμά τε Diels (*ibid.*) coll. Simplic. 615. 26 ἔλεγον γὰρ ἐκεῖνοι τὸ κενὸν ἐπεισιέναι τῷ κόσμῳ οἷον ἀναπνέοντι ἤτοι εἰσπνέοντι αὐτῷ ὥσπερ πνεῦμα ἀπὸ τοῦ ἔξωθεν περικεχυμένου, Philop. 615. 22. Fort. πνεῦμά τι.—C.]

[2] [τῆς del. Bonitz.—C.]

CHAPTER VII

ARGUMENT

Repetition and development of the theme that opened the preceding chapter. Implied definition of the 'void' as 'dimensional interval containing no heavy or buoyant substance.' Is this equivalent to not containing anything at all? (213 b 30–214 a 11).

Another definition, at once more explicit and more guarded, defines it as dimensionality with no compositum *of any kind in it; which condition would be met, if it were the ultimate material of all things, or matter devoid of form. But this is to make it the Platonic* (*not our*) *'place'; and we have seen that it is fundamental to 'place,' however conceived, that the body 'in a place' should be able to leave it, which is not true of its 'matter.' Indeed the definition of 'void' as 'place with nothing in it' makes it unnecessary to repeat*

of the limitless breath—regarding the heavens as breathing the very vacancy—which vacancy 'distinguishes' natural objects, as constituting a kind of separation and division between things next to each other, its prime seat being in numbers, since it is this void that delimits their nature.[a]

This then is a fair summary of the number and character of the grounds alleged for believing or disbelieving in the existence of the void.

[a] Apparently the essential discreteness of number was emphasized by the Pythagoreans under this figure of vacancies so separating numbers that they never could become continuous. See Gen. Introd. p. lxxxvi.

CHAPTER VII

ARGUMENT (*continued*)

the arguments already urged in dealing with 'place,' which showed both how the conception of self-existing dimensionality arises and why it is an illusion (214 a 11–26).

It remains to show that the alleged proofs of the existence of void are inconclusive. It is not implied in motion, whether 'motion' includes every kind of change or only local movement; for a ring or a sphere may move round its centre, every part following simultaneously in the track of its precursor. A vortex in an otherwise undifferentiated medium, aeriform or liquid, is another case of the same principle. Nor do the phenomena of expansion and contraction or of growth justify the hypothesis of the void, and the real difficulties of that hypothesis are only aggravated by it (214 a 26–b 11).

Πρὸς δὲ τὸ ποτέρως ἔχει δεῖ λαβεῖν τί σημαίνει τοὔνομα. δοκεῖ δὴ τὸ κενὸν τόπος εἶναι ἐν ᾧ μηδέν ἐστιν. τούτου δ' αἴτιον ὅτι τὸ ὂν σῶμα οἴονται εἶναι, πᾶν δὲ σῶμα ἐν τόπῳ, κενὸν δ' ἐν ᾧ τόπῳ μηδέν ἐστι σῶμα, ὥστ' εἴ που μὴ ἔστι σῶμα, κενὸν εἶναι ἐνταῦθα. σῶμα δὲ πάλιν ἅπαν οἴονται εἶναι ἁπτόν· τοιοῦτο δὲ ὃ ἂν ἔχῃ βάρος ἢ κουφότητα. συμβαίνει οὖν ἐκ συλλογισμοῦ τοῦτο εἶναι κενόν, ἐν ᾧ μηδέν ἐστι βαρὺ ἢ κοῦφον. ταῦτα μὲν οὖν, ὥσπερ εἴπομεν καὶ πρότερον, ἐκ συλλογισμοῦ συμβαίνει· ἄτοπον δὲ εἰ ἡ στιγμὴ κενόν· δεῖ γὰρ τόπον εἶναι, ἐν ᾧ σώματός ἐστι διάστημα ἁπτοῦ. ἀλλ' οὖν φαίνεται λέγεσθαι τὸ κενὸν ἕνα μὲν τρόπον τὸ μὴ πλῆρες αἰσθητοῦ σώματος κατὰ τὴν ἁφήν· αἰσθητὸν δ' ἐστὶ κατὰ τὴν ἁφὴν τὸ βάρος ἔχον καὶ κουφότητα. διὸ κἂν ἀπορήσειέ τις, τί ἂν φαῖεν, εἰ ἔχοι τὸ διάστημα χρῶμα ἢ ψόφον, πότερον κενὸν ἢ οὔ; ἢ δῆλον ὅτι εἰ μὲν δέχοιτο σῶμα ἁπτόν, κενόν,[1] εἰ δὲ μή, οὔ.

[1] [κενὸν E, Philop. 621. 9 (lemma), Simplic. 654. 30, 655. 13: κενὸν εἶναι cett.—C.]

---

[a] If we pin them down (perhaps to alternatives) we shall be able to draw logical conclusions which will put its reality to the test.

[b] *i.e.* no qualitive experiences which are not things are recognized by them as existing.

[c] [δεῖ γὰρ κτλ. is obscure in construction and interpretation. (1) Carteron translates: '*en effet il faut que le vide soit un lieu où il y ait extension d'un corps tangible.*' (2) Perhaps, 'for that in which (as in the void) there is extension of tangible body (*i.e.* room, whether empty or full, for an extended body) must be a place (and a point is not a place),' or (3) 'for it (the point, if it were a void) must be a place in which there is extension of tangible body (but a point is not that).'—C.]

To determine whether the void exists or not, we must know what they who use the word really mean by it.[a] The current answer is, 'a place in which there is nothing.' But that is in the mouths of people who consider that nothing 'exists' except material substances ;[b] so they mean that a body must be in a place, but a place in which there is not a body must be absolutely empty ; so that wherever there is not 'body' there is vacancy. Again they regard all corporeal substances as 'tangible' ; and this they regard as identical with 'possessed of gravity or levity.' Thus we may logically expand their definition of the void as 'that in which there is nothing that is either heavy or buoyant.' For this, as we have just said, is implicit in their actual words ; but no one could suppose them to mean that a point is a void, though it complies with the definition just given. We must add that in defining a void as a place they assign to it dimensionality as of a tangible body.[c] However, this seems to be one conception of the void : 'that which is empty of all tangibly perceptible body'—'tangibly perceptible' being equated with 'possessed of gravity or levity.' So, on this line of inquiry, if we go on to ask whether dimensional extension would be empty if it were coloured or sonant, presumably the answer would be, 'If it were still capable of receiving corporeal bulk, yes ; if otherwise, no.'[d]

[d] If light and sound can be propagated over void spaces, are those spaces still empty while they are passing through them ? A curious anticipation of the question raised in connexion with the hypothesis of the 'luminiferous aether.'

214 a Ἄλλον δὲ τρόπον, ἐν ᾧ μὴ τόδε τι μηδ' οὐσία τις σωματική. διό φασί τινες εἶναι τὸ κενὸν τὴν τοῦ σώματος ὕλην, οἵπερ καὶ τὸν τόπον, τὸ αὐτὸ τοῦτο λέγοντες οὐ καλῶς· ἡ μὲν γὰρ ὕλη οὐ χωριστὴ τῶν πραγμάτων, τὸ δὲ κενὸν ζητοῦσιν ὡς χωριστόν.

Ἐπεὶ δὲ περὶ τόπου διώρισται, καὶ τὸ κενὸν ἀνάγκη τόπον εἶναι (εἰ ἔστιν) ἐστερημένον σώματος, τόπος δὲ καὶ πῶς ἔστι καὶ πῶς οὐκ ἔστιν εἴρηται, φανερὸν ὅτι οὕτω μὲν κενὸν οὐκ ἔστιν, οὔτε ἀχώριστον οὔτε κεχωρισμένον· τὸ γὰρ κενὸν οὐ σῶμα ἀλλὰ σώματος διάστημα βούλεται εἶναι. διὸ καὶ τὸ κενὸν δοκεῖ τι εἶναι, ὅτι καὶ ὁ τόπος, καὶ διὰ ταὐτά. ἥκει γὰρ δὴ ἡ κίνησις ἡ κατὰ τόπον καὶ τοῖς τὸν τόπον φάσκουσιν εἶναί τι παρὰ τὰ σώματα τὰ ἐμπίπτοντα καὶ τοῖς τὸ κενόν· αἴτιον δὲ κινήσεως οἴονται εἶναι τὸ κενὸν οὕτως ὡς ἐν ᾧ κινεῖται· τοῦτο δ' ἂν εἴη οἷον τὸν τόπον φασί τινες εἶναι.

---

[a] The identification of matter and place is attributed to Plato, 209 b 11.

[b] It is difficult to be sure what 'inseparable' and 'separable' here refer to; and the commentators are not convincing. I take it to mean: 'If we mean by "place" the inner surface of the containing body which is inseparably inherent in that body itself, it is clear that the void cannot be that kind of place. If, on the other hand, we mean the intramural space apart from any body contained, we have seen that neither has that any independent and detachable existence.' [Aristotle argues against the existence of (1) a void '*separate from* bodies,' in the sense of a spatial frame containing bodies (*e.g.* atoms) moving about in it, with empty intervals between them, chap. viii. to 216 a 26; (2) a void *occupied by* bodies, 216 a 26–b 21; (3) void *interstices* inside bodies, chap. ix.—C.]

[c] Because, as we saw, the dimensions of the retreating

According to another representation, however, the void is ' that in which there is no concrete *compositum* uniting form and matter, or corporeal entity.' It is from this point of view that some thinkers regard the void as the ' matter ' of corporeal entities, regarding it as identical with ' place,' to which they assign the same rôle.[a] But this will not do, for things cannot detach themselves from their material, and what the believers in the void are in quest of is something from which things can separate themselves and come away.

Now since we have already determined the nature of ' place ' and the way in which it does or does not exist, and since the void (if there is such a thing) must be conceived as place in which there might be body but is not, it is clear that, so conceived, the void cannot exist at all, either as inseparable or separable ; [b] for what is meant by it is not body but the bodiless dimensions of body.[c] So we see that vacuity was supposed to have an independent existence of its own because ' place ' was so conceived, and for the same reasons. For those who say that ' place ' has an existence of its own, apart from the corporeally extended bodies that successively occupy it, do so in view of the phenomenon of local movement,[d] and so too do those who say the like of vacuity, the essential function assigned to vacuity in this connexion being that of the medium ' in which ' the motion occurs, which is just what certain other thinkers would say of place.

body that are inherent in it are not left behind it as a ' place ' ; whereas, if we suppose it to have dimensions that can be left behind, we are led into many self-contradictions. *Cf.* p. 309.

[d] [ἥκει, local movement ' *results*,' ' *is obtained*.' Bonitz (Index, 316 b 44) compares the phrase ὁ αὐτὸς ἥξει λόγος.—C.]

214 a Οὐδεμία δ' ἀνάγκη, εἰ κίνησις ἔστι, εἶναι κενόν. ὅλως μὲν οὖν πάσης κινήσεως οὐδαμῶς, διὸ καὶ Μέλισσον ἔλαθεν· ἀλλοιοῦσθαι γὰρ τὸ πλῆρες ἐνδέχεται. ἀλλὰ δὴ οὐδὲ τὴν κατὰ τόπον κίνησιν· ἅμα γὰρ ἐνδέχεται ὑπεξιέναι ἀλλήλοις, οὐδενὸς ὄντος διαστήματος χωριστοῦ παρὰ τὰ σώματα τὰ κινούμενα· καὶ τοῦτο δῆλον καὶ ἐν ταῖς τῶν συνεχῶν δίναις, ὥσπερ καὶ ἐν ταῖς τῶν ὑγρῶν.

Ἐνδέχεται δὲ καὶ πυκνοῦσθαι μὴ εἰς τὸ κενὸν
214 b ἀλλὰ διὰ τὸ τὰ ἐνόντα ἐκπυρηνίζειν (οἷον ὕδατος συνθλιβομένου τὸν ἐνόντα ἀέρα), καὶ αὐξάνεσθαι οὐ μόνον εἰσιόντος τινὸς ἀλλὰ καὶ ἀλλοιώσει, οἷον εἰ ἐξ ὕδατος γίνοιτο ἀήρ.

Ὅλως δὲ ὅ τε περὶ τῆς αὐξήσεως λόγος καὶ τοῦ εἰς τὴν τέφραν ἐγχεομένου ὕδατος αὐτὸς αὐτὸν ἐμποδίζει. ἢ γὰρ οὐκ αὐξάνεται ὁτιοῦν, ἢ οὐ σώματι, ἢ ἐνδέχεται δύο σώματα ἐν ταὐτῷ

---

[a] *δίνη* may mean either the revolution of a ring or of a wheel round its centre or the swirl of a vortex within a medium only differentiated from it by not sharing the movement.

[b] One explanation of contraction is given and one of expansion, but both are evidently meant to apply to both expansions and contractions. The expulsion and intrusion hypothesis is probably introduced with reference to the supposed case of the wine and wine-skins (213 b 16), though we are not told how to apply it. In any case it is of minor importance. On the other hand, the theory that matter is intrinsically 'contractile' (*i.e.* capable of dimensional expansion and contraction while preserving its continuity) is of fundamental importance to Aristotle's physical philosophy.

[c] 'Water' = representative liquid. 'Air' = representative gas.

[d] In this passage Aristotle shows neither how the dilemma

But the truth is that the self-existence of vacuity is not involved in any way whatever in the fact of motion. This needs no proof, if ' motion ' be taken as the typical and inclusive expression for every kind of ' passing from this to that,' for a *plenum* might undergo qualitive modification without the aid of any *vacuum*—though this obvious fact escaped Melissus. But, further, even for motion in the restricted sense of vection and rotation vacuity is not necessary ; for things can simultaneously give place to each other without there being any intervenient dimensionality sejunct from the moving bodies, as is evident in the case of the rotation of continuously coherent bodies and also in the vortices of liquids through liquids.[a]

Nor does condensation necessarily imply the entrance of particles into vacant dimensional interstices ; for we may suppose the condensing body to be squeezing out some foreign substance that it contained, as water for instance when pressed together extrudes the air that is in it. Or we may suppose a body to expand under the influence of qualitive modification without anything else entering into it,[b] as in the case of water becoming air.[c]

To put the case more generally, the arguments derived from the phenomenon of growth or expansion and from the case of the water poured into the ashes, refute themselves by proving too much. For the problem of growth lands us in the dilemma :[d] Either the growth does not occur in any and every part of the growing body, or things grow without the access of any corporeal substance, or two bodies

rises nor how it can be solved. But the question is treated at length in *De gen. et cor.* i. 5.

214 b εἶναι (ἀπορίαν οὖν κοινὴν ἀξιοῦσι λύειν, ἀλλ' οὐ κενὸν δεικνύουσιν ὡς ἔστιν), ἢ πᾶν εἶναι ἀναγκαῖον τὸ σῶμα κενόν, εἰ πάντῃ αὐξάνεται καὶ αὐξάνεται διὰ κενοῦ. ὁ δ' αὐτὸς λόγος καὶ ἐπὶ τῆς τέφρας.

Ὅτι μὲν οὖν ἐξ ὧν δεικνύουσιν εἶναι τὸ κενὸν λύειν ῥᾴδιον, φανερόν.

CHAPTER VIII

ARGUMENT

Let us now rehearse our reasons for denying the reality of a self-existent and self-constituted 'void' (in the sense of abstract or absolute 'space') such as some thinkers have assumed.

Mere limitless vacancy cannot be, as is alleged, the cause of motion either in the proper sense of vection and rotation or as the type of all 'change' or 'passing.' Indeed it is only in the narrower sense that anyone makes the claim for it. And in this sense we see that the elementary bodies have each their specific direction (fire 'up' from *the centre and earth 'down'* towards *the centre, for example) and obviously mere 'vacancy' cannot differentiate its causal action so as to produce these different effects* (214 b 12–17).

Again, if the void be conceived as the potential recipient of body that lacks the actuality of having received it, then, if we start with it as an undifferentiated emptiness outside all body and with a body that has no differentiated trend, how could there be anywhere in the vacancy to which the body should go rather than anywhere else or rather than remain motionless? For it cannot go into the void every way at once (b 17–19).

The same argument applies to those who allege an unoccupied void not as the cause but the condition of movement, on the ground that a body cannot move into a place except if

can occupy the same space—our friends, then, are attempting to solve a problem common to us and to them, but they throw no light on the nature of the 'vacuity' they wish to establish—or the whole growing body is vacuity and nothing else, if the body is growing or expanding at every point and can only do so where there is vacuity to grow or expand into. The same argument applies to the 'ashes' case.

Obviously, then, there is no difficulty in refuting the arguments urged to prove the existence of vacuity.

CHAPTER VIII

ARGUMENT (*continued*)

that place be a something which is detached from body and unoccupied (b 19–22).

It is quite natural that the same reasoning should be repeated in dealing with the void and with 'place'; for it applies to the void, and whenever the champions of the 'void' try to define it, they can only do so by making it (with whatever qualification) a kind of place (b 22–24).

And then how can any determined or defined place or void be 'in' place or void at large? For it is one thing for a given undifferentiated mass to be 'in' a body differentiated from it that abides unaffected by its departure, and another thing for a given 'space' to be in abstract or undifferentiated 'space.' This latter could but be 'in' as the part is 'in' the whole, not as the content locally 'in' a continent differentiated from it (b 24–27).

[*Void, so far from being necessary to motion, makes motion impossible* (b 28–215 a 1).

The natural motion of the simple bodies is prior to all other motions, but could not occur in void (a 1–14).

A projectile could not move in void (a 14–22).

ARGUMENT (*continued*)

The speed of a moving body varies either (1) *with the resistance of the medium traversed or* (2) *with its own weight or lightness.* (1) *If water offers* (*say*) *twice the resistance of air, the times taken to traverse equal distances in water and air will be as* 2 *to* 1. *But void offers a zero resistance,*

214 b Ὅτι δ' οὐκ ἔστι κενὸν οὕτω κεχωρισμένον ὡς ἔνιοί φασι, λέγωμεν πάλιν.

Εἰ γάρ ἐστιν ἑκάστου φορά τις τῶν ἁπλῶν σωμάτων φύσει (οἷον τῷ πυρὶ μὲν ἄνω, τῇ δὲ γῇ κάτω καὶ πρὸς τὸ μέσον), δῆλον ὅτι οὐκ ἂν τὸ κενὸν αἴτιον εἴη τῆς φορᾶς. τίνος οὖν αἴτιον ἔσται τὸ κενόν; δοκεῖ γὰρ αἴτιον εἶναι κινήσεως τῆς κατὰ τόπον, ταύτης δ' οὐκ ἔστιν.

Ἔτι εἰ ἔστι τι οἷον τόπος ἐστερημένος σώματος, ὅταν ᾖ κενόν, ποῦ οἰσθήσεται τὸ εἰστεθὲν εἰς αὐτὸ σῶμα; οὐ γὰρ δὴ εἰς ἅπαν. ὁ δ' αὐτὸς λόγος καὶ πρὸς τοὺς τὸν τόπον οἰομένους εἶναί τι κεχωρισμένον, εἰς ὃν φέρεται· πῶς γὰρ οἰσθήσεται τὸ ἐντεθὲν ἢ μενεῖ; καὶ περὶ τοῦ ἄνω καὶ κάτω καὶ περὶ τοῦ κενοῦ ὁ αὐτὸς ἁρμόσει λόγος εἰκότως·

[a] 'Return,' because the treatment of *topos* has already covered from one point of view what will now be re-investigated from another, namely in relation to motion.

[b] When you suppose the *kenon* to be emptiness within a *topos*, or the *topos* itself to be abstract dimensionality, this is to say abstract space; in either case it can exercise no influence on the entrant body to determine its motion or rest. We must remember that our present hypothesis is that bodies have no inherent trend of their own; but if they had, the direction in which their trend would manifest itself would depend at any moment on the relative gravity or levity of the immediately adjacent material surrounding—a function which vacancy could not perform.

ARGUMENT (*continued*)

and should be traversed in no time at all (a 24–b 22). *If it took any time, it can be shown that it would travel through a plenum and a void in the same time; which is absurd* (b 22–216 a 11). (2) *In a void all bodies, however light or heavy, would move at the same rate; which is impossible* (a 11–23).

A void occupied *by bodies is superfluous, if we conceive the bulk of a body as distinct from its qualities* (a 23–b 21). —*C.*]

Let us return [a] to the demonstration that vacuum with the independent existence that some assign to it does not really exist.

For if each of the elementary bodies has a natural trend of its own—fire upwards, and earth downwards and towards the centre—it is clear that vacancy cannot be the cause of these trends. Of what kind of motion-change, then, can it be the cause? For it is just of this local movement that it is supposed to be the cause and that it is not.

Again, if, when there is vacancy, there is something that might be called a place bereft of any material content, whither shall the body placed in it move? Clearly not anywhere-or-everywhere. And the same argument applies to those who hold that 'place' is itself something that exists apart from matter, into which the travelling body is carried. For how can the body placed in it either move or stay still? [b] And it is nothing astonishing that the same argument should occur positively with respect to the up-and-downness of place as we conceive it (which makes vacancy superfluous as the condition of movement of the elements) and with reference to the supposed nature of vacancy

214 b τὸ γὰρ κενὸν τόπον ποιοῦσιν οἱ εἶναι φάσκοντες. καὶ πῶς δὴ ἐνέσται ἢ ἐν τῷ τόπῳ ἢ ἐν τῷ κενῷ; οὐ γὰρ συμβαίνει, ὅταν ὅλον τεθῇ ὡς ἐν κεχωρισμένῳ τόπῳ καὶ ὑπομένοντι σῶμά τι[1]· τὸ γὰρ μέρος ἂν μὴ χωρὶς τιθῆται, οὐκ ἔσται ἐν τόπῳ ἀλλ' ἐν τῷ ὅλῳ. ἔτι εἰ μὴ τόπος, οὐδὲ κενὸν ἔσται.

Συμβαίνει δὲ τοῖς λέγουσιν εἶναι κενὸν ὡς ἀναγκαῖον, εἴπερ ἔσται κίνησις, τοὐναντίον μᾶλλον, ἂν

[1] [σώματι codd. contradicts the assumption above, εἰ ἔστι τι οἷον τόπος ἐστερημένος σώματος. σῶμά τι (as read in the Tauchnitz ed. 1881) can be supported by Simplic. 665. 32 οὐ γὰρ συμβαίνει, τουτέστιν οὐ κατὰ λόγον ἀπαντᾷ, ἂν ᾖ ὁ τόπος ὑπομένουσά τις φύσις τριχῇ διαστατὴ καὶ ἐν αὐτῷ ὑποτεθῇ σῶμά τι . . . 35 ὅταν ὅλον τι σῶμα τεθῇ ἐν τόπῳ, Philop. 636. 11 οὐ συμβαίνει τοῦτο, ὅταν σῶμά τι ἐν κεχωρισμένῳ καὶ διαστηματικῷ τόπῳ ἐντεθῇ.—C.]

---

[a] We have shown that the vacuum cannot be the cause of the up-and-down motion natural to the several elements, because that is already determined by their relation to the supreme cosmic site and by the nature of the several media in which they may find themselves, causing them to rise through media less buoyant than themselves and to sink through media less dense than themselves but to rest in those of the same gravity as themselves. We are now to prove further that, even if the motion of the natural bodies were not already accounted for without recourse to the idea of vacuity, it would still be vain to introduce vacuity as the cause of these movements. For the movements are differentiated from each other and vacancy as such is undifferentiated, and if this is nothing but the same argument taken from the negative side, this is nothing surprising. For those we are arguing against, by defining the *kenon* as a kind of *topos* and *topos* itself as bare dimensionality, reduce *topos* and *kenon* to identity.

[b] For in a vacuum, supposing the body in question to be 'in' it as determined or located by it, any portion of that

itself (which precludes all possibility of its being such a cause in any case); for our adversaries reduce both place and void to mere abstract dimensionality.[a] And indeed how can anything be *in* either 'place' so conceived or void, in the proper local sense of 'in'?[b] But when (as with us) the place of a body is external to itself and has material existence of its own as its contents come and go, this is not what happens but the parts of it exist in it as parts exist in the whole, not as the content in the continent.[c] Finally, in whatever sense the void is identified with 'place,' when 'place' in that sense is shown not to exist the void vanishes with it.[d]

But indeed on reflection we see that the assertion 'motion is impossible except into vacancy' is not only untrue but the contradiction of the truth, for

body would equally be determined by the special space assigned to it and occupied by it.

[c] This passage is lightly passed over by Themistius as identical in purport with 211 b 19 ff. which has already been dealt with. Philoponus further remarks on the relative lucidity of this former passage, contrasted with its obscure and truncated form here. [Simplicius, 665. 26, says that *οὐ γὰρ συμβαίνει . . . ἐν τῷ ὅλῳ* is added in certain copies after *καὶ πῶς δὴ ἐνέσται ἐν τόπῳ* (*sic*) *ἢ ἐν τῷ κενῷ*; which would be plain enough without the addition.—C.] I take it to be an interpolation. In any case it is difficult to get any sense out of it and it has no independent value. [Simplicius explains: When a whole body is put in a place, it is held that the whole occupies the place, whereas the parts, if continuous, are *in* that whole which is in a place, but not in a place on their own account unless they are separated from the whole and their continuity with respect to it is dissolved. The resulting absurdities are set forth at 211 b 19.—C.]

[d] The boldness of the explanation and general interpretation of this section is supported by all the old commentators, Greek, Arabic, and mediaeval Latin. Averroes is more than usually helpful here.

214 b τις ἐπισκοπῇ, μὴ ἐνδέχεσθαι μηδὲ ἓν κινεῖσθαι, ἐὰν ᾖ κενόν. ὥσπερ γὰρ οἱ διὰ τὸ ὅμοιον φάμενοι τὴν γῆν ἠρεμεῖν, οὕτως καὶ ἐν τῷ κενῷ ἀνάγκη ἠρεμεῖν· οὐ γὰρ ἔστιν οὗ μᾶλλον ἢ ἧττον κινηθή-
215 a σεται· ᾗ γὰρ κενόν, οὐκ ἔχει διαφοράν.

Πρῶτον μὲν οὖν ὅτι πᾶσα κίνησις ἢ βίᾳ ἢ κατὰ φύσιν· ἀνάγκη δ', ἄνπερ ᾖ βίαιος, εἶναι καὶ τὴν κατὰ φύσιν (ἡ μὲν γὰρ βίαιος παρὰ φύσιν ἐστίν, ἡ δὲ παρὰ φύσιν ὑστέρα τῆς κατὰ φύσιν)· ὥστ' εἰ μὴ κατὰ φύσιν ἐστὶν ἑκάστῳ τῶν φυσικῶν σωμάτων κίνησις, οὐδὲ τῶν ἄλλων ἔσται κινήσεων οὐδεμία. ἀλλὰ μὴν φύσει γε πῶς ἔσται μηδεμιᾶς οὔσης διαφορᾶς κατὰ τὸ κενὸν καὶ τὸ ἄπειρον; ᾗ μὲν γὰρ ἄπειρον, οὐδὲν ἔσται ἄνω οὐδὲ κάτω οὐδὲ μέσον, ᾗ δὲ κενόν, οὐδὲν διαφέρει τὸ ἄνω τοῦ κάτω (ὥσπερ γὰρ τοῦ μηδενὸς οὐδεμία ἔστι διαφορά, οὕτως καὶ τοῦ μὴ ὄντος· τὸ δὲ κενὸν μὴ ὄν τι καὶ στέρησις δοκεῖ εἶναι)· ἡ δὲ φύσει φορὰ διάφορος, ὥστ' ἔσται τὰ φύσει διάφορα. ἢ οὖν οὐκ ἔστι φύσει οὐδαμοῦ οὐδενὶ φορά, ἢ εἰ τοῦτ' ἔστιν, οὐκ ἔστι κενόν.

Ἔτι νῦν μὲν κινεῖται τὰ ῥιπτούμενα τοῦ ὤσαντος οὐχ ἁπτομένου, ἢ δι' ἀντιπερίστασιν (ὥσπερ ἔνιοί

---

[a] [Anaximander first declared that the earth was freely suspended and was not held in place by anything, but stayed where it was διὰ τὴν ὁμοίαν πάντων ἀπόστασιν, Hippol. *Ref.* i. 6. 3, Aristot. *De caelo* 295 b 11 διὰ τὴν ὁμοιότητα μένειν, Plato, *Phaedo* 109 A.—C.]

[b] Plato's "matter" is not conceived by him as equivalent to "nothing," but at the same time it is "not anything"

it is impossible for there to be any motion at all if there be a circumambient void. Just as it has been asserted that the earth is at rest by the law of symmetry,[a] so must it be in vacancy, since no preference can be given to one line of motion more than to another, inasmuch as the void, as such, is incapable of differentiation.

To begin with, every motion must be either natural or forced, and there can be no such thing as forced movement if there is no natural movement (for forced movement is movement counter to that which is natural, and the unnatural presupposes the natural); so that, unless every natural body has a natural movement, there cannot be any other movement at all. But how can there be any natural movement in the undifferentiated limitless void? For *qua* limitless it can have no top or bottom or middle, and *qua* vacancy it can have no differentiated directions of up and down (since the non-existent can no more be differentiated than 'nothing' can, and the void is conceived as not being a thing, but as mere shortage[b]); whereas natural trends are differentiated and differentiate the substances that manifest them. So it follows either that nothing has any natural trend in any direction or (since this is not so) that the supposed vacancy-as-conditioning-movement is a fiction.

In the second place, projectiles move when the body that impelled them is no longer in contact with them, whether this be due (as some suppose)

till it is differentiated by "form." [Or, 'the void is regarded as a sort of "nothing"—a *no-thing* in the privative sense.' The Atomists called the void 'nothing,' meaning not that it was *non-existent* (μὴ ὂν ἁπλῶς), but that it was *not any thing* (μὴ ὄν τι)—the only 'things' being solid atoms.—C.]

215 a φασιν) ἢ διὰ τὸ ὠθεῖν τὸν ὠσθέντα ἀέρα θάττω κίνησιν τῆς τοῦ ὠσθέντος φορᾶς ἣν φέρεται εἰς τὸν οἰκεῖον τόπον. ἐν δὲ τῷ κενῷ οὐδὲν τούτων ἐνδέχεται ὑπάρχειν, οὐδ' ἔσται φέρεσθαι ἀλλ' ἢ ὡς τὸ ὀχούμενον. ἔτι οὐδεὶς ἂν ἔχοι εἰπεῖν διὰ τί κινηθὲν στήσεταί που· τί γὰρ μᾶλλον ἐνταῦθα ἢ ἐνταῦθα; ὥστ' ἢ ἠρεμήσει ἢ εἰς ἄπειρον ἀνάγκη φέρεσθαι, ἐὰν μή τι ἐμποδίσῃ κρεῖττον.

Ἔτι νῦν μὲν εἰς τὸ κενὸν διὰ τὸ ὑπείκειν φέρεσθαι δοκεῖ· ἐν δὲ τῷ κενῷ πάντῃ ὁμοίως τὸ τοιοῦτον, ὥστε πάντῃ οἰσθήσεται.

Ἔτι δὲ καὶ ἐκ τῶνδε φανερὸν τὸ λεγόμενον. ὁρῶμεν γὰρ τὸ αὐτὸ βάρος καὶ σῶμα θᾶττον φερόμενον διὰ δύο αἰτίας, ἢ τῷ διαφέρειν τὸ δι' οὗ (οἷον δι' ὕδατος ἢ γῆς ἢ ἀέρος) ἢ τῷ διαφέρειν τὸ φερόμενον, ἐὰν τἆλλα ταὐτὰ ὑπάρχῃ, διὰ τὴν ὑπεροχὴν τοῦ βάρους ἢ τῆς κουφότητος.

Τὸ μὲν οὖν δι' οὗ φέρεται αἴτιον ὅτι ἐμποδίζει, μάλιστα μὲν ἀντιφερόμενον, ἔπειτα καὶ μένον· μᾶλλον δὲ τὸ μὴ εὐδιαίρετον· τοιοῦτο δὲ τὸ παχύ-
215 b τερον. τὸ δὴ ἐφ' οὗ Α οἰσθήσεται διὰ τοῦ Β τὸν ἐφ' ᾧ Γ χρόνον, διὰ δὲ τοῦ Δ λεπτομεροῦς ὄντος τὸν ἐφ' ᾧ Ε, εἰ ἴσον τὸ μῆκος τὸ τοῦ Β τῷ Δ,

---

[a] [ἀντιπερίστασις is defined by Simplicius (on 267 a 12 ff.) as the interchange of places, when a body A thrusts a body B out of its place and occupies that place, and B similarly displaces C, and so on until the last displaced body occupies the place of A.—C.]

[b] Here we should suppose Aristotle to have in his mind, not the "vast inane" in general, but either a vacant space *across* which, or unlimited space *into* which, the projectile is launched. For the impact itself must have a direction relatively to the movement of the impelling body and the

to a circulating thrust,[a] or to the air being set by the original impact in more rapid motion than that of the natural movement of the missile towards the place proper to it. But in vacancy neither of these agencies would be in operation, so that nothing could go on moving unless it were carried. Nor (if it did move) could a reason be assigned why the projectile should ever stop—for why here more than there? It must therefore either not move at all, or continue its movement without limit, unless some stronger force impedes it.

Yet again, we are told that things go *into* vacant spaces because they offer no resistance; but *in* vacancy there is no resistance in any direction, so that, as to direction of movement, it would be a case of one-as-good-as-another.[b]

A further proof of our thesis is this: We see that the velocity of a moving weight or mass depends on two conditions: (1) the distinctive nature of the medium—water, earth, or air—through which the motion occurs, and (2) the comparative gravity or levity of the moving body itself, other conditions being equal.

(1) Now the medium reduces velocity all the more if it is itself moving in the opposite direction, but it also reduces it, though in a lesser degree, if it is quiescent; and this impedium of motion is proportional to the resistance the medium offers to cleavage, which is to say its density. Thus, if one medium is easier to cleave than another, the time taken in

centre of gravity of the body impinged upon. A frame of reference of some sort, therefore, is implicit in the hypothesis of any impact and projection at all. But it does not look as if Aristotle had thought this out.

215 b κατὰ τὴν ἀναλογίαν τοῦ ἐμποδίζοντος σώματος. ἔστω γὰρ τὸ μὲν Β ὕδωρ, τὸ δὲ Δ ἀήρ· ὅσῳ δὴ λεπτότερον ἀὴρ ὕδατος καὶ ἀσωματώτερον, τοσούτῳ θᾶττον τὸ Α διὰ τοῦ Δ οἰσθήσεται ἢ διὰ τοῦ Β. ἐχέτω δὴ τὸν αὐτὸν λόγον ὅνπερ διέστηκεν ἀὴρ πρὸς ὕδωρ, τὸ τάχος πρὸς τὸ τάχος. ὥστ' εἰ διπλασίως λεπτόν, ἐν διπλασίῳ χρόνῳ τὴν τὸ Β δίεισιν ἢ τὴν τὸ Δ, καὶ ἔσται ὁ ἐφ' ᾧ Γ χρόνος διπλάσιος τοῦ ἐφ' ᾧ Ε. καὶ ἀεὶ δὴ ὅσῳ ἂν ᾖ ἀσωματώτερον καὶ ἧττον ἐμποδιστικὸν καὶ εὐδιαιρετώτερον δι' οὗ φέρεται, θᾶττον οἰσθήσεται.

Τὸ δὲ κενὸν οὐδένα ἔχει λόγον ᾧ ὑπερέχεται ὑπὸ τοῦ σώματος, ὥσπερ οὐδὲ τὸ μηδὲν πρὸς ἀριθμόν. εἰ γὰρ τὰ τέτταρα τῶν τριῶν ὑπερέχει ἑνί, πλείονι δὲ τοῖν δυοῖν, καὶ ἔτι πλείονι τοῦ ἑνὸς ἢ τοῖν δυοῖν, τοῦ δὲ[1] μηδενὸς οὐκέτι ἔχει λόγον ᾧ ὑπερέχει· ἀνάγκη γὰρ τὸ ὑπερέχον διαιρεῖσθαι εἴς τε τὴν ὑπεροχὴν καὶ τὸ ὑπερεχόμενον, ὥστε ἔσται τὰ τέτταρα ὅσῳ τε ὑπερέχει καὶ οὐδέν (διὸ οὐδὲ γραμμὴ στιγμῆς ὑπερέχει, εἰ μὴ σύγκειται ἐκ στιγμῶν). ὁμοίως δὲ καὶ τὸ κενὸν πρὸς τὸ πλῆρες

[1] [δὲ om. I. It is doubtful where the apodosis begins; possibly at 19 ὁμοίως δὲ (δὴ ?), 16 ἀνάγκη . . . 19 ἐκ στιγμῶν being a parenthesis.—C.]

[a] Here and elsewhere I follow the usual custom of simplifying the cumbrous mathematical methods often employed by Greek writers. The paraphrase, I trust, accurately represents the reasoning of the original.

[b] And however often you 'don't take any,' you will never have taken all, whereas if you take some (however little) often enough, you will eventually have taken all. [17 ὥστε ἔσται, κτλ. literally, 'consequently four will have to consist of the amount by which it exceeds (zero, namely 4) and nothing' (but it is absurd to speak of 'nothing' as a constituent of four).—C.]

travelling a given distance through it will be proportionately less.[a] *E.g.*, if the media are water and air, the ratio of air's cleavableness and unsubstantiality to that of water will give the ratio of the velocity of the passage through air to that through water.

Velocity in air : velocity in water =
Cleavability of air : cleavability of water.

So, if air is twice as easy to cleave as water, the passage through air will be twice as swift, and the time taken in covering a given distance half as long in air as in water. According to this universal principle, then, the velocity will in every case be greater in proportion to the unsubstantiality and diminished power of impeding and easier cleavability of the medium.

But the nonexistent substantiality of vacuity cannot bear any ratio whatever to the substantiality of any material substance, any more than zero can bear a ratio to a number. For if we divide a constant quantity c (that which exceeds) into two variable parts, a (the excess) and b (the exceeded), then, as a increases, b will decrease and the ratio $a : b$ will increase; but when the whole of c is in section a there will be none of c for section b; and it is absurd to speak of 'none of c' as 'a part of c.' So the ratio $a : b$ will cease to exist, because b has ceased to exist and only a is left, and there is no proportion between something and nothing.[b] (And in the same way there is no such thing as the proportion between a line and a point, because, since a point is no part of a line, taking a point is not taking any of the line.) And in like manner there is not any ratio that can

215 b οὐδένα οἷόν τε ἔχειν λόγον· ὥστ' οὐδὲ τὴν κίνησιν, ἀλλ' εἰ διὰ τοῦ λεπτοτάτου ἐν τοσῳδὶ τὴν τοσήνδε φέρεται, διὰ τοῦ κενοῦ παντὸς ὑπερβάλλει λόγου.

Ἔστω γὰρ τὸ Ζ κενόν, ἴσον δὲ τῷ μεγέθει τοῖς Β καὶ Δ. τὸ δὴ Α εἰ δίεισι καὶ κινηθήσεται ἐν τινὶ μὲν χρόνῳ, τῷ ἐφ' οὗ Η, ἐν ἐλάττονι δὲ ἢ τῷ ἐφ' οὗ Ε, τοῦτον ἕξει τὸν λόγον τὸ κενὸν πρὸς τὸ πλῆρες. ἀλλ' ἐν τοσούτῳ χρόνῳ ὅσος ἐφ' οὗ τὸ Η, τοῦ Δ τὸ Α δίεισι τὴν τὸ Θ. δίεισι δέ γε κἂν ᾖ τι λεπτότητι διαφέρον τοῦ ἀέρος ἐφ' ᾧ τὸ Ζ, ταύτην τὴν ἀναλογίαν ἣν ἔχει ὁ χρόνος ἐφ' ᾧ Ε πρὸς τὸν ἐφ' ᾧ Η. ἂν γὰρ ᾖ τοσούτῳ λεπτότερον τὸ ἐφ' ᾧ Ζ σῶμα τοῦ Δ, ὅσῳ ὑπερέχει τὸ Ε τοῦ Η, ἀντεστραμ-
216 a μένως δίεισι τῷ τάχει ἐν τῷ τοσούτῳ, ὅσον τὸ Η, τὴν τὸ Ζ τὸ ἐφ' οὗ Α, ἐὰν φέρηται. ἐὰν τοίνυν μηδὲν ᾖ σῶμα ἐν τῷ Ζ, ἔτι θᾶττον. ἀλλ' ἦν ἐν τῷ Η. ὥστ' ἐν ἴσῳ χρόνῳ δίεισι πλῆρές τε ὂν καὶ κενόν. ἀλλ' ἀδύνατον. φανερὸν τοίνυν ὅτι, εἰ ἔστι τις χρόνος ἐν ᾧ τοῦ κενοῦ ὁτιοῦν οἰσθήσεται, συμβήσεται τοῦτο τὸ ἀδύνατον· ἐν ἴσῳ γὰρ ληφθήσεται πλῆρές τε ὂν[1] διεξιέναι τι καὶ κενόν· ἔσται γάρ τι ἀνάλογον σῶμα ἕτερον πρὸς ἕτερον ὡς χρόνος πρὸς χρόνον.

[1] [ὂν om. E, Simplic. 675. 15.—C.]

express the excess of the resistance of a plenum over that of a vacuum, or of the consequential velocity in the vacuum over that in the plenum ; but if movement through the finest medium covers a given distance in a given time, movement through the void is out of all proportion.

For if we take D_1 and D_2 to represent the densities of any two media, and T_1 and T_2 to represent the length of time occupied by a body under otherwise similar conditions in passing through equal spaces in the respective media, then if there be any proportion that expresses the ratio of the velocity through vacancy, let it be that which covers the constant distance in T_3, less than T_2 but standing to it in a certain ratio. Now take a medium with a density D_3 that stands to D_2 in the inverse proportion of T_3 to T_2

$$D_3 : D_2 :: T_2 : T_3$$

T_3 will represent the time taken to traverse the constant distance through the medium of D_3. But the time occupied if there were no medium at all should be less than T_3, whereas T_3 was taken as that time itself. Thus a body would pass through a plenum and a vacuum with equal velocity. Which is impossible. Evidently, then, if any period of time be assigned for the passage of any selected distance you may choose through vacancy, we shall be landed in this impossible consequence ; for the assumption will involve the passage of a body through a material medium and through vacancy at the same velocity, since one may always take a medium whose density bears such a proportion to that of a standard medium as will correspond to the proportion of the supposed transit through vacancy to that of the standard transit.

216 a Ὡς δ' ἐν κεφαλαίῳ εἰπεῖν, δῆλον τὸ τοῦ συμβαίνοντος αἴτιον, ὅτι κινήσεως μὲν πρὸς κίνησιν πάσης ἔστι λόγος (ἐν χρόνῳ γάρ ἐστι, χρόνου δὲ παντὸς ἔστι πρὸς χρόνον, πεπερασμένων ἀμφοῖν), κενοῦ δὲ πρὸς πλῆρες οὐκ ἔστιν.

Ἧι μὲν οὖν διαφέρουσι δι' ὧν φέρονται, ταῦτα συμβαίνει, κατὰ δὲ τὴν τῶν φερομένων ὑπεροχὴν τάδε· ὁρῶμεν γὰρ τὰ μείζω ῥοπὴν ἔχοντα ἢ βάρους ἢ κουφότητος (ἐὰν τἆλλα ὁμοίως ἔχῃ τοῖς σχήμασι) θᾶττον φερόμενα τὸ ἴσον χωρίον καὶ κατὰ λόγον ὃν ἔχουσι τὰ μεγέθη πρὸς ἄλληλα. ὥστε καὶ διὰ τοῦ κενοῦ. ἀλλ' ἀδύνατον· διὰ τίνα γὰρ αἰτίαν οἰσθήσεται θᾶττον; ἐν μὲν γὰρ τοῖς πλήρεσιν ἐξ ἀνάγκης· θᾶττον γὰρ διαιρεῖ τῇ ἰσχύι τὸ μεῖζον· ἢ γὰρ σχήματι διαιρεῖ, ἢ ῥοπῇ ἣν ἔχει τὸ φερόμενον ἢ τὸ ἀφεθέν. ἰσοταχῆ ἄρα πάντ' ἔσται. ἀλλ' ἀδύνατον.

Ὅτι μὲν οὖν εἰ ἔστι κενόν, συμβαίνει τοὐναντίον ἢ δι' ὃ κατασκευάζουσιν οἱ φάσκοντες εἶναι κενόν, φανερὸν ἐκ τῶν εἰρημένων.

---

[a] The word ῥοπή is difficult to translate without giving Aristotle credit for a closer approach to the conceptions of modern science than would be justified ; for ' virtual velocity ' is the term we should use to express the thrust or pressure by which the falling stone makes its way downwards through the water or a rising bubble upwards through it, and Aristotle had no such precise conception of the phenomenon as would be implied in these words. He uses ῥοπή pretty much as a term applicable alike to the downward trend of a body heavier than its medium and to the upward trend of one lighter. Aquinas seems to me entirely correct in explaining that this greater ῥοπή may be due either to the greater size of the one body if they are of the same texture or to its greater gravity or levity if they are of the same size. On the radical fallacies of this whole section consult page lxiv.

[b] It is tantalizing to find Aristotle actually arriving at the fact, familiar in modern laboratories, that a feather and a

To sum up. The principle that leads to our conclusion is that there must always be a ratio between any two velocities (for they are both measured by time, and between two determinate periods of time there must be ratio); but there is no such ratio of densities between vacancy and any kind of fullness.

This is how velocities are affected by the media through which the movement occurs. But (2) as to differences that depend on the moving bodies themselves, we see that of two bodies of similar formation the one that has the stronger trend [a] downward by weight or upward by buoyancy, as the case may be, will be carried more quickly than the other through a given space in proportion to the greater strength of this trend. And this should hold in vacancy as elsewhere. But it cannot; for what reason can be assigned for this greater velocity? If the passage is through a medium, there must be such a difference; for when there is anything there to cleave, the body superior in force of its thrust will necessarily cleave the medium faster, since either its more suitable shape or the natural thrust it exercises, whether following its natural movement or being thrown, makes it cleave the better. Where there is nothing to cleave, therefore, all bodies will move at the same velocity; which is impossible.[b]

It is obvious, then, that the hypothesis of 'the void' leads to conclusions the precise opposite of those in support of which it was invented by its advocates.

guinea, to take the classical example, will fall at the same pace through a vacuum, but treating it as a *reductio ad absurdum*. [This truth was divined, without experiment, by Epicurus, *Ep*. i. 61 ἰσοταχεῖς ἀναγκαῖον τὰς ἀτόμους εἶναι, ὅταν διὰ τοῦ κενοῦ εἰσφέρωνται μηθενὸς ἀντικόπτοντος. *Cf*. Lucr. ii. 238.—C.]

216 a Οἱ μὲν οὖν οἴονται τὸ κενὸν εἶναι, εἴπερ ἔσται ἡ κατὰ τόπον κίνησις, ἀποκεκριμένον καθ' αὑτό· τοῦτο δὲ ταὐτόν ἐστι τῷ τὸν τόπον φάναι εἶναί τι κεχωρισμένον· τοῦτο δ' ὅτι ἀδύνατον, εἴρηται πρότερον. καὶ καθ' αὑτὸ δὲ σκοποῦσι φανείη ἂν τὸ λεγόμενον κενὸν ὡς ἀληθῶς κενόν. ὥσπερ γὰρ ἐὰν ἐν ὕδατι τιθῇ τις κύβον, ἐκστήσεται τοσοῦτον ὕδωρ ὅσος ὁ κύβος, οὕτω καὶ ἐν ἀέρι· ἀλλὰ τῇ αἰσθήσει ἄδηλον. καὶ ἀεὶ δὴ ἐν παντὶ σώματι ἔχοντι μετάστασιν, ἐφ' ὃ πέφυκε μεθίστασθαι ἀνάγκη (ἂν μὴ συμπιλῆται) μεθίστασθαι, ἢ κάτω ἀεί, εἰ κάτω ἡ φορὰ ὥσπερ γῆς, ἢ ἄνω, εἰ πῦρ, ἢ ἐπ' ἄμφω, ἢ[1] ὁποῖον ἄν τι ᾖ τὸ ἐντιθέμενον. ἐν δὲ δὴ τῷ κενῷ τοῦτο μὲν ἀδύνατον—οὐδὲ γὰρ σῶμα—διὰ δὲ τοῦ κύβου τὸ ἴσον διάστημα διεληλυθέναι 216 b δόξειεν ⟨ἂν⟩[2] ὅπερ ἦν καὶ πρότερον ἐν τῷ κενῷ, ὥσπερ ἂν εἰ τὸ ὕδωρ μὴ μεθίστατο τῷ ξυλίνῳ κύβῳ μηδ' ὁ ἀὴρ ἀλλὰ πάντῃ διῄεσαν δι' αὐτοῦ. ἀλλὰ μὴν καὶ ὁ κύβος ἔχει τοσοῦτον μέγεθος ὅσον κατέχει τὸ κενόν· ὃ εἰ καὶ θερμὸν ἢ ψυχρόν ἐστιν ἢ βαρὺ ἢ κοῦφον, οὐδὲν ἧττον ἕτερον τῷ εἶναι

[1] [ἢ om. Prantl. Philop. 665. 1 records the reading ἢ ὁποῖον ἄν τι ῃ, but questions it, adding εἰ δὲ ἔστιν (ἡ γραφὴ) 'ὅποι ἂν ᾖ,' τουτέστιν κτλ.—C.]

[2] [⟨ἂν⟩ supplevi.—C.]

The hypothesis of the void, then, was introduced under the belief that the phenomenon of local movement involved the reality of self-existing vacancy, which is equivalent to the reality of ' space ' sejunct from material content or continent; and this we have already shown to be impossible. But a consideration of the question on its own merits will also demonstrate the inanity, in very truth, of this doctrine of the ' inane.' For just as a cube displaces its own bulk of water if immersed in it, so does it also if the medium be air, only that the displacement is not perceptible to the senses. So, whatever the yielding medium may be, it must (unless it condenses) yield in whatever direction is determined by the natural movement of the intrusive body, always making way underneath it if it be earth, above it if it be fire, or below it or above it according as the medium may be lighter or heavier than the moving body.[a] Now this yielding is impossible in vacuity, which is not a material entity at all, and one must suppose that the dimensionality already there in the place before it was occupied must interpenetrate the equal dimensionality of the intrusive cube when it enters ; just as if the water or air should not make way for the wooden cube but should permeate it all through. But then the cube itself has its own bulk equal to that of the vacancy that now permeates it, which bulk, whether it be hot or cold or heavy or buoyant, is nevertheless different

[a] If the medium be air, for example, it will make way for water on the water's under side (and well up over it), but to fire on the fire's upper side, etc.—one way or the other, therefore, as determined by the natural movement of the intrusive body.

216 b πάντων τῶν παθημάτων ἐστί, καὶ εἰ μὴ χωριστόν· λέγω δὲ τὸν ὄγκον τοῦ ξυλίνου κύβου. ὥστ' εἰ καὶ χωρισθείη τῶν ἄλλων πάντων καὶ μήτε βαρὺ μήτε κοῦφον εἴη, καθέξει τὸ ἴσον κενὸν καὶ ἐν τῷ αὐτῷ ἔσται τῷ τοῦ τόπου καὶ τῷ τοῦ κενοῦ μέρει ἴσῳ αὐτῷ. τί οὖν διοίσει τὸ τοῦ κύβου σῶμα τοῦ ἴσου κενοῦ καὶ τόπου; καὶ εἰ δύο τοιαῦτα, διὰ τί οὐ καὶ ὁποσαοῦν ἐν τῷ αὐτῷ ἔσται; ἓν μὲν δὴ τοῦτο ἄτοπον καὶ ἀδύνατον.

Ἔτι δὲ φανερὸν ὅτι τοῦτο ὁ κύβος ἕξει καὶ μεθιστάμενος, ὃ καὶ τὰ ἄλλα σώματα πάντ' ἔχει. ὥστ' εἰ τοῦ τόπου μηδὲν διαφέρει, τί δεῖ ποιεῖν τόπον τοῖς σώμασι παρὰ τὸν ἑκάστου ὄγκον, εἰ ἀπαθὲς ὁ ὄγκος; οὐδὲν γὰρ συμβάλλεται, εἰ ἕτερον περὶ αὐτὸν ἴσον διάστημα τοιοῦτον εἴη.

[Ἔτι δεῖ δῆλον εἶναι οἷον κενὸν ἐν τοῖς κινουμένοις. νῦν δ' οὐδαμοῦ ἐντὸς τοῦ κόσμου· ὁ γὰρ ἀὴρ ἔστι τι, οὐ δοκεῖ δέ γε. οὐδὲ τὸ ὕδωρ, εἰ ἦσαν οἱ ἰχθύες σιδηροῖ· τῇ ἀφῇ γὰρ ἡ κρίσις τοῦ ἁπτοῦ.]

Ὅτι μὲν τοίνυν οὐκ ἔστι κεχωρισμένον κενόν, ἐκ τούτων ἐστὶ δῆλον.

---

[a] [On this 'bulk' as the 'matter of growth and diminution,' logically (though not actually) distinguishable from the 'matter of qualitative change,' see *De gen. et corr.* 320 b 16 ff. and Joachim *ad loc.*—C.]

[b] The old commentators add the corollary that, if the dimensions of a body *qua* dimensions require a place to be in or to go into, so must the dimensions of a *vacuum*, and so on *ad infinitum*.

[c] The bracketed passage (which was unknown to the Greek commentators) is evidently and admittedly spurious. What it means is that, if fishes were made of iron, they would not be able to feel the water and would be as apt to equate

in its nature from all these qualities, even though it may not be separable from them—I mean the bulk, as such, of the wooden cube itself.[a] So that, even if this bulk could be isolated from all these other qualities and be neither heavy nor buoyant, it would still embrace an equal measure of vacancy, and would coincide with a portion of 'space' and 'vacuity' equal to itself. How, then, would the material bulk of the cube differ from an equal portion of vacancy or space? And if there could be two such coincident entities, why not any number you choose to name? So this is one paradoxical and absurd implication of the doctrine of the void.

But further the cube in question, if it should shift, will still have its own bulk, just as every other body has, as such. And if this differs in no respect from 'place,' why imagine a place for a body to go into in addition to its own bulk, if that bulk as such has no physical properties? For if another equal dimensionality were to permeate it, it would make no difference.[b]

[Further, vacancy (if it exist) should manifest its distinctive character amongst the transmutable substances that constitute the realm of nature. But in fact it is nowhere to be discovered in the cosmos. For air is substantial, though it does not appear so, any more than water would to fishes if they were made of iron; for touch is the test of existence for the tangible.] [c]

From all this it is clear that there is no such thing as a self-existing void.

water with nothing as we are to equate air with it. Prantl appositely refers us to the myth in Plato's *Phaedo*, 109 c.

CHAPTER IX

ARGUMENT

[Having disposed of the argument that motion requires a void 'separate from bodies,' in and through which bodies may move, Aristotle turns to the supposed necessity of empty interstices inside *bodies, into which they may contract. This internal void had been invoked to account for differences of specific density, of lightness or heaviness, and of hardness or softness. It was argued that without it no compression was possible, and hence either* (1) *there could be no movement at all, or* (2) *movement must cause the whole material universe to bulge, since nothing can contract to make room for it, or* (3) *whenever a given amount of air passes into water (with a reduction of volume), a corresponding amount of water must simultaneously expand, somewhere else, into air* (216 b 22–30).

The alleged internal void, accounting for tenuity and contraction, might be conceived in two ways :

(*a*) *As sporadic cavities into which particles of the contracting body can draw together. But our arguments against self-determined 'places' exclude the existence of such 'voids'* (b 30–33).

216 b Εἰσὶ δέ τινες οἳ διὰ τοῦ μανοῦ καὶ πυκνοῦ οἴονται φανερὸν εἶναι ὅτι ἔστι κενόν. εἰ μὲν γὰρ μὴ ἔστι μανὸν καὶ πυκνόν, οὐδὲ συνιέναι καὶ πιλεῖσθαι οἷόν τε· εἰ δὲ τοῦτο μὴ εἴη, ἢ ὅλως κίνησις οὐκ ἔσται, ἢ κυμανεῖ τὸ ὅλον (ὥσπερ ἔφη Ξοῦθος), ἢ εἰς ἴσον ἀεὶ[1] μεταβάλλειν ἀέρα καὶ

[1] [ἀεὶ <δεῖ> Bonitz, Prantl.—C.]

[a] [A Pythagorean, Diels, *Vors.* 23.—C.]

CHAPTER IX

ARGUMENT (*continued*)

(b) As somehow diffused throughout the tenuous substance. In this case its function is changed: it is no longer a vacancy in which *movement takes place, but could only account for the rising upward of buoyant things. But this is open to the objection that void cannot move or have a place* (b 33–217 a 10).

Since, however, not all movement is re-entrant, contraction and expansion must be possible, or one of the three alternatives above stated cannot be escaped (a 10–20).

The true explanation is that, just as the same matter passes from hot to cold without any accession or loss, so the same matter serves for a large and for a small body without accession or loss, without any need of empty interstices (a 20–b 11).

The internal void has been invoked to explain both lightness and soft texture. But, if both qualities were due to more void in the body, they ought always to go together, whereas in fact iron is lighter than lead but harder (b 11–20).

Conclusion: the void does not exist, either as separate from bodies or as involved in tenuous substances (b 20–28).—C.]

But other thinkers hold that it is tenuity and density that prove the existence of the void. For they can only be caused by things drawing together and being compressed, which involves the existence of vacant spaces through which their particles can approach or recede from each other. And if there were no such drawing of things together to make room, then either (1) there would be no movement at all, or (2) the universe would (to use Xuthus's [a] expression) be thrust up into a wave, or (3) a balance must always be preserved, for instance, between the transformation of water into air and air into water.

216 b ὕδωρ. λέγω δ' οἷον εἰ ἐξ ὕδατος κυάθου γέγονεν ἀήρ, ἅμα ἐξ ἴσου ἀέρος ὕδωρ τοσοῦτον γεγενῆσθαι, ἢ κενὸν εἶναι ἐξ ἀνάγκης· συμπιλεῖσθαι γὰρ καὶ συνεπεκτείνεσθαι οὐκ ἐνδέχεται ἄλλως.

Εἰ μὲν οὖν τὸ μανὸν λέγουσι τὸ πολλὰ κενὰ κεχωρισμένα ἔχον, φανερὸν ὡς, εἰ μηδὲ κενὸν ἐνδέχεται εἶναι χωριστὸν ὥσπερ μηδὲ τόπον ἔχοντα διάστημα αὑτοῦ, οὐδὲ μανὸν οὕτως.

Εἰ δὲ μὴ χωριστόν, ἀλλ' ὅμως ἐνεῖναί τι κενόν, ἧττον μὲν ἀδύνατον, συμβαίνει δὲ πρῶτον μὲν οὐ πάσης κινήσεως αἴτιον τὸ κενὸν ἀλλὰ τῆς ἄνω
217 a (τὸ γὰρ μανὸν κοῦφον, διὸ καὶ τὸ πῦρ μανὸν εἶναί φασιν)· ἔπειτα κινήσεως αἴτιον οὐχ οὕτω τὸ κενὸν ὡς ἐν ᾧ, ἀλλ' ὥσπερ οἱ ἀσκοὶ τῷ φέρεσθαι αὐτοὶ ἄνω φέρουσι τὸ συνεχές, οὕτω τὸ κενὸν ἀνωφερές. καίτοι πῶς οἷόν τε φορὰν εἶναι κενοῦ ἢ τόπον κενοῦ; κενοῦ γὰρ γίγνεται κενόν, εἰς ὃ φέρεται. ἔτι δὲ πῶς ἐπὶ τοῦ βαρέος ἀποδώσουσι τὸ φέρεσθαι κάτω; καὶ δῆλον ὅτι εἰ ὅσῳ ἂν μανότερον καὶ κενώτερον ᾖ ἄνω οἰσθήσεται, εἰ ὅλως εἴη κενόν, τάχιστ' ἂν φέροιτο. ἴσως δὲ καὶ τοῦτ' ἀδύνατον κινηθῆναι· λόγος δ' ὁ αὐτός, ὥσπερ ὅτι ἐν τῷ

[a] [By ‘self-determined (or ‘separate,’ *cf.* ἀποκεκριμένον, 217 b 20) voids’ Philoponus understands empty interstices occurring here and there in bodies (like pores in a sponge), and large enough for other bodies to move into them; whereas the alternative (*b*) in the next paragraph he takes to mean: interstices so small and diffused that nothing can intrude (as in a body composed of spherical atoms all in contact with one another). Alternative (*a*) professes to explain compression by particles drawing together into the empty pores (still regarding void as that ‘in which’ movement can take place, 217 a 1). Alternative (*b*), Aristotle says, could profess to explain only movement of buoyancy

I mean, for example, that if a cupful of water becomes air, at the same time a corresponding amount of air must become water, unless there is such a thing as void; for without void nothing can be pressed together or stretched out.

Well, but if (*a*) they mean by 'tenuous' that which has a great number of self-determined 'voids'[a] in it, we have only to point out that there can be no such things, any more than there can be a 'place' with self-determined dimensions. So neither can the tenuous so conceived exist.

But, if (*b*) they mean that the vacuum is not self-determined, but nevertheless is there in the tenuous body, then they are talking more reasonably. But, in the first place, this means that we are no longer speaking of the void as the condition of all movement, but only of the movement of ascending (for the tenuous is light, which is why tenuity is ascribed to fire); and, in the second place, that we are no longer supposing it to be the necessary condition of movement as being 'that in which the movement takes place,' but as being the efficient cause of a movement in the capacity of an 'elevator,' just as bladders elevate whatever is attached to them. But how can a void move or have a place? The void would require another void to move into. Again, how do they account for the downward trend of the weighty? Moreover, if things tend upwards in proportion to their rarity and emptiness, that which is pure emptiness should move upwards quickest of all. Or perhaps we should say that even this will not do; for the same reasoning which

(for nothing can get into these diffused interstices), and cannot really explain that.—C.]

κενῷ ἀκίνητα πάντα, οὕτω καὶ τὸ κενὸν ὅτι ἀκίνητον· ἀσύμβλητα γὰρ τὰ τάχη.

Ἐπεὶ δὲ κενὸν μὲν οὔ φαμεν εἶναι, τἆλλα δ' ἠπόρηται ἀληθῶς—ὅτι ἢ κίνησις οὐκ ἔσται, εἰ μὴ ἔσται πύκνωσις καὶ μάνωσις, ἢ κυμανεῖ ὁ οὐρανός, ἢ ἀεὶ ἴσον ὕδωρ ἐξ ἀέρος ἔσται καὶ ἀὴρ ἐξ ὕδατος (δῆλον γὰρ ὅτι πλείων ἀὴρ ἐξ ὕδατος γίνεται)—ἀνάγκη τοίνυν, εἰ μὴ ἔστι πίλησις, ἢ ἐξωθούμενον τὸ ἐχόμενον τὸ ἔσχατον κυμαίνειν ποιεῖν, ἢ ἄλλοθί που ἴσον μεταβάλλειν ἐξ ἀέρος ὕδωρ, ἵν' ὁ πᾶς ὄγκος τοῦ ὅλου ἴσος ᾖ, ἢ μηδὲν κινεῖσθαι. ἀεὶ γὰρ μεθισταμένου τοῦτο συμβήσεται, ἂν μὴ κύκλῳ περιίστηται· οὐκ ἀεὶ δ' εἰς τὸ κύκλῳ ἡ φορά, ἀλλὰ καὶ εἰς εὐθύ.

Οἱ μὲν δὴ διὰ ταῦτα κενόν τι φαῖεν ἂν εἶναι· ἡμεῖς δὲ λέγομεν ἐκ τῶν ὑποκειμένων ὅτι ἐστὶν ὕλη μία τῶν ἐναντίων (θερμοῦ καὶ ψυχροῦ καὶ τῶν ἄλλων τῶν φυσικῶν ἐναντιώσεων), καὶ ἐκ δυνάμει ὄντος ἐνεργείᾳ ὂν γίνεται, καὶ οὐ χωριστὴ μὲν ἡ ὕλη τῷ δ' εἶναι ἕτερον, καὶ μία τῷ ἀριθμῷ (εἰ ἔτυχε) χροιᾶς καὶ θερμοῦ καὶ ψυχροῦ. ἔστι δὲ καὶ σώματος ὕλη καὶ μεγάλου καὶ μικροῦ ἡ

---

[a] The moving body must thrust out some other body, and the thrust must be passed on till something is thrust up beyond the previous limits of the universe.

showed that nothing can move *in vacuo* would also show that vacuum itself cannot move either, for its velocity would be out of all proportion to that of anything else.

Since, then, we deny the existence of the void and the other alternatives are really exhaustive—namely that, on the supposition of there being no possibility of densification and tenuification, either there can be no movement at all, or else the universe must fluctuate, or when water turns into air a corresponding amount of air must change into water (since the bulk of the air is obviously greater than that of the water)—it does follow necessarily that, if no condensation has taken place, either the continuously transmitted thrust must cause the boundary of the universe to bulge[a] or somewhere else an equivalent amount of air must be changed into water so that the whole bulk of the universe shall remain the same, or nothing must move at all. For a choice between these alternatives is all that is left when anything occupies a new place otherwise than in a re-entrant cycle; and such movements straight on, and not re-entrant, do actually occur.

Now, whereas our opponents would find in this dilemma a ground for asserting the existence of a void, our own explanation is based on the established principle that it is the same matter which experiences the contrasted affections of heat and cold and the other physical opposites, passing either way from a potential to an actual affection, never existing in separation from all attributes, but maintaining its identity through the changing modes of its existence—for instance its colour or its temperature. Similarly the matter of a body may also remain identical when

217 a αὐτή. δῆλον δέ· ὅταν γὰρ ἐξ ὕδατος ἀὴρ γένηται, ἡ αὐτὴ ὕλη οὐ προσλαβοῦσά τι ἄλλο ἐγένετο, ἀλλ' ὃ ἦν δυνάμει, ἐνεργείᾳ ἐγένετο· καὶ πάλιν ὕδωρ ἐξ ἀέρος ὡσαύτως—ὁτὲ μὲν εἰς μέγεθος ἐκ μικρότητος, ὁτὲ δ' εἰς μικρότητα ἐκ μεγέθους.

Ὁμοίως τοίνυν κἂν ἀὴρ πολὺς ὢν ἐν ἐλάττονι γίγνηται ὄγκῳ καὶ ἐξ ἐλάττονος μείζων, ἡ δυνάμει οὖσα γίνεται ὕλη ἄμφω. ὥσπερ γὰρ καὶ ἐκ ψυχροῦ θερμὸν καὶ ἐκ θερμοῦ ψυχρὸν ἡ αὐτή, ὅτι 217 b ἦν δυνάμει, οὕτω καὶ ἐκ θερμοῦ μᾶλλον θερμόν, οὐδενὸς γενομένου ἐν τῇ ὕλῃ θερμοῦ ὃ οὐκ ἦν θερμὸν ὅτε ἧττον ἦν θερμόν. ὥσπερ γε οὐδ' ἡ τοῦ μείζονος κύκλου περιφέρεια καὶ κυρτότης ἐὰν γίνηται ἐλάττονος κύκλου (ἡ αὐτὴ οὖσα ἢ ἄλλη), ἐν οὐθενὶ ἐγγέγονε τὸ κυρτὸν ὃ ἦν οὐ κυρτὸν ἀλλ' εὐθύ—οὐ γὰρ τῷ διαλείπειν τὸ ἧττον ἢ τὸ μᾶλλόν ἐστιν—οὐδ' ἔστι τῆς φλογὸς λαβεῖν τι μέγεθος ἐν ᾧ οὐ καὶ θερμότης καὶ λευκότης ἔνεστιν. οὕτω τοίνυν καὶ ἡ πρότερον θερμότης τῇ ὕστερον. ὥστε καὶ τὸ μέγεθος καὶ ἡ μικρότης τοῦ αἰσθητοῦ ὄγκου οὐ προσλαβούσης τι τῆς ὕλης ἐπεκτείνεται, ἀλλ' ὅτι δυνάμει ἐστὶν ἡ ὕλη ἀμφοῖν·

---

[a] [*Or* 'without any part of the matter becoming hot, which (part) was not hot when the whole was less hot.' Both the original and the increased heat can be equably diffused throughout the whole. We need not suppose that some cold *part* has got warmer.—C.]

[b] [τῷ διαλείπειν. The lesser heat is not to be explained by the *failure* or *omission* of heat in some part of the body, nor the subsequent greater heat by the making up of the omission. No more need change of bulk be explained by addition or subtraction of matter.—C.]

it becomes greater or smaller in bulk. This is manifestly the case ; for when water is transformed into air the same matter, without taking on anything additional, is transformed from what it was, by passing into the actuality of that which before was only a potentiality to it. And it is just the same when air is transformed into water, the one transition being from smaller to greater bulk, and the other from greater to smaller.

And, again on just the same principle, when a greater mass of air contracts into a smaller bulk, or the reverse, it is the same matter which, having the potentiality of either bulk, realizes one or the other alternatively. For just as it is the same matter that passes from cold to hot and from hot to cold, because in either case it was already in potentiality what it now becomes in actuality, so also the same matter can pass from hot to hotter, without any other hot thing coming into it that was not in it when less hot.[a] Just as, when the circumference and curvature of the greater circle becomes that of a lesser one (whether you regard it as retaining its identity or not), in no part of it does something acquire curvature that was not curved but straight before—for the more or less of things does not come about by discontinuity and interpolation [b]—so neither can you find in a flame any tract which is not both hot and bright. It is in this way, then, and not by interpolations, that a former heat becomes subsequently greater. So, too, the smallness of a physical bulk is not stretched into largeness by the matter of it annexing anything to itself, but is possible because that matter had itself the potentiality of

217 b ὥστ' ἐστὶ τὸ αὐτὸ πυκνὸν καὶ μανόν, καὶ μία ὕλη αὐτῶν.

Ἔστι δὲ τὸ μὲν πυκνὸν βαρύ, τὸ δὲ μανὸν κοῦφον.[1] δύο γὰρ ἔστιν ἐφ' ἑκατέρου, τοῦ τε πυκνοῦ καὶ τοῦ μανοῦ· τό τε γὰρ βαρὺ καὶ τὸ σκληρὸν πυκνὰ δοκεῖ εἶναι, καὶ τἀναντία μανά, τό τε κοῦφον καὶ τὸ μαλακόν. διαφωνεῖ δὲ τὸ βαρὺ καὶ τὸ σκληρὸν ἐπὶ μολίβδου καὶ σιδήρου.

Ἐκ δὴ τῶν εἰρημένων φανερὸν ὡς οὔτ' ἀποκεκριμένον κενὸν ἔστιν (οὔθ' ἁπλῶς οὔτ' ἐν τῷ μανῷ) οὔτε δυνάμει, εἰ μή τις βούλεται πάντως καλεῖν κενὸν τὸ αἴτιον τοῦ φέρεσθαι. οὕτω δ' ἡ τοῦ βαρέος καὶ κούφου ὕλη, ᾗ τοιαύτη, εἴη ἂν τὸ κενόν· τὸ γὰρ πυκνὸν καὶ τὸ μανὸν κατὰ ταύτην τὴν ἐναντίωσιν φορᾶς ποιητικά, κατὰ δὲ τὸ σκληρὸν καὶ μαλακὸν πάθους καὶ ἀπαθείας, καὶ οὐ φορᾶς ἀλλ' ἑτεροιώσεως μᾶλλον.

Καὶ περὶ μὲν κενοῦ, πῶς ἔστι καὶ πῶς οὐκ ἔστι, διωρίσθω τὸν τρόπον τοῦτον.

[1] [After *κοῦφον* our MSS. read: *ἔτι ὥσπερ ἡ τοῦ κύκλου περιφέρεια συναγομένη εἰς ἔλαττον οὐκ ἄλλο τι λαμβάνει τὸ κοῖλον, ἀλλ' ὃ ἦν συνήχθη, καὶ τοῦ πυρὸς ὅ τι ἂν τις λάβῃ πᾶν ἔσται θερμόν, οὕτω καὶ τὸ πᾶν συναγωγῇ καὶ διαστολῇ τῆς αὐτῆς ὕλης.* This interpolation, omitted by Themistius, 139. 23, 'not in certain copies' (Simpl. 691. 2), 'regarded as spurious' (Philop. 701. 4), is a doublet of the illustrations in the preceding paragraph.—C.]

---

[a] [*Cf.* Joachim on *De gen. et corr.* p. 124: 'We must not think of a dense body as one in which there are few or small "pores," and of a "rare" body as one with large or many

either bulk. So it is the very same thing that is now dense and now tenuous, and the matter of both is identically the same.[a]

And note that the dense is heavy, and the tenuous light. For density and tenuity are accompanied by two characteristics each, inasmuch as heavy and hard things are held to be dense, and light and soft ones tenuous. But in the case of iron and lead, the superior heaviness does not go with superior hardness.

Our conclusion is that the void does not exist either in separation (whether considered by itself or as implicated in tenuous substances) or even potentially ;[b] unless anyone should choose to *call* that which causes local movement ' vacancy.' For in that case the material components of the heavy and buoyant would, as such, *be* vacancy, for the dense and rare under that aspect are the causes of movement, whereas under their aspect of hard and soft they are the cause of readiness or unreadiness to yield, and so of qualitive modification rather than of local movement.

Let this stand then as our conclusion about how the void does or does not exist.

gaps interspacing its corporeal particles. We must rather conceive of ὕλη as a material capable of filling space with all possible degrees of intensity, or capable of expanding and contracting without a break in its continuity.'—C.]

[b] On the hypothesis that the void receives that which moves into it and is vacated by that which moves out of it, it will be void in actuality when it is unoccupied, and in potentiality when it is occupied by what may move out of it. Hence the potentiality of the occupied vacancy is contrasted with the two supposed forms of actual vacancy, namely the vast inane outside the material universe and the interspersed inane amongst its particles.

CHAPTER X

ARGUMENT

Difficulties presented by the conception of time. No light on the subject to be found in the speculations of earlier thinkers (217 b 29–218 b 9).

(*At this point Aristotle makes a fresh start, and the remaining portion properly belongs to Chapter XI., the subject being an examination of our general impression that there is a close connexion between time and movement or change.*)

217 b Ἐχόμενον δὲ τῶν εἰρημένων ἐστὶν ἐπελθεῖν περὶ χρόνου. πρῶτον δὲ καλῶς ἔχει διαπορῆσαι περὶ αὐτοῦ καὶ διὰ τῶν ἐξωτερικῶν λόγων, πότερον τῶν ὄντων ἐστὶν ἢ τῶν μὴ ὄντων, εἶτα τίς ἡ φύσις αὐτοῦ.

Ὅτι μὲν οὖν ἢ ὅλως οὐκ ἔστιν ἢ μόλις καὶ ἀμυδρῶς, ἐκ τῶνδέ τις ἂν ὑποπτεύσειεν.

Τὸ μὲν γὰρ αὐτοῦ γέγονε καὶ οὐκ ἔστι, τὸ δὲ
218 a μέλλει καὶ οὔπω ἔστιν· ἐκ δὲ τούτων καὶ ὁ ἄπειρος καὶ ὁ ἀεὶ λαμβανόμενος χρόνος σύγκειται. τὸ δ' ἐκ μὴ ὄντων συγκείμενον ἀδύνατον ἂν εἶναι δόξειε μετέχειν οὐσίας.

Πρὸς δὲ τούτοις παντὸς μεριστοῦ, ἐάνπερ ᾖ, ἀνάγκη, ὅτε ἔστιν, ἤτοι πάντα τὰ μέρη εἶναι ἢ ἔνια. τοῦ δὲ χρόνου τὰ μὲν γέγονε τὰ δὲ μέλλει, ἔστι δ' οὐδέν, ὄντος μεριστοῦ. τὸ δὲ νῦν οὐ μέρος· μετρεῖ τε γὰρ τὸ μέρος, καὶ συγκεῖσθαι δεῖ τὸ

CHAPTER X

ARGUMENT (*continued*)

Time is evidently connected with change and movement, but not identical with them. It determines 'fast' and 'slow' because it determines in how much of itself a movement covers a given distance (which distance is a spatial magnitude), or a change effects a given qualitive modification. But it cannot measure itself, for it presents no spatial magnitude and no change of quality for itself to measure (b 9–20).

The subject of inquiry next in succession is 'time.' It will be well to begin with the questions which general reflections suggest as to its existence or non-existence and its nature.

The following considerations might make one suspect either that there is really no such thing as time, or at least that it has only an equivocal and obscure existence.

(1) Some of it is past and no longer exists, and the rest is future and does not yet exist; and time, whether limitless or any given length of time we take, is entirely made up of the no-longer and not-yet; and how can we conceive of that which is composed of non-existents sharing in existence in any way?

(2) Moreover, if anything divisible exists, then, so long as it is in existence, either all its parts or some of them must exist. Now time is divisible into parts, and some of these were in the past and some will be in the future, but none of them exists. The present 'now' is not part of time at all, for a part measures the whole, and the whole must be made up of the

218 a ὅλον ἐκ τῶν μερῶν· ὁ δὲ χρόνος οὐ δοκεῖ συγκεῖσθαι ἐκ τῶν νῦν.

Ἔτι δὲ τὸ νῦν, ὃ φαίνεται διορίζειν τὸ παρελθὸν καὶ τὸ μέλλον, πότερον ἓν καὶ ταὐτὸν ἀεὶ διαμένει ἢ ἄλλο καὶ ἄλλο, οὐ ῥᾴδιον ἰδεῖν.

Εἰ μὲν γὰρ ἀεὶ ἕτερον καὶ ἕτερον, μηδὲν δ' ἐστὶ τῶν ἐν τῷ χρόνῳ ἄλλο καὶ ἄλλο μέρος ἅμα (ὃ μὴ περιέχει, τὸ δὲ περιέχεται, ὥσπερ ὁ ἐλάττων χρόνος ὑπὸ τοῦ πλείονος), τὸ δὲ νῦν μὴ ὂν πρότερον δὲ ὂν ἀνάγκη ἐφθάρθαι ποτέ, καὶ τὰ νῦν ἅμα μὲν ἀλλήλοις οὐκ ἔσται, ἐφθάρθαι δὲ ἀνάγκη ἀεὶ τὸ πρότερον. ἐν ἑαυτῷ μὲν οὖν ἐφθάρθαι οὐχ οἷόν τε, διὰ τὸ εἶναι τότε, ἐν ἄλλῳ δὲ νῦν ἐφθάρθαι τὸ πρότερον νῦν οὐκ ἐνδέχεται. ἔστω γὰρ ἀδύνατον ἐχόμενα εἶναι ἀλλήλων τὰ νῦν ὥσπερ στιγμὴ στιγμῆς. εἴπερ οὖν ἐν τῷ ἐφεξῆς οὐκ ἔφθαρται ἀλλ' ἐν ἄλλῳ, ἐν τοῖς μεταξὺ τοῖς νῦν ἀπείροις οὖσιν ἅμα ἂν εἴη· τοῦτο δ' ἀδύνατον.

Ἀλλὰ μὴν οὐδ' ἀεὶ τὸ αὐτὸ διαμένειν δυνατόν. οὐδενὸς γὰρ διαιρετοῦ πεπερασμένου ἓν πέρας ἐστίν, οὔτ' ἂν ἐφ' ἓν ᾖ συνεχὲς οὔτ' ἂν ἐπὶ πλείω· τὸ δὲ νῦν πέρας ἐστί, καὶ χρόνον ἔστι λαβεῖν πεπερασμένον. ἔτι εἰ τὸ ἅμα εἶναι κατὰ χρόνον καὶ μήτε πρότερον μήτε ὕστερον τὸ ἐν τῷ αὐτῷ εἶναι καὶ ἐν τῷ νῦν ἐστιν, εἰ τά τε πρότερον καὶ τὰ ὕστερον ἐν τῷ νῦν τῳδί ἐστιν, ἅμα ἂν εἴη

---

[a] [*Or* 'and if anything that does not exist now, but formerly did exist, must have perished at some moment.'—C.]

[b] This is proved in Book VI. chap. i.

parts, but we cannot say that time is made up of 'nows.'

(3) Nor is it easy to see whether the 'now' that appears to divide the past and the future (*a*) is always one and the same or (*b*) is perpetually different.

(*b*) For if it is perpetually different, and if no two sectional parts of time can exist at once (unless one includes the other, the longer the shorter), and if the 'now' that is not, but was, must have ceased to be at *some* time or other,[a] so also no two 'nows' can exist together, but the past 'now' must have perished before there was any other 'now.' Now it cannot have ceased to be when it was itself the 'now,' for that is just when it existed ; but it is impossible that the past 'now' should have perished in any other 'now' but itself. For we must lay it down as an axiom that there can be no next 'now' to a given 'now', any more than a next point to a given point.[b] So that if it did not perish in the next 'now,' but in some subsequent one, it would have been in existence coincidently with the countless 'nows' that lie between the 'now' in which it was and the subsequent 'now' in which we are supposing it to perish ; which is impossible.

(*a*) But neither can it continuously persist in its identity. For nothing which is finite and divisible is bounded by a single limit, whether it be continuous in one dimension only or in more than one ; [c] but the 'now' is a time limit, and if we take any limited period of time, it must be determined by two limits, which cannot be identical. Again, if simultaneity in time, and not being before or after, means coinciding and being in the very 'now' wherein they coincide, then, if the before and the after were both in the persistently identical 'now' we are discussing, what

[c] See p. lxxxiv.

218 a τὰ εἰς ἔτος γενόμενα μυριοστὸν τοῖς γενομένοις τήμερον, καὶ οὔτε πρότερον οὔθ' ὕστερον οὐδὲν ἄλλο ἄλλου.

Περὶ μὲν οὖν τῶν ὑπαρχόντων αὐτῷ τοσαῦτ' ἔστω διηπορημένα.

Τί δ' ἐστὶν ὁ χρόνος καὶ τίς αὐτοῦ ἡ φύσις, ὁμοίως ἔκ τε τῶν παραδεδομένων ἄδηλόν ἐστι καὶ περὶ ὧν τυγχάνομεν διεληλυθότες πρότερον.
218 b οἱ μὲν γὰρ τὴν τοῦ ὅλου κίνησιν εἶναί φασιν, οἱ δὲ τὴν σφαῖραν αὐτήν. καίτοι τῆς περιφορᾶς καὶ τὸ μέρος χρόνος τίς ἐστι, περιφορὰ δέ γε οὔ· μέρος γὰρ περιφορᾶς τὸ ληφθέν, ἀλλ' οὐ περιφορά. ἔτι δ' εἰ πλείους ἦσαν οἱ οὐρανοί, ὁμοίως ἂν ἦν ὁ χρόνος ἡ ὁτουοῦν αὐτῶν κίνησις, ὥστε πολλοὶ χρόνοι ἅμα. ἡ δὲ τοῦ ὅλου σφαῖρα ἔδοξε μὲν τοῖς εἰποῦσιν εἶναι ὁ χρόνος ὅτι ἔν τε τῷ χρόνῳ πάντα ἐστὶ καὶ ἐν τῇ τοῦ ὅλου σφαίρᾳ· ἔστι δ' εὐηθικώτερον τὸ εἰρημένον ἢ ὥστε περὶ αὐτοῦ τὰ ἀδύνατα ἐπισκοπεῖν.

Ἐπεὶ δὲ δοκεῖ μάλιστα κίνησις εἶναι καὶ μεταβολή τις ὁ χρόνος, τοῦτ' ἂν εἴη σκεπτέον. ἡ μὲν οὖν ἑκάστου μεταβολὴ καὶ κίνησις ἐν αὐτῷ τῷ μεταβάλλοντι μόνον ἐστίν, ἢ οὗ ἂν τύχῃ ὂν αὐτὸ τὸ κινούμενον καὶ μεταβάλλον· ὁ δὲ χρόνος ὁμοίως

---

[a] [Plato, according to Eudemus and Theophrastus (Simpl. 700. 18 ; Diels, *Dox*. 492).—C.]

[b] [Pythagoreans, Diels, *Vors*. 45 B 33.—C.]

[c] The characteristic motion of the heavens is re-entrant 'circumlation.' But part of a re-entrant circumlation is not re-entrant, whereas time (in addition to not being re-entrant, it might have been added, and therefore having no

happened ten thousand years ago would be simultaneous with what is happening to-day, and nothing would be before or after anything else.

Let this suffice as to the problems raised by considering the properties of time.

But what time really is and under what category it falls, is no more revealed by anything that has come down to us from earlier thinkers than it is by the considerations that have just been urged. For (*a*) some [a] have identified time with the revolution of the all-embracing heaven, and (*b*) some [b] with that heavenly sphere itself. But (*a*) a partial revolution is time just as much as a whole one is, but it is not just as much a revolution ; for any finite portion of time is a portion of a revolution, but is not a revolution.[c] Moreover, if there were more universes than one, the re-entrant circumlation of each of them would be time, so that several different times would exist at once. And (*b*) as to those who declare the heavenly sphere itself to be time, their only reason was that all things are contained 'in the celestial sphere' and also occur 'in time,' which is too childish to be worth reducing to absurdities more obvious than itself.

Now the most obvious thing about time is that it strikes us as some kind of 'passing along' and changing ; but if we follow this clue, we find that, when any particular thing changes or moves, the movement or change is in the moving or changing thing itself or occurs only where that thing is ; whereas 'the

natural unit) is such that any portion of it (a day, for instance) is time just as truly and completely as any other portion (a day and night, for instance). So the Greek commentators understand this obscure and apparently not very significant passage. [If time is actually *identified* with the diurnal revolution, there is no time of less duration than a day.—C.]

218 b καὶ πανταχοῦ καὶ παρὰ πᾶσιν. ἔτι δὲ μεταβολὴ μὲν ἔστι πᾶσα θάττων καὶ βραδυτέρα, χρόνος δ' οὐκ ἔστιν· τὸ γὰρ βραδὺ καὶ ταχὺ χρόνῳ ὥρισται, ταχὺ μὲν τὸ ἐν ὀλίγῳ πολὺ κινούμενον, βραδὺ δὲ τὸ ἐν πολλῷ ὀλίγον· ὁ δὲ χρόνος οὐχ ὥρισται χρόνῳ, οὔτε τῷ ποσός τις εἶναι οὔτε τῷ ποιός. ὅτι μὲν τοίνυν οὐκ ἔστι κίνησις, φανερόν· μηδὲν δὲ διαφερέτω λέγειν ἡμῖν ἐν τῷ παρόντι κίνησιν ἢ μεταβολήν.

---

[a] If more or less change can take place in the same time, time itself cannot be identical with change. For if it were, equal portions of time would coincide with more or less of time itself.

CHAPTER XI

ARGUMENT

As to time, Aristotle enters into no profound metaphysical speculations as to its essential nature. He is content with attempting to bring precision into the thoughts of the plain man, who may easily fall into confusions and contradictions when he tries to give himself an account of what he really means by it.

He finds that we are aware of what we call the passage of time if, and only if, we are aware of some kind of change or movement ; but it need not be spatial *movement or change ; a mental experience, though it came to us in darkness and silence, when we were aware of nothing outside our own consciousness, would make us aware of the passage of time, no less than an objective change which came to us through the senses.*

Now movement or change implies interval, and therefore some kind of magnitude. And magnitude is continuous,

passage of time ' is current everywhere alike and is in relation with everything. And further, all changes may be faster or slower, but not so time ; for fast and slow are defined by time, ' faster ' being more change in less time, and ' slower ' less in more. But time cannot measure time thus, as though it were a distance (like the space passed through in motion) or a qualitive modification, as in other kinds of change.[a] It is evident, therefore, that time is not identical with movement ; nor, in this connexion, need we distinguish between movement and other kinds of change.

CHAPTER XI

ARGUMENT (*continued*)

therefore movement or change, and time (which keeps pace with change) must also be continuous.

As local movement, or change of local position, is the simplest and most obvious form of change, we will begin by examining it. Suppose A, B, C, D, etc. to be successive points on a line. If we say that C comes after *B and* before *D we mean that the distance from A to C is* more *than the distance from A to B and* less *than the distance from A to D. If P has moved along the line (starting at A) we say that if it reached C it had moved* more *than if it reached B and* less *than if it reached D because C comes* after *B and* before *D.*

This more-or-lessness, or before-or-afterness is primarily a measure of space, and derivatively of movement ; for it is the change of relative position that makes us aware of movement. Thus the standard unit for the measurement of movement would naturally be : that movement which covers the standard unit of length.

ARGUMENT (*continued*)

Now let us examine spatial movement in its relation to time. Time is the same everywhere, and more than one movement goes on at the same time, and the relation of time to movement is different in different cases.

For suppose P and P′ to be two points which begin to move from A simultaneously, and suppose P reaches B and P′ reaches D simultaneously, then P′ has moved more *than P in the interval between the initial and the final ' now ' of the interval.*

This leads us to another kind of more-or-lessness in movement : the more-or-lessness of intensity or concentration of the motion itself, or, as we should say, the speed of the motion. For if C comes after *B* in space, *but the arrival of P′ at C comes* before *the arrival of P at B* in time, *then the motion of P′ is faster* (i.e. *more concentrated*) *than the motion of P.*

Now that by which we count the more-or-lessness of motion itself (i.e. *its rapidity or slowness*) *is precisely what we mean by Time ; and if we fix a unit of time, we shall say that the* faster *movement covers a given distance in* less *time than the* slower *one.*

The precise relation, then, of time to movement is that Time is the measure ; the counting, or the dimension, of movement which is primary to motion itself. *It determines the speed of a movement with a test that is not identical with the statical or local more-or-lessness of the distance covered. Aristotle expresses this by saying that time is the ' number ' (that is, system of units) which measures the kind of more-or-lessness which is peculiar to motion, as distinct from that which is determined by distances covered.*

Again, if the moving point P makes a pause at the point B and then continues its movement, then the ' station ' divides the motion, as a point divides a line. The whole of the movement takes place either before or after the ' station ' ; just as the whole of the line comes either before or after the point of division B, the point itself being no part of the line. The point of division B has a double function : it marks the

ARGUMENT (*continued*)

end of ' the before ' and the beginning of ' the after.' It is therefore called a ' double point ' ; not in the sense of being two separate points, but as having a double function. Similarly the ' station ' has the double function of being the final limit of the-movement-before *and the initial limit of* the-movement-after, *but is not itself any part of the motion. And with respect to space ; the-movement-after-the-pause began* where *the movement-before-the-pause ended ; there was no space between them, and the same* where *has the double function of marking the beginning of the one and the end of the other. But with respect to time, the case is different : for the* when *which marks the beginning of the-movement-after-the-pause is not the same* when *as that which marked the end of the-movement-before-the-pause : for time flowed on continuously during the pause, and other changes were going on.*

So that, though the ' station ' is marked in Space by one point, which is itself no part of space, it is marked in Time by an interval, which is itself a part of time, and whose initial and final limits are marked by two separate ' nows ' (analogous to two distinct points).

When we come to the question how the units of movement in this sense are to be established, we shall find that it can only be done by fixing on some standard of re-entrant movement assumed to be uniform.[a]

[a] This preliminary discussion of *motion* leads up to the full exposition of *movement* in Vol. II. Bks. V.-VIII.

[*Continued on next page.*]

ARGUMENT (*continued*)

All this can be expressed in more modern language by saying that time is the non-spatial dimension of movement. Ever since Fourrier this has been symbolically expressed by saying that the dimensions of rate of movement are space entering positively and time entering negatively, or the dimensions of R are ST^{-1}. To realize this is to realize the greatness of Aristotle's achievement, and to see how directly it opens the way to the four-dimensional algebra to which so much attention has recently been directed in connexion with the doctrine of relativity.

218 b Ἀλλὰ μὴν οὐδ' ἄνευ γε μεταβολῆς· ὅταν γὰρ μηδὲν αὐτοὶ μεταβάλλωμεν τὴν διάνοιαν ἢ λάθωμεν μεταβάλλοντες, οὐ δοκεῖ ἡμῖν γεγονέναι χρόνος, καθάπερ οὐδὲ τοῖς ἐν Σαρδοῖ μυθολογουμένοις καθεύδειν παρὰ τοῖς ἥρωσιν, ὅταν ἐγερθῶσιν· συνάπτουσι γὰρ τὸ πρότερον νῦν τῷ ὕστερον νῦν καὶ ἓν ποιοῦσιν, ἐξαιροῦντες διὰ τὴν ἀναισθησίαν τὸ μεταξύ. ὥσπερ οὖν εἰ μὴ ἦν ἕτερον τὸ νῦν ἀλλὰ ταὐτὸ καὶ ἕν, οὐκ ἂν ἦν χρόνος, οὕτως καὶ ἐπεὶ λανθάνει ἕτερον ὄν, οὐ δοκεῖ εἶναι τὸ μεταξὺ χρόνος. εἰ δὴ τὸ μὴ οἴεσθαι εἶναι χρόνον τότε συμβαίνει ἡμῖν ὅταν μὴ ὁρίζωμεν μηδεμίαν μεταβολὴν ἀλλ' ἐν ἑνὶ καὶ ἀδιαιρέτῳ φαίνηται ἡ ψυχὴ μένειν, ὅταν δ' αἰσθώμεθα καὶ ὁρίσωμεν, τότε φαμὲν γεγονέναι χρόνον, φανερὸν ὅτι οὐκ ἔστιν
219 a ἄνευ κινήσεως καὶ μεταβολῆς χρόνος.

Ὅτι μὲν οὖν οὔτε κίνησις οὔτ' ἄνευ κινήσεως ὁ χρόνος ἐστί, φανερόν.

Ληπτέον δέ, ἐπεὶ ζητοῦμεν τί ἐστιν ὁ χρόνος, ἐντεῦθεν ἀρχομένοις, τί τῆς κινήσεώς ἐστιν. ἅμα

ARGUMENT (*continued*)

Aristotle, however, carries his investigation no further than the comparison of movements, each of which is uniform with itself. He is aware of the fact of acceleration (in which T enters twice as a dimension symbolized as T^{-2}), but he does not carry his investigations into this further region.

On the other hand, time cannot be disconnected from change; for when we experience no changes of consciousness, or, if we do, are not aware of them, no time seems to have passed, any more than it did to the men in the fable who 'slept with the heroes' [a] in Sardinia, when they awoke; for under such circumstances we fit the former 'now' on to the later, making them one and the same and eliminating the interval between them, because we did not perceive it. So, just as there would be no time if there were no distinction between this 'now' and that 'now,' but it were always the same 'now'; in the same way there appears to be no time between two 'nows' when we fail to distinguish between them. Since, then, we are not aware of time when we do not distinguish any change (the mind appearing to abide in a single indivisible and undifferentiated state), whereas if we perceive and distinguish changes, then we say that time has elapsed, it is clear that time cannot be disconnected from motion and change.

Plainly, then, time is neither identical with movement nor capable of being separated from it.

In our attempt to find out what time is, therefore, we must start from the question, in what way it pertains to movement. For when we are aware of

[a] [Sons of Herakles and of the daughters of Thespius, who were said to have colonized Sardinia (see Frazer on Apollodorus ii. 7. 6).—C.]

219 a γὰρ κινήσεως αἰσθανόμεθα καὶ χρόνου· καὶ γὰρ ἐὰν ᾖ σκότος καὶ μηδὲν διὰ τοῦ σώματος πάσχωμεν, κίνησις δέ τις ἐν τῇ ψυχῇ ἐνῇ, εὐθὺς ἅμα δοκεῖ τις γεγονέναι καὶ χρόνος. ἀλλὰ μὴν καὶ ὅταν γε χρόνος δοκῇ γεγονέναι τις, ἅμα καὶ κίνησίς τις φαίνεται γεγονέναι. ὥστε ἤτοι κίνησις ἢ τῆς κινήσεώς τί ἐστιν ὁ χρόνος· ἐπεὶ οὖν οὐ κίνησις, ἀνάγκη τῆς κινήσεώς τι εἶναι αὐτόν.

Ἐπεὶ δὲ τὸ κινούμενον κινεῖται ἔκ τινος εἴς τι καὶ πᾶν μέγεθος συνεχές, ἀκολουθεῖ τῷ μεγέθει ἡ κίνησις· διὰ γὰρ τὸ τὸ μέγεθος εἶναι συνεχὲς καὶ ἡ κίνησίς ἐστι συνεχής· διὰ δὲ τὴν κίνησιν ὁ χρόνος· ὅση γὰρ ἡ κίνησις, τοσοῦτος καὶ ὁ χρόνος ἀεὶ δοκεῖ γεγονέναι. τὸ δὲ δὴ πρότερον καὶ ὕστερον ἐν τόπῳ πρῶτόν ἐστιν. ἐνταῦθα μὲν δὴ τῇ θέσει· ἐπεὶ δ' ἐν τῷ μεγέθει ἐστὶ τὸ πρότερον καὶ ὕστερον, ἀνάγκη καὶ ἐν κινήσει εἶναι τὸ πρότερον καὶ ὕστερον, ἀνάλογον τοῖς ἐκεῖ. ἀλλὰ μὴν καὶ ἐν χρόνῳ ἐστὶ τὸ πρότερον καὶ ὕστερον διὰ τὸ ἀκολουθεῖν ἀεὶ θατέρῳ θάτερον αὐτῶν. ἔστι δὲ τὸ πρότερον καὶ ὕστερον αὐτῶν[1] ἐν τῇ κινήσει, ὃ μέν ποτε ὂν κίνησίς ἐστιν· τὸ μέντοι εἶναι αὐτῷ ἕτερον καὶ οὐ κίνησις. ἀλλὰ μὴν καὶ τὸν χρόνον γε

[1] [αὐτῶν om. H: αὐτῶν ἐν τῇ κινήσει om. Philop. 720. 24 (lemma).—C.]

---

[a] 'From here to there' implies an interval between them, and this implies some kind of magnitude.

[b] [Philop. 720. 26 paraphrases: 'This before-and-after, when observed in motion and in time, in respect of the substrate (ὑποκείμενον)—that is the meaning of ὃ μέν ποτε ὄν—is nothing but the motion, but in respect of its definition is distinct: as ascent and descent in respect of their substrate are a ladder, but in definition distinct from it.' *Cf. De gen. et corr.* 319 b 2 ᾗ ἔστι μὲν ὡς (*sc.* ἡ ὕλη, the matter under-

movement we are thereby aware of time, since, even if it were dark and we were conscious of no bodily sensations, but something were 'going on' in our minds, we should, from that very experience, recognize the passage of time. And conversely, whenever we recognize that there has been a lapse of time, we by that act recognize that something 'has been going on.' So time must either itself be movement, or if not, must pertain to movement; and since we have seen that it is not identical with movement, it must pertain to it in some way.

Well then, since anything that moves moves from a 'here' to a 'there,'[a] and magnitude as such is continuous, movement is dependent on magnitude; for it is because magnitude is continuous that movement is so also, and because movement is continuous so is time; for (excluding differences of velocity) the time occupied is conceived as proportionate to the distance moved over. Now, the primary significance of before-and-afterness is the local one of 'in front of' and 'behind.' There it is applied to order of position; but since there is a before-and-after in magnitude, there must also be a before-and-after in movement in analogy with them. But there is also a before-and-after in time, in virtue of the dependence of time upon motion. Motion, then, is the objective seat of before-and-afterness both in movement and in time; but conceptually the before-and-afterness is distinguishable from movement.[b] Now, when we determine a movement by

lying the four simple bodies) ἡ αὐτή, ἔστι δὲ ὡς ἑτέρα; ὃ μὲν γάρ ποτε ὂν ὑπόκειται, τὸ αὐτό, τὸ δ' εἶναι οὐ τὸ αὐτό, 'for that which underlies them, whatever its nature may be *qua* underlying them, is the same, but its actual being is not the same' (Joachim).—C.] See pp. 209 and 213.

219 a γνωρίζομεν, ὅταν ὁρίσωμεν τὴν κίνησιν τὸ πρότερον καὶ ὕστερον ὁρίζοντες· καὶ τότε φαμὲν γεγονέναι χρόνον, ὅταν τοῦ προτέρου καὶ ὑστέρου ἐν τῇ κινήσει αἴσθησιν λάβωμεν. ὁρίζομεν δὲ τῷ ἄλλο καὶ ἄλλο ὑπολαβεῖν αὐτὰ καὶ μεταξύ τι αὐτῶν ἕτερον· ὅταν γὰρ ἕτερα τὰ ἄκρα τοῦ μέσου νοήσωμεν καὶ δύο εἴπῃ ἡ ψυχὴ τὰ νῦν—τὸ μὲν πρότερον τὸ δ' ὕστερον—τότε καὶ τοῦτό φαμεν εἶναι χρόνον· τὸ γὰρ ὁριζόμενον τῷ νῦν χρόνος εἶναι δοκεῖ· καὶ ὑποκείσθω.

Ὅταν μὲν οὖν ὡς ἓν τὸ νῦν αἰσθανώμεθα καὶ μὴ ἤτοι ὡς πρότερον καὶ ὕστερον ἐν τῇ κινήσει ἢ ὡς τὸ αὐτὸ μὲν προτέρου δὲ καὶ ὑστέρου τινός, οὐ δοκεῖ χρόνος γεγονέναι οὐθείς, ὅτι οὐδὲ κίνησις.
219 b ὅταν δὲ τὸ πρότερον καὶ ὕστερον, τότε λέγομεν χρόνον· τοῦτο γάρ ἐστιν ὁ χρόνος, ἀριθμὸς κινήσεως κατὰ τὸ πρότερον καὶ ὕστερον.

Οὐκ ἄρα κίνησις ὁ χρόνος, ἀλλ' ᾗ ἀριθμὸν ἔχει ἡ κίνησις. σημεῖον δέ· τὸ μὲν γὰρ πλεῖον καὶ ἔλαττον κρίνομεν ἀριθμῷ, κίνησιν δὲ πλείω καὶ ἐλάττω χρόνῳ· ἀριθμὸς ἄρα τις ὁ χρόνος. ἐπεὶ δ' ἀριθμός ἐστι διχῶς (καὶ γὰρ τὸ ἀριθμούμενον καὶ τὸ ἀριθ-

---

[a] That is, 'whose beginning and end are each a "now."'

defining its first and last limit, we also recognize a lapse of time ; for it is when we are aware of the measuring of motion by a prior and posterior limit that we may say time has passed. And our determination consists in distinguishing between the initial limit and the final one, and seeing that what lies between them is distinct from both ; for when we distinguish between the extremes and what is between them, and the mind pronounces the 'nows' to be two—an initial and a final one—it is then that we say that a certain time has passed ; for that which is determined either way by a 'now'[a] seems to be what we mean by time. And let this be accepted and laid down.

Accordingly, when we perceive a 'now' in isolation, that is to say not as one of two, an initial and a final one in the motion, nor yet as being a final 'now' of one period and at the same time the initial 'now' of a succeeding period, then no time seems to have elapsed, for neither has there been any corresponding motion. But when we perceive a distinct before and after, then we speak of time ; for this is just what time is, the calculable measure or dimension of motion with respect to before-and-afterness.

Time, then, is not movement, but that by which movement can be numerically estimated. To see this, reflect that we estimate any kind of more-and-lessness by number ; so, since we estimate all more-or-lessness on some numerical scale and estimate the more-or-lessness of motion by time, time is a scale on which something (to wit movement) can be numerically estimated. But now, since 'number' has two meanings (for we speak of the 'numbers'

219 b μητὸν ἀριθμὸν λέγομεν, καὶ ᾧ ἀριθμοῦμεν), ὁ δὲ[1] χρόνος ἐστὶ τὸ ἀριθμούμενον καὶ οὐχ ᾧ ἀριθμοῦμεν. ἔστι δ' ἕτερον ᾧ ἀριθμοῦμεν καὶ τὸ ἀριθμούμενον.

Καὶ ὥσπερ ἡ κίνησις ἀεὶ ἄλλη καὶ ἄλλη, καὶ ὁ χρόνος. ὁ δ' ἅμα πᾶς χρόνος ὁ αὐτός· τὸ γὰρ νῦν τὸ αὐτὸ ὅ ποτ' ἦν, τὸ δ' εἶναι αὐτῷ ἕτερον· τὸ δὲ νῦν τὸν χρόνον μετρεῖ, ᾗ πρότερον καὶ ὕστερον. τὸ δὲ νῦν ἔστι μὲν ὡς τὸ αὐτό, ἔστι δ' ὡς οὐ τὸ αὐτό· ᾗ μὲν γὰρ ἐν ἄλλῳ καὶ ἄλλῳ, ἕτερον (τοῦτο δ' ἦν αὐτῷ τὸ νῦν[2]), ᾗ δὲ ὅ ποτε ὄν ἐστι τὸ νῦν, τὸ αὐτό. ἀκολουθεῖ γάρ, ὡς ἐλέχθη, τῷ μὲν μεγέθει ἡ κίνησις, ταύτῃ δ' ὁ χρόνος, ὥς φαμεν· καὶ ὁμοίως δὴ τῇ στιγμῇ τὸ φερόμενον, ᾧ τὴν κίνησιν γνωρίζομεν καὶ τὸ πρότερον ἐν αὐτῇ καὶ τὸ ὕστερον. τοῦτο δὲ ὃ μέν ποτε ὂν τὸ αὐτό (ἢ στιγμὴ γὰρ ἢ λίθος ἤ τι ἄλλο τοιοῦτόν ἐστι), τῷ λόγῳ δὲ ἄλλο· ὥσπερ οἱ σοφισταὶ λαμβάνουσιν ἕτερον τὸ Κορίσκον ἐν Λυκείῳ εἶναι καὶ τὸ Κορίσκον ἐν ἀγορᾷ, καὶ τοῦτο δὴ τῷ ἄλλοθι καὶ ἄλλοθι εἶναι ἕτερον. τῷ δὲ φερομένῳ ἀκολουθεῖ τὸ νῦν, ὥσπερ ὁ χρόνος τῇ

[1] [δὲ: δὴ FG.—C.]

[2] [τὸ νῦν εἶναι Philop. 726. 21 (paraphr.), Bonitz, Prantl.—C.]

[a] The contrast is between the *numeri numerati* and the *numeri numerantes* and between the 'concrete' and 'abstract' of recent arithmetical terminology. In counting the successive 'nows,' we are counting sections of continuous time; but we are counting them in abstract numbers. Time, then, is a concrete numerable, not an abstract numerator.

[b] [The following paragraph points out 'that as movement is recognized by observing a single moving body successively at different points, the passage of time is recognized by noting that the single character of "nowness" has been

that are counted in the thing in question, and also of the 'numbers' by which we count them and in which we calculate), we are to note that time is the countable thing that we are counting, not the numbers we count in—which two things are different.[a]

And [b] as motion is a continuous flux, so is time; but at any given moment time is the same everywhere, for the 'now' itself is identical in its essence, but the relations into which it enters differ in different connexions, and it is the 'now' that marks off time as before and after. But this 'now,' which is identical everywhere, itself retains its identity in one sense, but does not in another; for inasmuch as the point in the flux of time which it marks is changing (and so to mark it is its essential function) the 'now' too differs perpetually, but inasmuch as at every moment it is performing its essential function of dividing the past and future it retains its identity. For there is a dependent sequence, as we have shown, of movement upon magnitude and (we may add) of time upon movement; and the moving object, by which we become aware of movement and its before-and-afterness, may be regarded as a point;[c] and throughout its course this—whether point or stone or what you like—retains its identity, but its relations alter: as the Sophists distinguish between Koriscos in the Lyceum and Koriscos in the market-place, so this moving object also is different in so far as it is perpetually marking a different position. And as time follows the analogy of movement, so does the 'now' of time follow the analogy of the moving

attached to more than one experienced event' (Ross, *Aristotle*, p. 90).—C.]

[c] 'As a point,' that is, as a mere indicator of motion, in abstraction from all its other properties.

κινήσει· τῷ γὰρ φερομένῳ γνωρίζομεν τὸ πρότερον καὶ ὕστερον ἐν κινήσει· ᾗ δ' ἀριθμητὸν τὸ πρότερον καὶ ὕστερον, τὸ νῦν ἐστιν· ὥστε καὶ ἐν τούτοις, ὃ μέν ποτε ὂν νῦν ἐστι, τὸ αὐτό (τὸ πρότερον γὰρ καὶ ὕστερόν ἐστιν ἐν κινήσει), τὸ δ' εἶναι ἕτερον· ᾗ ἀριθμητὸν γὰρ τὸ πρότερον καὶ ὕστερον, τὸ νῦν ἐστιν. καὶ γνώριμον δὲ μάλιστα τοῦτ' ἔστιν· καὶ γὰρ ἡ κίνησις διὰ τὸ κινούμενον καὶ ἡ φορὰ διὰ τὸ φερόμενον· τόδε γάρ τι τὸ φερόμενον, ἡ δὲ κίνησις οὔ. ἔστι μὲν οὖν ὡς τὸ αὐτὸ τὸ νῦν ἀεί, ἔστι δ' ὡς οὐ τὸ αὐτό· καὶ γὰρ τὸ φερόμενον.

Φανερὸν δὲ καὶ ὅτι εἴτε χρόνος μὴ εἴη, τὸ νῦν οὐκ ἂν εἴη, εἴτε τὸ νῦν μὴ εἴη, χρόνος οὐκ ἂν εἴη· ἅμα γὰρ ὥσπερ τὸ φερόμενον καὶ ἡ φορά, οὕτως καὶ ὁ ἀριθμὸς ὁ τοῦ φερομένου καὶ ὁ τῆς φορᾶς. χρόνος μὲν γὰρ ὁ τῆς φορᾶς ἀριθμός, τὸ νῦν δὲ ὡς τὸ φερόμενον οἷον μονὰς ἀριθμοῦ.

Καὶ συνεχής τε δὴ ὁ χρόνος τῷ νῦν, καὶ διῄρηται

---

[a] [*Or*, 'the "now" is the before and after, *qua* countable.' Themist. 150. 23 *τὸ δὲ ἀριθμούμενον πρότερόν τε καὶ ὕστερον τὸ νῦν ἐστιν*, Simplic. 723. 29 *καθόσον δὲ ἀριθμητὸν τὸ πρότερον τοῦτο καὶ ὕστερον, κατὰ τοσοῦτον, φησί, νῦν δείκνυται*.—C.]

[b] I understand this to be a reference to the general principle of the more easy apprehensibility of the concrete and the greater intellectual luminosity of the abstract. [Themistius, 150. 27 explains: The 'now,' as a sort of particular existent, is more *cognizable* than time (*γνωριμώτερον τοῦ χρόνου*), just as the moving body is more so than its motion.—C.]

[c] The moving object (whatever it may be) is apprehended through the senses. It is related to the flux of Motion (apprehended through the intelligence) as 'now' is related to the flux of Time. It is no more a *part* of Motion than 'now' is a *part* of Time. It divides perpetually different past and future Motion just as 'now' divides perpetually different past and future Time.

object, since it is by the moving object that we come to know the before-and-after in motion, and it is in virtue of the countableness of its before-and-afters that the 'now' exists;[a] so that the 'now,' wherever found in the before-and-afters, is identical (for it is simply the mark of the before-and-afters in motion), but the before-and-afternesses it marks differ; though the nature of the 'now' depends on the markableness of any before-and-after in general, not on the specific before-and-after marked by it. And it is this specifically related 'now' that is nearest to our apprehension,[b] just as motion-change is apprehended through the changing object, and translation through the translated object, for this object is a concrete thing, which motion is not. There is a sense, then, in which what we mean when we say 'now' is always the same, and a sense in which it is not, just as is the case with anything that is in motion.

It is evident, too, that neither would time be if there were no 'now,' nor would 'now' be if there were no time; for they belong to each other as the moving thing and the motion do, so that whatever ticks off the position of the one ticks off the other. For time is the dimension proper to motion, and the 'now' corresponds to the moving object as the numerical monad.[c]

So, too, time owes its continuity to the 'now,' and yet is divided by reference to it, since in this

Time, Distance and Motion are all divisible into innumerable parts or units. The 'numbers' by which Time units are counted are static 'nows,' the 'numbers' which count Distance units are static points, and the 'numbers' which count Motion units are static objects. None of these numerators are *parts* of the numerables which they count.

220 a κατὰ τὸ νῦν· ἀκολουθεῖ γὰρ καὶ τοῦτο τῇ φορᾷ καὶ τῷ φερομένῳ· καὶ γὰρ ἡ κίνησις καὶ ἡ φορὰ μία τῷ φερομένῳ, ὅτι ἕν, καὶ οὐχ ὅ ποτε ὄν (καὶ γὰρ ἂν διαλίποι) ἀλλὰ τῷ λόγῳ· καὶ ὁρίζει δὴ[1] τὴν πρότερον καὶ ὕστερον κίνησιν τοῦτο. ἀκολουθεῖ δὲ καὶ τοῦτό πως τῇ στιγμῇ· καὶ γὰρ ἡ στιγμὴ καὶ συνέχει τὸ μῆκος καὶ ὁρίζει· ἔστι γὰρ τοῦ μὲν ἀρχὴ τοῦ δὲ τελευτή. ἀλλ' ὅταν μὲν οὕτω λαμβάνῃ τις ὡς δυσὶ χρώμενος τῇ μιᾷ, ἀνάγκη ἵστασθαι, εἰ ἔσται ἀρχὴ καὶ τελευτὴ ἡ αὐτὴ στιγμή· τὸ δὲ νῦν, διὰ τὸ κινεῖσθαι τὸ φερόμενον, ἀεὶ ἕτερον· ὥσθ' ὁ χρόνος ἀριθμὸς οὐχ ὡς τῆς αὐτῆς στιγμῆς, ὅτι ἀρχὴ καὶ τελευτή, ἀλλ' ὡς τὰ ἔσχατα τῆς γραμμῆς μᾶλλον, καὶ οὐχ ὡς τὰ μέρη, διά τε τὸ εἰρημένον (τῇ γὰρ μέσῃ στιγμῇ ὡς δυσὶ χρήσεται, ὥστε ἠρεμεῖν συμβήσεται), καὶ ἔτι φανερὸν ὅτι οὐδὲ μόριον τὸ νῦν τοῦ χρόνου, οὐδ' ἡ διαίρεσις τῆς κινήσεως, ὥσπερ οὐδ' αἱ στιγμαὶ τῆς γραμμῆς· αἱ δὲ γραμμαὶ αἱ δύο τῆς μιᾶς μόρια. ᾗ μὲν οὖν

[1] [καὶ ὁρίζει δὴ HI, Philop. 734. 12 (lemma): καὶ γὰρ ὁρίζει cett. The moving body *both* constitutes the *continuity* of the motion *and* at any moment marks the boundary (*division*) between the motion that has taken place and that which will follow. (*Cf.* Themist. 151. 11 (τὸ φερόμενον) καὶ διορίζει τὴν προτέραν καὶ ὑστέραν κίνησιν καὶ πάλιν συνέχει.)—C.]

---

[a] [The 'now' is analogous to the dividing point in a bisected line, in so far as either can be regarded as the end-point of what comes before and also the starting-point of what comes after. But the 'now' is not stationary; so when time is *counted*, we do not take the same 'nows' twice as we took the dividing point), but different 'nows,' like the two extremeties of the line—two different points. —C.)

[b] If the line AB is divided at the point C, the two lines

respect also the analogy with the translation and the object translated holds good; for the movement or translation is one-and-continuous in virtue of the identity of the translated object—not its identity *qua* object (for it would preserve that if it stopped) but its unbroken identity *qua* 'the thing that is being moved'; and it is this that also marks the division between the movement before and the movement after. And there is an analogy also between such a 'body that is being moved' and a point; for it is a point that both constitutes (by its movement) the continuity of the line it traces and also marks the end of the line that is behind and the beginning of the line in front. If, however, one ascribes the latter function to it, regarding the one point in two capacities—as the end of one section of the line and the beginning of another—it must have been arrested, since its identity in this 'statical' relation must be preserved. But the 'now,' as it follows the object in motion, marks a perpetually different position, so that time is not counted as if by one and the same point,—since each point in it so counted is a double point, being end and beginning at once,—but rather as the two extremities of the line,[a] and not as *parts* of it, for the reason already stated (that, if one were to count the dividing point in its two capacities that would involve a pause), and because it is obvious that the 'now' is not a portion of time, just as the division of motion is not part of motion any more than points are of a line; it is the two sections that are *parts* of the one line.[b] The 'now,' therefore, as a limit is not

AC, CB are *parts* of AB; the point C is *not* a part of AB. —C.]

220 a πέρας τὸ νῦν, οὐ χρόνος ἀλλὰ συμβέβηκεν, ᾗ δ' ἀριθμεῖ, ἀριθμός· τὰ μὲν γὰρ πέρατα ἐκείνου μόνον ἐστὶν οὗ ἐστι πέρατα, ὁ δ' ἀριθμὸς ὁ τῶνδε τῶν ἵππων—ἡ δεκάς—καὶ ἄλλοθι.

Ὅτι μὲν τοίνυν ὁ χρόνος ἀριθμός ἐστι κινήσεως κατὰ τὸ πρότερον καὶ ὕστερον, καὶ συνεχής (συνεχοῦς γάρ), φανερόν.

CHAPTER XII

ARGUMENT

There is no smallest unit of a continuous magnitude, and therefore not of time (220 a 27–32).

Time is ' much ' or ' little ' according to the number of its units, and ' long ' or ' short ' as a continuous magnitude. But it is not ' quick ' or ' slow,' any more than other magnitudes, or than numbers, are in themselves quick or slow (a 32–b 5).

Any given period of time is the same everywhere, and any given ' now ' cuts across all changes ' everywhere ' and ' at once,' for the ' now ' of each change is the same ' now ' and is the ' now ' of time. And though every ' now ' divides the past and the future, whether of time or change, the past and the future that it divides are always changing, so that although ' now ' is the same everywhere taken ' across ' changes, it is always different at every-when taken ' along ' a change or changes or time as a dimension, and any period of time is different from one that comes before or after it (b 5–14).

Mutual determinations of time and motion (b 14-32).

Time is a dimension of motion itself (since one mobile moving faster than another means that it covers the same

Ἐλάχιστος δὲ ἀριθμὸς ὁ μὲν ἁπλῶς ἐστιν ἡ δυάς· τὶς δ' ἀριθμὸς ἔστι μὲν ὡς ἔστιν, ἔστι δ' ὡς

time, but is incidental to time, while as the numerator it is a number ; for limits are limits only of the particular thing they limit, whereas the number 10, for instance, pertains equally to the ten horses (say) the sum of which it has defined, and to anything else numerable.

That time, then, is the dimension of movement in its before-and-afterness, and is continuous (because movement is so), is evident.

CHAPTER XII

ARGUMENT (*continued*)

distance in less time) and also of the duration of movement. But in this second sense it also measures the duration of everything else that begins and ceases to be. And it is in this sense of being embraced or bounded by it that things in general (including movements) are most properly said to be 'in time,' on the analogy of the wine which, being embraced or bounded by the flask (in space), is said to be 'in the flask.' Another sense of 'having existence in time,' that must not be allowed to mislead us (b 32–221 a 26).

Development of the meaning of being 'embraced by time,' and being 'affected by time' (a 26–b 3).

Things eternal (b 3–7).

Time, in directly measuring the duration of motion, incidentally measures the duration of cessation of motion, but not of 'all that does not move' ; for what cannot move cannot cease to move and so has no beginning or end of its non-movement (b 7–23).

Of existents and non-existents as subject or not to time (b 23–222 a 9).

The dyad is the smallest possible *abstract number*. In one sense there is no smallest possible number,

220 a οὐκ ἔστιν· οἷον γραμμῆς ἐλάχιστος πλήθει μὲν ἔστιν αἱ δύο ἢ ἡ μία, μεγέθει δ' οὐκ ἔστιν ἐλάχιστος· ἀεὶ γὰρ διαιρεῖται πᾶσα γραμμή. ὥσθ' ὁμοίως καὶ ὁ χρόνος· ἐλάχιστος γὰρ κατὰ μὲν ἀριθμόν ἐστιν ὁ εἷς ἢ οἱ δύο, κατὰ μέγεθος δ' οὐκ ἔστιν.

220 b Φανερὸν δὲ καὶ ὅτι ταχὺς μὲν καὶ βραδὺς οὐ λέγεται, πολὺς δὲ καὶ ὀλίγος καὶ μακρὸς καὶ βραχύς. ᾗ μὲν γὰρ συνεχής, μακρὸς καὶ βραχύς, ᾗ δ' ἀριθμός, πολὺς καὶ ὀλίγος· ταχὺς δὲ καὶ βραδὺς οὐκ ἔστιν· οὐδὲ γὰρ ἀριθμὸς ᾧ ἀριθμοῦμεν ταχὺς καὶ βραδὺς οὐδείς.

Καὶ ὁ αὐτὸς δὴ πανταχοῦ ἅμα· πρότερον δὲ καὶ ὕστερον οὐχ ὁ αὐτός, ὅτι καὶ ἡ μεταβολὴ ἡ μὲν παροῦσα μία, ἡ δὲ γεγενημένη καὶ ἡ μέλλουσα ἑτέρα. ὁ δὲ χρόνος ἀριθμός ἐστιν οὐχ ᾧ ἀριθμοῦμεν ἀλλ' ὁ ἀριθμούμενος· οὗτος δὲ συμβαίνει πρότερον καὶ ὕστερον ἀεὶ ἕτερος· τὰ γὰρ νῦν ἕτερα·

[a] In multiplicity there is a minimum in magnitude, there is not of numerables. The Greek text gives 'two or one' as the smallest number, though it gives no alternative to the dyad as the smallest abstract number (abstract number being defined as a 'plurality of ones,' the unit itself is not a number but a numerable). Themistius does not seem to have read ἢ ἡ μία (or ὁ εἷς ἢ in l. 32). Philoponus finds it embarrassing. [Alexander, quoted with approval by Simplic. 730. 13, suggests that Aristotle might speak of 'one line' or 'one unit of time' as a minimum *number*, if using words popularly.—C.]

[b] Which small lines could be taken for the units instead of the larger one, and these again could be divided up into still smaller units, and so on without end, for there is no indivisible line. The smaller the unit is, the smaller will a

but in another sense there is ;[a] for, whatever line you take for the unit, two is the smallest number of such units, but in magnitude there is no minimum, for any line whatever may itself be divided into smaller lines.[b] So too with time, 'two' is the smallest possible number of time units, but there is no smallest possible time unit itself that may be selected.

Observe too that we do not speak of time itself as 'swift or slow,' but as consisting of 'many or few' of the units in which it is counted, or as 'long and short' when we regard it as a continuum. It would not be swift or slow, even if we supposed it to be the counter that counts, not the dimension that is counted (which it really is) ; for abstract numbers are in no case swift or slow, though the counting of them may be.

Moreover, though time is identical everywhere simultaneously, it is not identical if taken twice successively ;[c] for the change it measures, likewise, is one when considered as present, but not one if considered as partly past and partly future.[d] And time considered numerically is concrete, not abstract ; whereby follows that it changes from the former to the latter 'now,'[e] inasmuch as these 'nows' them-

given aggregate of those units be, and as there is no limit to the conceptual smallness of a unit, so neither is there a limit to the conceptual smallness of a given aggregate of them.

[c] What time measures is not the nature of the change, but its before-and-afterness. Therefore it is not differentiated by the specific nature of the several contemporaneous changes, but it is by its very definition differentiated by the before-and-afterness that it counts.

[d] For the change which has already occurred is different from that which is about to occur.

[e] πρότερον καὶ ὕστερον here (rather misleadingly) used for γεγενημένον καὶ μέλλον.

220 b ἔστι δὲ ὁ ἀριθμὸς εἷς μὲν καὶ ὁ αὐτὸς ὁ τῶν ἑκατὸν ἵππων καὶ ὁ τῶν ἑκατὸν ἀνθρώπων, ὧν δ' ἀριθμὸς ἕτερα, οἱ ἵπποι τῶν ἀνθρώπων. ἔτι ὡς ἐνδέχεται κίνησιν εἶναι τὴν αὐτὴν καὶ μίαν πάλιν καὶ πάλιν, οὕτω καὶ χρόνον, οἷον ἐνιαυτὸν ἢ ἔαρ ἢ μετόπωρον.

Οὐ μόνον δὲ τὴν κίνησιν τῷ χρόνῳ μετροῦμεν, ἀλλὰ καὶ τῇ κινήσει τὸν χρόνον, διὰ τὸ ὁρίζεσθαι ὑπ' ἀλλήλων· ὁ μὲν γὰρ χρόνος ὁρίζει τὴν κίνησιν ἀριθμὸς ὢν αὐτῆς, ἡ δὲ κίνησις τὸν χρόνον. καὶ λέγομεν πολὺν ἢ ὀλίγον χρόνον τῇ κινήσει μετροῦντες, καθάπερ καὶ τῷ ἀριθμητῷ τὸν ἀριθμόν, οἷον τῷ ἑνὶ ἵππῳ τὸν τῶν ἵππων ἀριθμόν. τῷ μὲν γὰρ ἀριθμῷ τὸ τῶν ἵππων πλῆθος γνωρίζομεν, πάλιν δὲ τῷ ἑνὶ ἵππῳ τὸν τῶν ἵππων ἀριθμὸν αὐτόν. ὁμοίως δὲ καὶ ἐπὶ τοῦ χρόνου καὶ τῆς κινήσεως· τῷ μὲν γὰρ χρόνῳ τὴν κίνησιν, τῇ δὲ κινήσει τὸν χρόνον μετροῦμεν. καὶ τοῦτ' εὐλόγως συμβέβηκεν· ἀκολουθεῖ γὰρ τῷ μὲν μεγέθει ἡ κίνησις, τῇ δὲ κινήσει ὁ χρόνος, τῷ καὶ ποσὰ καὶ συνεχῆ καὶ διαιρετὰ εἶναι· διὰ μὲν γὰρ τὸ τὸ μέγεθος εἶναι τοιοῦτον ἡ κίνησις ταῦτα πέπονθεν, διὰ δὲ τὴν κίνησιν ὁ χρόνος. καὶ μετροῦμεν καὶ τὸ μέγεθος τῇ κινήσει καὶ τὴν κίνησιν τῷ μεγέθει· πολλὴν γὰρ εἶναί φαμεν τὴν ὁδόν, ἂν ᾖ ἡ πορεία πολλή, καὶ ταύτην

[a] This leads up to the selection of the rotation of the starry sphere as the standard for measuring time. It indicates that evenly flowing time can be cut up into presumably equal periods (years), or may present us with a recognizable recurrence of experiences or phenomena. But time itself must not be thought of as 'moving in a circle' any more than 'in a line.' This and the following parallels are technically imperfect, for in one case the unit of what

selves are different; just as the number of a hundred horses is identical with that of a hundred men, but the horses enumerated are different from the men enumerated. Now note further that as there may be movement (of rotation to wit) that covers the same course over and over again, in like manner we mark off time by the year or by spring or autumn.[a]

And not only do we measure the length of uniform movement by time, but also the length of time by uniform movement, since they mutually determine each other[b]; for the time taken determines the length moved over (the time units corresponding to the space units), and the length moved over determines the time taken. And when we call time 'much' or 'little' we are estimating it in units of uniform motion, as we measure the 'number' of anything we count by the units we count it in—the number of horses, for example, by taking one horse as our unit. For when we are told the number of horses, we know how many there are in the troop; and by counting how many there are, horse by horse, we know their number. And so too with time and uniform motion, for we measure them by each other either way. And this is only natural, for movement corresponds to linear magnitude, and time to movement, in being a quantity, in being continuous, and in being divisible; for it is from linear magnitude that motion takes on these qualities, and from motion that time does. That we do measure linear magnitude by movement, and *vice versa*, is evidenced from our saying that it is

is measured is itself a magnitude of the same order as what it measures (line measuring line), in the other it is proportional to it but not identical with it.

[b] V being countant, S (which is proportional to V) and T determine each other.

220 b πολλήν, ἂν ἡ ὁδὸς ᾖ πολλή. καὶ τὸν χρόνον, ἂν ἡ κίνησις, καὶ τὴν κίνησιν, ἂν ὁ χρόνος.

221 a Ἐπεὶ δ' ἐστὶν ὁ χρόνος μέτρον κινήσεως καὶ τοῦ κινεῖσθαι, μετρεῖ δ' οὗτος τὴν κίνησιν τῷ ὁρίσαι τινὰ κίνησιν ἣ καταμετρήσει τὴν ὅλην (ὥσπερ καὶ τὸ μῆκος ὁ πῆχυς τῷ ὡρίσθαι τι μέγεθος ὃ ἀναμετρήσει τὸ ὅλον), καὶ ἔστι τῇ κινήσει τὸ ἐν χρόνῳ εἶναι τὸ μετρεῖσθαι τῷ χρόνῳ καὶ αὐτὴν καὶ τὸ εἶναι αὐτῆς—ἅμα γὰρ τὴν κίνησιν καὶ τὸ εἶναι τῇ κινήσει μετρεῖ, καὶ τοῦτ' ἔστιν αὐτῇ τὸ ἐν χρόνῳ εἶναι, τὸ μετρεῖσθαι αὐτῆς τὸ εἶναι—δῆλον[1] ὅτι καὶ τοῖς ἄλλοις τοῦτ' ἔστι τὸ ἐν χρόνῳ εἶναι, τὸ μετρεῖσθαι αὐτῶν τὸ εἶναι ὑπὸ τοῦ χρόνου.

Τὸ γὰρ ἐν χρόνῳ εἶναι δυοῖν ἐστι θάτερον, ἓν μὲν τὸ εἶναι τότε ὅτε ὁ χρόνος ἔστιν, ἓν δὲ τὸ ὥσπερ ἔνια λέγομεν ὅτι ἐν ἀριθμῷ ἐστιν· τοῦτο δὲ σημαίνει ἤτοι ὡς μέρος ἀριθμοῦ καὶ πάθος, καὶ ὅλως ὅτι τοῦ ἀριθμοῦ τι, ἢ ὅτι ἔστιν αὐτοῦ ἀριθμός.

Ἐπεὶ δ' ἀριθμὸς ὁ χρόνος, τὸ μὲν νῦν καὶ τὸ πρότερον καὶ ὅσα τοιαῦτα οὕτως ἐν χρόνῳ ὡς ἐν ἀριθμῷ μονὰς καὶ τὸ περιττὸν καὶ ἄρτιον (τὰ μὲν

[1] [δῆλον Themist. 154. 10, Simplic. 735. 8, Philop. 749. 20: δῆλον δὲ EFGI, Philop. 749. 24: δὲ δῆλον H.—C.]

---

[a] [*Literally*, ' Since time is a measure of movement and of the movement's actually taking place, and this measurement of movement by time is effected by determining a certain unit of movement which shall serve to measure off the whole movement (as the cubit serves to measure length by being fixed upon as a unit of magnitude which will serve to measure off the whole length), and for movement to be ' in time ' means that both the movement itself and its existence (or taking place) are measured by time—for in measuring the

a great 'way' if it is a great 'walk,' or *vice versa*. So too with time and movement: we speak of a 'long walk' taking a 'long time,' or *vice versa*.

It [a] is by reference to the standard unit of time that we determine the relative velocity of two several motions. For we ask what distance either motion has covered during the lapse of the standard unit of time, and pronounce the motion itself fast or slow in proportion as that distance is great or small. But that same standard unit of time measures the duration of a motion. So the way in which a motion exists in time is by both itself and its duration being measured by time. For time measures both the motion and its duration by the same act, and its duration being so measured constitutes it as existing in time. But it is obvious that other things as well as motion exist in time because their existence too is measured by time.

For this phrase 'existing in time' is ambiguous. It may mean (1) existing when time also exists, or (2) it may mean 'in time' in a sense analogous to that in which we say that certain things exist 'in number'; and this phrase again is ambiguous, for it may mean (*a*) that they exist in number as parts or affections of it, or generally that they pertain to it in some way or other, or (*b*) it may mean that they themselves can be counted.

Now (2) taking time as a number scale, (*a*) the 'now' and the 'before' and suchlike exist in time as the monad and the odd and even exist in number

movement time does also measure the movement's existence, and this fact that its existence is measured is what it means for a movement to be 'in time'—evidently for all other things also to be 'in time' means that their existence is measured by time.'—C.]

221 a γὰρ τοῦ ἀριθμοῦ τι, τὰ δὲ τοῦ χρόνου τί ἐστι), τὰ δὲ πράγματα ὡς ἐν ἀριθμῷ τῷ χρόνῳ ἐστίν· εἰ δὲ τοῦτο, περιέχεται ὑπ᾿ ἀριθμοῦ ὥσπερ καὶ τὰ ἐν τόπῳ ὑπὸ τόπου.

Φανερὸν δὲ καὶ ὅτι οὐκ ἔστι τὸ ἐν χρόνῳ εἶναι τὸ εἶναι ὅτε ὁ χρόνος ἔστιν, ὥσπερ οὐδὲ τὸ ἐν κινήσει εἶναι οὐδὲ τὸ ἐν τόπῳ ὅτε ἡ κίνησις καὶ ὁ τόπος ἔστιν. εἰ γὰρ ἔσται τὸ ἔν τινι οὕτως, πάντα τὰ πράγματα ἐν ὁτῳοῦν ἔσται, καὶ ὁ οὐρανὸς ἐν τῇ κέγχρῳ· ὅτε γὰρ ἡ κέγχρος ἔστιν ἔστι καὶ ὁ οὐρανός. ἀλλὰ τοῦτο μὲν συμβέβηκεν, ἐκεῖνο δ᾿ ἀνάγκη παρακολουθεῖν—καὶ τῷ ὄντι ἐν χρόνῳ εἶναί τινα χρόνον ὅτε κἀκεῖνο ἔστι, καὶ τῷ ἐν κινήσει ὄντι εἶναι τότε κίνησιν.

Ἐπεὶ δ᾿ ἐστὶν ὡς ἐν ἀριθμῷ τὸ ἐν χρόνῳ, ληφθήσεταί τις πλείων χρόνος παντὸς τοῦ ἐν χρόνῳ ὄντος· διὸ ἀνάγκη πάντα τὰ ἐν χρόνῳ ὄντα περιέχεσθαι ὑπὸ χρόνου, ὥσπερ καὶ τἆλλα ὅσα ἔν τινί ἐστιν, οἷον τὰ ἐν τόπῳ ὑπὸ τοῦ τόπου. καὶ πάσχειν δή τι ὑπὸ τοῦ χρόνου, καθάπερ καὶ λέγειν εἰώθαμεν ὅτι κατατήκει ὁ χρόνος, καὶ γηράσκει πάνθ᾿ ὑπὸ τοῦ χρόνου καὶ ἐπιλανθάνεται διὰ τὸν
221 b χρόνον, ἀλλ᾿ οὐ μεμάθηκεν, οὐδὲ νέον γέγονεν οὐδὲ

(for these latter pertain to number just in the same way in which the former pertain to time); but (*b*) events have their places in time in a sense analogous to that in which any numbered group of things exist in number (*i.e.*, in such and such a definite number), and such things as these are *embraced* in number (*i.e.*, in time) as things that have locality are embraced in their places.

And it is further evident (1) that to be in existence while time is in existence does not constitute being 'in time,' just as neither is a thing constituted as in motion or in a place because a motion and place exist while it does. For if this constitutes being '*in*' a thing, everything would be in anything, and the universe in a grain of millet, because a grain of millet exists while the universe is in existence. But this latter is an incidental coincidence; whereas when a thing is said to exist in time, it follows of necessity that there should be time while this thing exists, and if it exists in motion, it follows of necessity that there should be motion while it exists.

And since what exists in time exists in it as number (that is to say, as countable), you can take a time longer than anything that exists in time. So we must add that for things to exist in time they must be embraced by time, just as with other cases of being 'in' something; for instance, things that are in places are embraced by place. And it will follow that they are in some respect affected by time, just as we are wont to say that time crumbles things, and that everything grows old under the power of time and is forgotten through the lapse of time. But we do not say that we have learnt, or that anything is made new or beautiful, by the mere lapse of time;

221 b καλόν· φθορᾶς γὰρ αἴτιος καθʼ αὑτὸν μᾶλλον ὁ χρόνος· ἀριθμὸς γὰρ κινήσεως, ἡ δὲ κίνησις ἐξίστησι τὸ ὑπάρχον. ὥστε φανερὸν ὅτι τὰ ἀεὶ ὄντα, ᾗ ἀεὶ ὄντα, οὐκ ἔστιν ἐν χρόνῳ· οὐ γὰρ περιέχεται ὑπὸ χρόνου, οὐδὲ μετρεῖται τὸ εἶναι αὐτῶν ὑπὸ τοῦ χρόνου· σημεῖον δὲ τούτου ὅτι οὐδὲ πάσχει οὐδὲν ὑπὸ τοῦ χρόνου ὡς οὐκ ὄντα ἐν χρόνῳ.

Ἐπεὶ δʼ ἐστὶν ὁ χρόνος μέτρον κινήσεως, ἔσται καὶ ἠρεμίας μέτρον κατὰ συμβεβηκός· πᾶσα γὰρ ἠρεμία ἐν χρόνῳ. οὐ γὰρ ὥσπερ τὸ ἐν κινήσει ὂν ἀνάγκη κινεῖσθαι, οὕτω καὶ τὸ ἐν χρόνῳ· οὐ γὰρ κίνησις ὁ χρόνος, ἀλλʼ ἀριθμὸς κινήσεως· ἐν ἀριθμῷ δὲ κινήσεως ἐνδέχεται εἶναι καὶ τὸ ἠρεμοῦν. οὐ γὰρ πᾶν τὸ ἀκίνητον ἠρεμεῖ, ἀλλὰ τὸ ἐστερημένον κινήσεως πεφυκὸς δὲ κινεῖσθαι, καθάπερ εἴρηται ἐν τοῖς πρότερον. τὸ δʼ εἶναι ἐν ἀριθμῷ ἐστι τὸ εἶναί τινα ἀριθμὸν τοῦ πράγματος καὶ μετρεῖσθαι τὸ εἶναι αὐτοῦ τῷ ἀριθμῷ ἐν ᾧ ἐστιν· ὥστʼ εἰ ἐν χρόνῳ, ὑπὸ χρόνου. μετρήσει δʼ ὁ χρόνος τὸ κινούμενον καὶ τὸ ἠρεμοῦν, ᾗ τὸ μὲν κινούμενον τὸ δὲ ἠρεμοῦν· τὴν γὰρ κίνησιν αὐτῶν μετρήσει καὶ τὴν ἠρεμίαν, πόση τις. ὥστε τὸ κινούμενον οὐχ ἁπλῶς ἔσται μετρητὸν ὑπὸ χρόνου ᾗ ποσόν τί ἐστιν, ἀλλʼ ᾗ ἡ κίνησις αὐτοῦ ποσή. ὥστʼ ὅσα μήτε κινεῖται μήτʼ ἠρεμεῖ οὐκ ἔστιν ἐν χρόνῳ· τὸ μὲν γὰρ ἐν χρόνῳ εἶναι τὸ μετρεῖσθαί ἐστι χρόνῳ, ὁ δὲ χρόνος κινήσεως καὶ ἠρεμίας μέτρον.

Φανερὸν οὖν ὅτι οὐδὲ τὸ μὴ ὂν ἔσται πᾶν ἐν

---

[a] [At 202 a 4.—C.]

for we regard time in itself as destroying rather than producing, for what is counted in time is movement, and movement dislodges whatever it affects from its present state. From all this it is clear that things which exist eternally, as such, are not in time; for they are not embraced by time, nor is their duration measured by time. This is indicated by their not suffering anything under the action of time as though they were within its scope.

And since time is the measure of a motion, it will also incidentally be the measure of rest; for all rest is in time. For a thing being in motion necessitates that it should be moving, but its being in time does not; for time is not identical with motion, but is that in terms of which motion is counted, and even if a thing is at rest, it may be countable by the same count as motion. For not everything that is unmoved is at rest, but that only which by its nature is capable of moving but now lacks its actual motion, as we have already noted.[a] But a thing existing in number means that it 'has' a number and that its existence is measured by that number; and so too in the case of time. And time will measure that which is in motion and that which is at rest, *as such*; for it is their motion and their rest of which it determines the amount. So that the thing in motion is not measured by time in all respects in its capacity of a quantum, but in so far as its motion is defined in quantity; hence that which is neither in motion nor at rest is not in time, since to be 'in time' means to be measured by time, and it is motion and rest of which time is the measure.

Clearly, then, not all non-existences are in time,

221 b χρόνῳ, οἷον ὅσα μὴ ἐνδέχεται ἄλλως, ὥσπερ τὸ τὴν διάμετρον εἶναι τῇ πλευρᾷ σύμμετρον. ὅλως γάρ, εἰ μέτρον μὲν ἔστι κινήσεως ὁ χρόνος καθ' αὑτό, τῶν δ' ἄλλων κατὰ συμβεβηκός, δῆλον ὅτι ὧν τὸ εἶναι μετρεῖ, τούτοις ἅπασιν ἔσται τὸ εἶναι ἐν τῷ ἠρεμεῖν ἢ κινεῖσθαι. ὅσα μὲν οὖν φθαρτὰ καὶ γενητὰ καὶ ὅλως ὁτὲ μὲν ὄντα ὁτὲ δὲ μή, ἀνάγκη ἐν χρόνῳ εἶναι· ἔστι γὰρ χρόνος τις πλείων, ὃς ὑπερέξει τοῦ τε εἶναι αὐτῶν καὶ τοῦ μετροῦντος τὴν οὐσίαν. τῶν δὲ μὴ ὄντων ὅσα μὲν περιέχει 222 a ὁ χρόνος, τὰ μὲν ἦν (οἷον Ὅμηρός ποτε ἦν) τὰ δὲ ἔσται (οἷον τῶν μελλόντων τι), ἐφ' ὁπότερα περιέχει, καὶ εἰ ἐπ' ἄμφω, ἀμφότερα καὶ ἦν καὶ ἔσται· ὅσα δὲ μὴ περιέχει μηδαμῇ, οὔτ' ἦν οὔτ' ἔστιν οὔτ' ἔσται. ἔστι δὲ τὰ τοιαῦτα τῶν μὴ ὄντων, ὅσων τἀντικείμενα ἀεὶ ἔστιν, οἷον τὸ ἀσύμμετρον εἶναι τὴν διάμετρον ἀεὶ ἔστι, καὶ οὐκ ἔσται τοῦτ' ἐν χρόνῳ. οὐ τοίνυν οὐδὲ τὸ σύμμετρον· διὸ ἀεὶ οὐκ ἔστιν, ὅτι ἐναντίον τῷ ἀεὶ ὄντι· ὅσων δὲ τὸ ἐναντίον μὴ ἀεί, ταῦτα δὲ δύναται καὶ εἶναι καὶ μή, καὶ ἔστι γένεσις καὶ φθορὰ αὐτῶν.

---

[a] *i.e.* there is no unit of length in terms of which both the diagonal and side can be precisely measured.

[b] [a 1, ἐφ' ὁπότερα περιέχει κτλ., 'according as they are embraced by a stretch of time lying wholly on one or the other side of the present moment ; or, if they are embraced

but only such as might exist; for instance, the commensurability of the diagonal and the side does not exist,[a] but its non-existence is not temporal. For, as a general proposition, if time is the measure of motion on its own account and of anything else only by incidental coincidence, obviously everything whose existence it measures must have its existence in rest or in motion. Accordingly, whatever is destructible or generable, or (more broadly) sometimes existing and sometimes not, must be embraced by time; for there must be some time great enough to exceed the time of their duration and therefore the time which measures their being. Among non-existents, on the other hand, those which are embraced by time either once were (as Homer once existed) or will be (*e.g.*, some future event), according as they are embraced by time on one or the other side of the present moment, or, if they are embraced in both directions, they can be either past or future[b]; whereas those which are not in any way embraced by time neither were nor are nor will be. Non-existents of this latter kind are all those things whose opposites eternally exist; for instance, the incommensurability of the diagonal eternally exists, and therefore is not in time. And it follows that neither is its commensurability in time; hence it is *eternally* non-existent, inasmuch as it is a contradiction of what is eternally existent; whereas things of which the opposite does not exist eternally may either be or not be, and so they can come into existence and vanish from it.

by time extending on both sides of the present moment, they can be either past or future.'—C.]

CHAPTER XIII

ARGUMENT

Analogy between a point and a 'now,' as uniters in their unity and dividers in their plurality (222 a 10–20).

Opening of an examination of vaguer meanings attached to the word 'now' and to other adverbs of time (a 20–28).

A discussion (episodical in the lexicographical context, but far outweighing it in importance) of what may be inferred from the measurability of any given distance either way in time. Is time itself limited? No, because motion, as will be shown more fully later on, is everlasting. Is time itself, then, a travelling thing analogous to the body in motion that remains the same, or is it a medium or dimension analogous to the supreme 'place' in which things move? Solution—that time (being, as defined, simply movement-change as measured and counted successionally) is analogous to the medium of motion rather than to the moving body—reached

Τὸ δὲ νῦν ἐστι συνέχεια χρόνου, ὥσπερ ἐλέχθη—συνέχει γὰρ τὸν χρόνον τὸν παρελθόντα καὶ ἐσόμενον—καὶ πέρας[1] χρόνου ἐστίν· ἔστι γὰρ τοῦ μὲν ἀρχή, τοῦ δὲ τελευτή. ἀλλὰ τοῦτ' οὐχ ὥσπερ ἐπὶ τῆς στιγμῆς μενούσης φανερόν. διαιρεῖ δὲ δυνάμει. καὶ ᾗ μὲν τοιοῦτο, ἀεὶ ἕτερον τὸ νῦν, ᾗ δὲ συνδεῖ, ἀεὶ τὸ αὐτό, ὥσπερ ἐπὶ τῶν μαθηματικῶν γραμμῶν· οὐ γὰρ ἡ αὐτὴ ἀεὶ μία στιγμὴ τῇ νοήσει—διαιρούντων γὰρ ἄλλη—ᾗ δὲ μία, ἡ αὐτὴ πάντῃ. οὕτω καὶ τὸ νῦν τὸ μὲν τοῦ χρόνου διαίρεσις κατὰ δύναμιν, τὸ δὲ πέρας ἀμφοῖν καὶ ἑνότης·

[1] [καὶ πέρας E, Simplic. 748. 18: καὶ ὅλως πέρας cett.—C.]

CHAPTER XIII

ARGUMENT (*continued*)

by renewed reference to the analogies between a ' point ' and a ' now ' (a 28–b 7).

Linguistic discussion completed. Provisional conclusion (b 7–29).

(Note.—*The linguistic notes in this chapter cannot be tortured into a discussion of English terms. While finding the best equivalents I could, I have in every case inserted the Greek words in brackets to remind the reader that the definitions (sometimes needing qualification in any case) refer to them and not to the English substitutes, for which latter they are naturally bad fits, by no fault of the tailor.*)

We have said that it is through the ' now ' that time is continuous, for it holds time past and future time together; and in its general character of ' limit ' it is at once the beginning of time to come and the end of time past. But in the case of the ' now ' this is not so obvious as in that of the stationary point; for, as well as actually continuing, it potentially divides time. And in this potentiality one ' now ' differs from another, but in its actual holding of time continuously together it always remains the same, as in the parallel case of mathematical lines traced by moving points, in which case the point too, if arrested as a divider, is not conceived as retaining its identity with the tracing point or another arrested point; for if we are dividing the line, the point differs at every division, but if we regard the line as a single undivided one, the point that traces it is the same all along. Thus too the ' now ' of time is a divider in mental potentiality, but a continuing unifier as the coincident end-term and

ἔστι δὲ ταὐτὸ καὶ κατὰ ταὐτὸ ἡ διαίρεσις καὶ ἡ ἕνωσις, τὸ δ' εἶναι οὐ ταὐτό.

Τὸ μὲν οὖν οὕτω λέγεται τῶν 'νῦν,' ἄλλο δ' ὅταν ὁ χρόνος ὁ τούτου ἐγγὺς ᾖ. 'ἥξει νῦν,' ὅτι τήμερον ἥξει· 'ἥκει νῦν,' ὅτι ἦλθε τήμερον. τὰ δ' ἐν Ἰλίῳ γέγονεν οὐ νῦν, οὐδ' ὁ κατακλυσμὸς γέγονε νῦν· καίτοι συνεχὴς χρόνος εἰς αὐτά, ἀλλ' ὅτι οὐκ ἐγγύς.

Τὸ δὲ 'ποτὲ' χρόνος ὡρισμένος πρὸς τὸ πρότερον νῦν, οἷον 'ποτὲ ἐλήφθη Τροία,' καὶ 'ποτὲ ἔσται κατακλυσμός'· δεῖ γὰρ πεπεράνθαι πρὸς τὸ νῦν. ἔσται ἄρα ποσός τις ἀπὸ τοῦδε χρόνος καὶ εἰς ἐκεῖνο, καὶ ἦν εἰς τὸ παρελθόν.

Εἰ δὲ μηδεὶς χρόνος ὃς οὐ ποτέ, πᾶς ἂν εἴη χρόνος πεπερασμένος. ἆρ' οὖν ὑπολείψει; ἢ οὔ, εἴπερ ἀεὶ ἔστι κίνησις. ἄλλος οὖν ἢ ὁ αὐτὸς πολλάκις; δῆλον ὅτι ὡς ἂν ἡ κίνησις, οὕτω καὶ ὁ χρόνος· εἰ μὲν γὰρ ἡ αὐτὴ καὶ μία γίνεταί ποτε, ἔσται καὶ χρόνος εἷς καὶ ὁ αὐτός, εἰ δὲ μή, οὐκ

---

[a] That is to say the 'now,' when arrested, divides the same specific past and present which it was dividing in its movement, but differs in the conditions of its functioning.

[b] [*Or* the *future* flood, mentioned in the next paragraph. *Cf.* Plato, *Timaeus* 22 c.—C.]

[c] [*Literally*, '" Some time " (or " one day ") means a time determined relatively to the " now " in the former

beginning-term of past and future time ; and these two capacities of potential divider and actual uniter pertain to the same actual ' now ' and on the same count of its being two limits at once, but its essential and defined functioning in the one capacity differs from that in the other.[a]

This is one of the meanings of ' now,' but it is also used for ' not far off in time.' ' He will come now,' if he will come to-day ; ' He has come but now,' if he came to-day. But we do not speak so of the Trojan war or Deucalion's flood[b] ; though time is continuous between us and these events, they are not near.

' Sometime ' is used when we wish to be no more definite than that the present ' now ' comes after it or the reverse.[c] When was Troy taken ? ' Sometime ' in the past.[d] When will the flood be ? ' Sometime ' in the future. There will be a measurable stretch of time from now onwards to that, or there has been one from that to now.

And since there is no time-ago or time-to-come that was not, or will not be, ' some time ' off, it would seem that all time is limited. Will it come to an end, then ? Surely not ; for if motion is everlasting, so is time. Is it, then, always a different stretch of time that continues the succession, or the same stretch of time taken repeatedly ? As to this, evidently it must conform to motion ; for whichever of these kinds of counting applies to motion

sense ' (*i.e.* the sense of the present moment, defined in the first paragraph).—C.]

[d] In the Greek idiom ' sometime ' (ποτέ) would not require the addition of past or future, the difference being sufficiently expressed by the tense of the verb.

222b ἔσται. ἐπεὶ δὲ τὸ νῦν τελευτὴ καὶ ἀρχὴ χρόνου, ἀλλ' οὐ τοῦ αὐτοῦ ἀλλὰ τοῦ μὲν παρήκοντος τελευτὴ ἀρχὴ δὲ τοῦ μέλλοντος, ἔχοι ἂν ὥσπερ ὁ κύκλος ἐν τῷ αὐτῷ πως τὸ κυρτὸν καὶ τὸ κοῖλον, οὕτω καὶ ὁ χρόνος ἀεὶ ἐν ἀρχῇ καὶ τελευτῇ. καὶ διὰ τοῦτο δοκεῖ ἀεὶ ἕτερος· οὐ γὰρ τοῦ αὐτοῦ ἀρχὴ καὶ τελευτὴ τὸ νῦν· ἅμα γὰρ ἂν καὶ κατὰ τὸ αὐτὸ τὰ ἀντικείμενα εἴη. καὶ οὐχ ὑπολείψει δή· ἀεὶ γὰρ ἐν ἀρχῇ.

Τὸ δ' 'ἤδη' τὸ ἐγγύς ἐστι τοῦ παρόντος νῦν ἀτόμου μέρος τοῦ μέλλοντος χρόνου—'πότε βαδίζεις;' 'ἤδη,' ὅτι ἐγγὺς ὁ χρόνος ἐν ᾧ μέλλει—καὶ τοῦ παρεληλυθότος χρόνου τὸ μὴ πόρρω τοῦ νῦν· 'πότε βαδίζεις;' 'ἤδη βεβάδικα.' τὸ δὲ Ἴλιον φάναι ἤδη ἑαλωκέναι οὐ λέγομεν, ὅτι πόρρω λίαν τοῦ νῦν. καὶ τὸ 'ἄρτι' τὸ ἐγγὺς τοῦ παρόντος νῦν μόριον[1] τοῦ παρελθόντος· 'πότε ἦλθες;' 'ἄρτι,' ἐὰν ᾖ ὁ χρόνος ἐγγὺς τοῦ ἐνεστῶτος νῦν. 'πάλαι' δὲ τὸ πόρρω. τὸ δ' 'ἐξαίφνης' τὸ ἐν ἀναισθήτῳ χρόνῳ διὰ μικρότητα ἐκστάν.

[1] [μόριον Themist. 158. 20, Simplic. 752. 11, Bonitz: τὸ μόριον codd.—C.]

---

[a] The successive repetitions of a recurrent movement are specifically the same but numerically different; the 'now' of one succeeds the 'now' of the other but does not repeat it.

[b] The analogy goes no further than to show that an identical line can be both concave and convex (though it has no breadth) at the same time, the one with reference to what is inside the circle, the other with reference to what is outside.

[c] [*Literally*, 'so also time is always both at a beginning and at an end. And for that reason it seems to be always different . . .'—C.]

must apply to time.[a] And besides, since the 'now' is the final limit and the initial limit of time, but not of the same time, but the final limit of time past and the initial limit of time to come, it must present a relation analogous to the kind of identity between the convexity and the concavity of the same circumference, which necessitates a difference between that with respect to which it bears the one character and that with respect to which it bears the other.[b] So too, since every 'now' is at once the initial limit and the final limit of a stretch of time, the two stretches must be different; [c] for the same 'now' cannot be both the beginning and the end of the same thing, for, if so, it would be both of two contradictories in the same subject at once. Neither, then, will time ever come to an end, for it is always at a beginning.

We say 'already' (*ēdē*) for any time close enough to the indivisible 'now' of the absolute present. 'When do you take your walk?' 'I am starting already,' or 'I have already taken it.' But the phrase only applies to the near future or past; so that we should not say 'Troy has already fallen.' Another term for the near past is 'just now' (*arti*). 'When did you arrive?' 'Just now,' if it was near to the instantaneous 'now' at which you speak; but 'some time ago' (*palai*) if the interval is considerable; and 'suddenly' [d] (*exaiphnēs*) if the passage of time is so short as to be imperceptible.

[d] In the Greek there is a rather faint etymological play upon the ἐξ ('out' or 'away from') of ἐξαίφνης, fairly represented by the etymological suggestion of 'sudden' (from the Latin *sub-it-aneus*, 'going stealthily.' *Cf.* Skeat).

222 b Μεταβολὴ δὲ πᾶσα φύσει ἐκστατικόν. ἐν δὲ τῷ χρόνῳ πάντα γίνεται καὶ φθείρεται· διὸ καὶ οἱ μὲν σοφώτατον ἔλεγον, ὁ δὲ Πυθαγόρειος Πάρων ἀμαθέστατον, ὅτι καὶ ἐπιλανθάνονται ἐν τούτῳ, λέγων ὀρθότερον. δῆλον οὖν ὅτι φθορᾶς μᾶλλον ἔσται καθ' αὑτὸν αἴτιος ἢ γενέσεως, καθάπερ ἐλέχθη καὶ πρότερον (ἐκστατικὸν γὰρ ἡ μεταβολὴ καθ' αὑτήν), γενέσεως δὲ καὶ τοῦ εἶναι κατὰ συμβεβηκός. σημεῖον δὲ ἱκανὸν ὅτι γίνεται μὲν οὐδὲν ἄνευ τοῦ κινεῖσθαί πως αὐτὸ καὶ πράττειν, φθείρεται δὲ καὶ μηδὲν κινούμενον· καὶ ταύτην μάλιστα λέγειν εἰώθαμεν ὑπὸ τοῦ χρόνου φθοράν. οὐ μὴν ἀλλ' οὐδὲ ταύτην ὁ χρόνος ποιεῖ, ἀλλὰ συμβαίνει ἐν χρόνῳ γίνεσθαι καὶ ταύτην τὴν μεταβολήν.

Ὅτι μὲν οὖν ἔστιν ὁ χρόνος καὶ τί, καὶ ποσαχῶς λέγομεν τὸ νῦν, καὶ τί τὸ ποτὲ καὶ τὸ ἄρτι καὶ τὸ ἤδη καὶ τὸ πάλαι καὶ τὸ ἐξαίφνης, εἴρηται.

---

[a] Nothing is known of any Paron, and Simplicius [754. 9] thinks the word is not a proper name at all, but the ordinary participle that means 'being present.' He quotes Eudemus as telling the tale that Simonides reciting at an Olympic festival extolled Time as the wisest of beings, since all learning and all memory are in him ; but a certain unnamed philosopher (παρόντα τινὰ τῶν σοφῶν) said : 'But look here, Simonides, don't we forget everything in time as well as learn it ?' Simplicius thinks that Aristotle is referring to this and speaks of the interrupter as 'The Pythagorean who was there.' [Diels, *Vors.* 16.—C.]

[b] [At 221 b 1.—C.]

All change is in its nature a 'passing away.' And it is 'in time' that everything begins and ceases to be; so some have called it the wisest of things, because it brings all knowledge, but the Pythagorean Paron [a] said it was the most ignorant, because it is in time too that everything is forgotten, and he was nearer the mark. Indeed, it is evident that the mere passage of time itself is destructive rather than generative, as we said earlier [b]; because change is primarily a 'passing away.' So it is only incidentally that time is the cause of things coming into being and existing. A sufficient indication of this may be found in the fact that nothing comes into being without being started by some cause and reacting to it,[c] but things perish without anything being stirred, and it is a kind of perishing without apparent provocation that we especially attribute to time.[d] But yet, after all, it is not really time itself that destroys things in this way, but the changes that do destroy take place concurrently with time.[e]

That there is such a thing as time, therefore, and what it is, and in how many senses we speak of 'now,' and what 'sometime' and 'but now' and 'already' and 'some time ago' and 'suddenly' mean, has now been said.

[c] [*καὶ πράττειν*. *Cf.* Themist. 159. 9 *προσδεῖταί τινος ἐνεργείας ὑφ' ἧς γίνεται, οἷον τέχνης ἢ φύσεως ἢ διδασκαλίας ἢ πράξεως*.—C.]

[d] Not for instance if a man was slain by a sword (Aquinas).

[e] [*Or*, 'but it is incidental that this change also occurs in time.'—C.]

CHAPTER XIV

ARGUMENT

Definition of velocity and examination of succession in time.

All change and every moving thing are in time (222 b 30–223 a 15).

Relation of time as numeration, potential or actual, to consciousness (a 16–29).

What kind of succession should time be standardized by? Time is identical as 'now' across *all changes everywhere and in every succession, but 'nows' and intercepted intervals between them are different from each other as taken* along *any succession whatever. But all such intervals in whatever*

Τούτων δ' ἡμῖν οὕτω διηριθμημένων φανερὸν ὅτι πᾶσα μεταβολὴ καὶ ἅπαν τὸ κινούμενον ἐν χρόνῳ. τὸ γὰρ θᾶττον καὶ βραδύτερον κατὰ πᾶσάν ἐστι μεταβολήν· ἐν πᾶσι γὰρ οὕτω φαίνεται. λέγω δὲ θᾶττον κινεῖσθαι τὸ πρότερον μεταβάλλον εἰς τὸ ὑποκείμενον κατὰ τὸ αὐτὸ διάστημα καὶ ὁμαλὴν κίνησιν κινούμενον· οἷον ἐπὶ τῆς φορᾶς, εἰ ἄμφω κατὰ τὴν περιφερῆ κινεῖται ἢ ἄμφω κατὰ τὴν εὐθεῖαν· ὁμοίως δὲ καὶ ἐπὶ τῶν ἄλλων. ἀλλὰ μὴν τό γε πρότερον ἐν χρόνῳ ἐστίν· πρότερον γὰρ καὶ ὕστερον λέγομεν κατὰ τὴν πρὸς τὸ νῦν ἀπόστασιν, τὸ δὲ νῦν ὅρος τοῦ παρήκοντος καὶ τοῦ μέλλοντος· ὥστ' ἐπεὶ τὰ νῦν ἐν χρόνῳ, καὶ τὸ πρότερον καὶ ὕστερον ἐν χρόνῳ ἔσται· ἐν ᾧ γὰρ

---

[a] ['homogeneous.' The two movements compared must be 'over the same course' (κατὰ τὸ αὐτὸ διάστημα) and *each* 'of uniform velocity' (ὁμαλὴν κίνησιν), for, if either varies in pace, it might be the quicker at some moments, the slower at others.—C.]

CHAPTER XIV

ARGUMENT (*continued*)

kind of succession run abreast and are measured by the same time as mutually determining and determined by any one of them (a 18–b 12).

Advantage of rotation as the form of movement by reference to which to standardize time. Loose ways of speaking of time and temporal things as ' circulating ' (b 12–224 a 2).

An interpolation of no value (a 2–15).

Conclusion (a 15–17).

All this being so established, it becomes clear that all changes and everything that moves are conditioned by time. For it is a patent fact that every change may be quicker or slower. And what I mean by one change being quicker than another is that, of two homogeneous [a] change-movements (either both on a periphery, for instance, or both on a straight line, if it be a local movement, and *mutatis mutandis* in other kinds of change), that one is the quicker which reaches a certain determined stage or point in its course ' before ' the other reaches the point at the same distance from the starting-point in its course. Now this ' before ' means before ' in time,' for both ' before ' and ' after ' are expressions of an interval between the ' nows ' of arrival ; [b] and since the ' now ' is a boundary between past and future, it follows that the two ' nows ' (of the former and latter arrival, namely) being both phenomena of time, so must their ' before ' and ' after ' be. For, whatever

[b] [*Literally*, ' we use the terms " earlier " and " later " with respect to the distance (from the event so described) to the present moment.'—C.]

223a τὸ νῦν, καὶ ἡ τοῦ νῦν ἀπόστασις. (ἐναντίως δὲ λέγεται τὸ πρότερον κατά τε τὸν παρεληλυθότα χρόνον καὶ τὸν μέλλοντα· ἐν μὲν γὰρ τῷ παρεληλυθότι πρότερον λέγομεν τὸ πορρώτερον τοῦ νῦν, ὕστερον δὲ τὸ ἐγγύτερον, ἐν δὲ τῷ μέλλοντι πρότερον μὲν τὸ ἐγγύτερον, ὕστερον δὲ τὸ πορρώτερον.) ὥστ᾽ ἐπεὶ τὸ μὲν πρότερον ἐν χρόνῳ, πάσῃ δ᾽ ἀκολουθεῖ κινήσει τὸ πρότερον, φανερὸν ὅτι πᾶσα μεταβολὴ καὶ πᾶσα κίνησις ἐν χρόνῳ ἐστίν.

Ἄξιον δ᾽ ἐπισκέψεως καὶ πῶς ποτε ἔχει ὁ χρόνος πρὸς τὴν ψυχήν, καὶ διὰ τί ἐν παντὶ δοκεῖ εἶναι ὁ χρόνος, καὶ ἐν γῇ καὶ ἐν θαλάττῃ καὶ ἐν οὐρανῷ. ἢ ὅτι κινήσεώς τι πάθος ἢ ἕξις, ἀριθμός γε ὤν, ταῦτα δὲ κινητὰ πάντα (ἐν τόπῳ γὰρ πάντα), ὁ δὲ χρόνος καὶ ἡ κίνησις ἅμα κατά τε δύναμιν καὶ κατ᾽ ἐνέργειαν.

Πότερον δὲ μὴ οὔσης ψυχῆς εἴη ἂν ὁ χρόνος ἢ οὔ, ἀπορήσειεν ἄν τις· ἀδυνάτου γὰρ ὄντος εἶναι τοῦ ἀριθμήσοντος ἀδύνατον καὶ ἀριθμητόν τι εἶναι, ὥστε δῆλον ὅτι οὐδ᾽ ἀριθμός· ἀριθμὸς γὰρ ἢ τὸ ἠριθμημένον ἢ τὸ ἀριθμητόν. εἰ δὲ μηδὲν ἄλλο πέφυκεν ἀριθμεῖν ἢ ψυχὴ καὶ ψυχῆς νοῦς, ἀδύνατον

it be that the 'now' pertains to, to that must the interval determined by it pertain. (But note that 'before' has opposite meanings according to whether it refers to past or future time; for in the past we regard the event that is farther from the present as 'before' the other and the nearer event as 'after' it, but in the future the nearer as 'before' and the farther as 'after' the other.) So, inasmuch as 'before' pertains to time, and may be a 'before' of arrival at a point of any kind of change-movement, it follows that every change or movement occurs 'in time.'

The relation of time to consciousness deserves examination, and so does the question why we conceive of time as immanent in everything in earth and sea and sky. As to the latter point, it is because time, being the numerator of motion, pertains to such motion wherever it exists, as an affection or disposition of it (namely, that it is either actually counted in units or potentially countable in such); and all things in the material universe are susceptible of motion (for they can all change their positions),[a] and time and movement run in pairs both potentially and actually.

The question remains, then, whether or not time would exist if there were no consciousness; for if it were impossible for there to be the factor that does the counting, it would be impossible that anything should be counted; so that evidently there could be no number, for a number is either that which has actually been counted or that which can be counted. And if nothing can count except consciousness, and consciousness only as intellect (not as sensation merely), it is impossible that time should exist if

[a] Which they can quit.

223 a εἶναι χρόνον ψυχῆς μὴ οὔσης, ἀλλ' ἢ τοῦτο ὅ ποτε ὂν ἔστιν ὁ χρόνος, οἷον εἰ ἐνδέχεται κίνησιν εἶναι ἄνευ ψυχῆς. τὸ δὲ πρότερον καὶ ὕστερον ἐν κινήσει ἐστίν· χρόνος δὲ ταῦτ' ἐστὶν ᾗ ἀριθμητά ἐστιν.

Ἀπορήσειε δ' ἄν τις καὶ ποίας κινήσεως ὁ χρόνος ἀριθμός. ἢ ὁποιασοῦν· καὶ γὰρ γίνεται ἐν χρόνῳ καὶ φθείρεται καὶ αὐξάνεται, καὶ ἀλλοιοῦται ἐν χρόνῳ καὶ φέρεται· ᾗ οὖν κίνησίς ἐστι, ταύτῃ ἐστὶν ἑκάστης κινήσεως ἀριθμός. διὸ κινήσεώς
223 b ἐστιν ἁπλῶς ἀριθμὸς συνεχοῦς, ἀλλ' οὐ τινός.

Ἀλλ' ἔστι νῦν κεκινῆσθαι καὶ ἄλλα, ὧν ἑκατέρας τῆς κινήσεως εἴη ἂν ἀριθμός. ἕτερος οὖν χρόνος ἔστι, καὶ ἅμα δύο ἴσοι χρόνοι ἂν εἶεν; ἢ οὔ· ὁ αὐτὸς[1] γὰρ χρόνος εἷς ὁμοίως καὶ ἅμα[2]· εἴδει δὲ καὶ οἱ μὴ ἅμα· εἰ γὰρ εἶεν κύνες, οἱ δ' ἵπποι, ἑκάτεροι δ' ἑπτά, ὁ αὐτὸς ἀριθμός. οὕτω καὶ τῶν κινήσεων τῶν ἅμα περαινομένων ὁ αὐτὸς χρόνος· ἀλλ' ἡ μὲν

[1] [αὐτὸς: ἅπας GI.—C.[

[2] [εἷς ὁμοίως καὶ ἅμα F, Bekker: εἷς καὶ ἴσος καὶ πᾶς ἅμα H: εἷς καὶ ἴσος καὶ ἅμα E.—C.]

---

[a] [τοῦτο ὅ ποτε ὂν ἐστιν ὁ χρόνος is explained by Philoponus (755. 17) as 'the substratum of time, namely movement.' Movement is the *objective* thing which contains before and after, whether we *count* these or not. If we count them, then we have time.—C.]

[b] [Either, ὁ αὐτὸς γὰρ χρόνος πᾶς καὶ εἷς ὁ ἴσος καὶ ἅμα (Torstrik, Prantl) or ὁ αὐτὸς γὰρ χρόνος πᾶς ὁ ἴσος καὶ ἅμα (Carteron) will yield the sense: 'For all stretches of time that are equal and simultaneous are numerically one and the same time, while (equal) stretches which are not simultaneous are the same conceptually (though not numerically),' *e.g.* two successive hours. *Cf.* Simplic. 764. 3 λύων οὖν τὴν ἀπορίαν οὐχ οὕτως ἔχει, φησίν· οὐ γὰρ πλείους ἅμα χρόνοι γίγνονται,

consciousness did not; unless as the 'objective thing' which is subjectively time to us,[a] if we may suppose that movement could thus objectively exist without there being any consciousness. For 'before' and 'after' are objectively involved in motion, and these, *qua* capable of numeration, constitute time.

It may be asked further to what kind of motion-change time does pertain. We may answer, 'It does not matter.' For things begin and cease to be, and grow, and change their qualities and their places 'in time'; so far, then, as change can be regarded as movement, so far time must be a numerator of every such kind of movement. We conclude, then, that time is the numeration of continuous movement, without any qualification, not only of some particular kind.

But if we take one kind of change and say 'now' with respect to it, other kinds of change, each of which has a specifically different unit to be counted in, will be at a certain stage of their change at this same 'now.' Can each of them have a different time, and must there be more than one time running concurrently? No; for it is the same lapse of time that is counted by two 'nows,' everywhere at once, whatever the units of movement or change; whereas the one-and-sameness of the units is determined by their kind and not by their 'at-once-ness'; [b] just as if there were dogs and horses, seven of each, the number would be the same, but the units numbered different. So, too, of all movement-changes determined simultaneously the time is the same; one may be quick and another

ἀλλ' ὁ ἴσος χρόνος πᾶς ἐν τῷ ἐνεστῶτι ὁ αὐτὸς εἷς ἅμα ἐστὶ τῷ ἀριθμῷ, εἴδει δὲ οἱ αὐτοί εἰσι καὶ οἱ μὴ ἅμα, τουτέστιν ὁ παρεληλυθὼς τῷ μέλλοντι.—C.]

223 b ταχεῖα ἴσως ἡ δ᾽ οὔ, καὶ ἡ μὲν φορὰ ἡ δ᾽ ἀλλοίωσις· ὁ μέντοι χρόνος ὁ αὐτός, εἴπερ καὶ ὁ ἀριθμὸς ἴσος καὶ ἅμα, τῆς τε ἀλλοιώσεως καὶ τῆς φορᾶς. καὶ διὰ τοῦτο αἱ μὲν κινήσεις ἕτεραι καὶ χωρίς, ὁ δὲ χρόνος πανταχοῦ ὁ αὐτός, ὅτι καὶ ὁ ἀριθμὸς εἷς καὶ ὁ αὐτὸς πανταχοῦ ὁ τῶν ἴσων καὶ ἅμα.

Ἐπεὶ δ᾽ ἔστι[1] φορὰ καὶ ταύτης ἡ κύκλῳ, ἀριθμεῖται δ᾽ ἕκαστον ἑνί τινι συγγενεῖ, μονάδες μονάδι, ἵπποι δ᾽ ἵππῳ, οὕτω καὶ ὁ χρόνος χρόνῳ τινὶ ὡρισμένῳ, μετρεῖται δὲ (ὥσπερ εἴπομεν) ὅ τε χρόνος κινήσει καὶ ἡ κίνησις χρόνῳ—τοῦτο δ᾽ ἐστίν, ὅτι ὑπὸ τῆς ὡρισμένης κινήσεως χρόνῳ μετρεῖται τῆς τε κινήσεως τὸ ποσὸν καὶ τοῦ χρόνου—εἰ οὖν τὸ πρῶτον μέτρον πάντων τῶν συγγενῶν, ἡ κυκλοφορία ἡ ὁμαλὴς μέτρον μάλιστα, ὅτι ὁ ἀριθμὸς ὁ ταύτης γνωριμώτατος.

[1] [ἔστι: πρώτη ἐστὶ Prantl.—C.]

---

[a] [ὑπὸ τῆς ὡρισμένης κινήσεως χρόνῳ. Taking χρόνῳ with ὡρισμένης (*cf.* 221 a 1 μετρεῖ δ᾽ οὗτος (ὁ χρόνος) τὴν κίνησιν τῷ ὁρίσαι τινὰ κίνησιν ἣ καταμετρήσει τὴν ὅλην), the meaning will be: 'that is because it is by the movement-determined-by-time that the quantity both of movement and of time is measured.' The unit of measurement (the day, for instance) is a movement determined by a certain time-length. Alexander, and some copies known to Simplicius (769. 15) read τῆς ὡρισμένης κινήσεως χρόνῳ. So also Philoponus, 782. 26 (lemma τῆς KM, ὑπὸ τῆς t), but he paraphrases: διότι ἡ ὑπὸ τοῦ χρόνου πρώτως ὁρισθεῖσα κίνησις, αὕτη καὶ τῆς πάσης ἐστὶ κινήσεως μέτρον καὶ τοῦ χρόνου.—C.]

[b] The unit fixed is the unit of time fixed in reference to some particular motion, and this becomes the measure of all time-lapses, whether of movement, modification, change of size, etc.; for the lapses of all these different manifestations of time are akin to each other with respect to before and after.

[c] The circumference of a circle or sphere, being a re-

slow, and one a change of place and the other of quality; the time, however, is the same, if the counting has reached the same number and been made simultaneously, whether of the qualitive modification or of the change of place. So the movements or changes are different and stand apart, but the time is the same everywhere, because the numeration, if made simultaneously and up to the same figure, is one and the same.

And now, keeping locomotion and especially rotation in mind, note that everything is counted by some unit of like nature to itself—monads monad by monad, for instance, and horses horse by horse—and so likewise time by some finite unit of time. But as we have said, motion and time mutually determine each other quantitively; and that because the standard of time established by the motion we select is the quantitive measure both of that motion and of time.[a] If, then, the standard once fixed measures all dimensionality of its own order,[b] a uniform rotation will be the best standard, since it is easiest to count.[c]

entrant curve, any part of which will fit any other part, is uniform all over; and motion upon it (supposing its velocity to be uniform) has a natural spatial unit in each completed circle. It is therefore easy to count, and since its natural unit is easy to divide or multiply into convenient secondary units, it furnishes a perfect standard by which to determine time. Aristotle does not say what particular circular motion we are to take as our standard, but the commentators are probably right in supposing that for practical purposes he accepted the the solar day, though on all theoretical grounds he should have taken the stellar day, as registering (in his astronomy) the prime heavenly movement and one that by the science of the day (which recognized differences in the length of the solar day, measured from one southing of the sun to another, at different seasons of the year) would seem to be demonstrably uniform.

223 b Ἀλλοίωσις μὲν οὖν οὐδ' αὔξησις οὐδὲ γένεσις οὐκ εἰσὶν ὁμαλεῖς, φορὰ δ' ἐστίν. διὸ καὶ δοκεῖ ὁ χρόνος εἶναι ἡ τῆς σφαίρας κίνησις, ὅτι ταύτῃ μετροῦνται αἱ ἄλλαι κινήσεις καὶ ὁ χρόνος ταύτῃ τῇ κινήσει. διὰ δὲ τοῦτο καὶ τὸ εἰωθὸς λέγεσθαι συμβαίνει· φασὶ γὰρ κύκλον εἶναι τὰ ἀνθρώπινα πράγματα καὶ τῶν ἄλλων τῶν κίνησιν ἐχόντων φυσικὴν καὶ γένεσιν καὶ φθοράν. τοῦτο δ', ὅτι ταῦτα πάντα τῷ χρόνῳ κρίνεται, καὶ λαμβάνει τελευτὴν καὶ ἀρχὴν ὥσπερ ἂν εἰ κατά τινα περίοδον· καὶ γὰρ ὁ χρόνος αὐτὸς εἶναι δοκεῖ κύκλος τις. τοῦτο δὲ πάλιν δοκεῖ διότι τοιαύτης ἐστὶ φορᾶς μέτρον καὶ μετρεῖται αὐτὸς ὑπὸ τοιαύτης. ὥστε τὸ λέγειν εἶναι τὰ γινόμενα τῶν πραγμάτων κύκλον τὸ λέγειν ἐστὶ τοῦ χρόνου εἶναί τινα κύκλον· τοῦτο δ', ὅτι μετρεῖται τῇ κυκλοφορίᾳ· παρὰ γὰρ τὸ
224 a μέτρον[1] οὐδὲν ἄλλο παρεμφαίνεται τὸ μετρούμενον, ἀλλ' ἢ πλείω μέτρα τὸ ὅλον.

[Λέγεται δ' ὀρθῶς καὶ ὅτι ἀριθμὸς μὲν ὁ αὐτὸς ὁ τῶν προβάτων καὶ τῶν κυνῶν, εἰ ἴσος ἑκάτερος·

[1] [μέτρον: μετροῦν F, Philop. 784. 15 (lemma).—C.]

---

[a] That is to say, one part of rotary motion can be fitted on to another and has no distinguishing quality, whereas in a given process of change or growth the stages differ from one another. ὁμαλής does not here mean ' uniform ' in the ordinary sense, *i.e.* a motion free from acceleration, positive or negative. Aristotle here assumes this kind of uniformity in the rotation which he is about to choose as the standard, and, that being assumed, there will be no distinction between two equal sections of the course.

[b] [*Cf.* 218 a 34 οἱ μὲν γὰρ τὴν τοῦ ὅλου κίνησιν εἶναί φασιν. —C.]

[c] Motion here includes every kind of successive passing from this to that, whether it be from this to that place, from this to that quality, or from this to that size.

Neither qualitive modification nor growth nor genesis has the kind of uniformity that rotation has;[a] and so time is regarded as the rotation of the sphere,[b] inasmuch as all other orders of motion[c] are measured by it, and time itself is standardized by reference to it. And this is the reason of our habitual way of speaking; for we say that human affairs and those of all other things that have natural movement and become and perish seem to be in a way circular, because all these things come to pass in time and have their beginning and end as it were 'periodically'[d]; for time itself is conceived as 'coming round'; and this again because time and such a standard rotation mutually determine each other. Hence, to call the happenings of a thing a circle is saying that there is a sort of circle of time; and that is because it is measured by a complete revolution, and the whole measurement of a thing is nought else but a defined number of the units of its measurements.[e]

[It is correct to say that the number of sheep and of dogs is the same number, if that of the sheep and that of the dogs are equal; but it is not the same

[d] 'Period' is a Greek word which means 'passage round.'

[e] Here in the time of Themistius (fourth century ?) the chapter and book seem to have ended, except perhaps the final summary. The interpolated passage is worthless. [Themistius, however, (at 162. 23) does mention the 'ten sheep' of this paragraph. It recurs to the statement (223 b 10) that one universal time is the measure or 'number' of movements which are distinct from one another and of different sorts (qualitative modification, growth, etc.), although these movements are not 'uniform,' whereas the rotation which measures time is uniform (b 12–21). A technical explanation is now given of how it is that *the same* 'number' can serve to measure things of different kinds.—C.]

224 a δεκὰς δὲ οὐχ ἡ αὐτὴ οὐδὲ δέκα ταὐτά, ὥσπερ οὐδὲ τρίγωνα τὰ αὐτὰ τὸ ἰσόπλευρον καὶ τὸ σκαληνές. καίτοι σχῆμά γε ταὐτό, ὅτι τρίγωνα ἄμφω· ταὐτὸ γὰρ λέγεται οὗ μὴ διαφέρει διαφορᾷ, ἀλλ' οὐχὶ οὗ διαφέρει· οἷον τρίγωνον τριγώνου διαφορᾷ διαφέρει —τοιγαροῦν ἕτερα τρίγωνα—σχήματος δὲ οὔ, ἀλλ' ἐν τῇ αὐτῇ διαιρέσει καὶ μιᾷ. σχῆμα γὰρ τὸ μὲν τοιόνδε κύκλος, τὸ δὲ τοιόνδε τρίγωνον, τούτου δὲ τὸ μὲν τοιόνδε ἰσόπλευρον, τὸ δὲ τοιόνδε σκαληνές. σχῆμα μὲν οὖν τὸ αὐτὸ καὶ τοῦτο (τρίγωνον γάρ), τρίγωνον δ' οὐ τὸ αὐτό. καὶ ὁ ἀριθμὸς δὴ ὁ αὐτός· οὐ γὰρ διαφέρει ἀριθμοῦ διαφορᾷ ὁ ἀριθμὸς αὐτῶν· δεκὰς δ' οὐχ ἡ αὐτή· ἐφ' ὧν γὰρ λέγεται, διαφέρει· τὰ μὲν γὰρ κύνες, τὰ δ' ἵπποι].

Καὶ περὶ μὲν χρόνου καὶ αὐτοῦ καὶ τῶν περὶ αὐτὸν οἰκείων τῇ σκέψει εἴρηται.

---

[a] [a 6 ταὐτὸ γὰρ λέγεται κτλ. *Literally*, '(A thing) is said to be " the same as " that from which it does not differ by a *differentia* (proper to that thing, οἰκείᾳ διαφορᾷ τούτου, Alex. *ap*. Simpl. 770. 23), and " not (the same as) " that from which it does (so) differ. For instance, a (scalene) triangle differs from an (isosceles) triangle by a *differentia*—so they are 'not the same' triangle—but does not so differ from

decad in each case, nor are the units of one the same as the units of the other, any more than a scalene and an isosceles are the same triangle, though they are the same figure, both being triangles ; for things bear the same name if they do not differ as to the characteristic in virtue of which that name is borne, in this case the differentia of triangle. They are different as triangles, therefore, but not different as figures, since they belong to one and the same figure denomination.[a] For, as a figure, this is a circle and that a triangle, but, as a triangle, this isosceles and that scalene. The two, then, have the same figure conformation (for both are triangular), but not the same triangle formation. So with the animals : the number of each ten-group is the same, for they do not differ in a numerical differentia ; but the ten-groups themselves are not the same, for the ten-ness is predicated of different subjects—dogs in the one case and horses in the other.]

This closes our investigation of time and its properties, in so far as they are germane to our inquiry.

' Figure,' but (both triangles) are in one and the same Table of Division (headed by the genus ' Figure,' which includes all sorts of triangles, as well as other figures).'—C.]